The Evolution of Phylogenetic Systematics

SPECIES AND SYSTEMATICS

www.ucpress.edu/go/spsy

The Species and Systematics series will investigate fundamental and practical aspects of systematics and taxonomy in a series of comprehensive volumes aimed at students and researchers in systematic biology and in the history and philosophy of biology. The book series will examine the role of descriptive taxonomy, its fusion with cyber-infrastructure, its future within biodiversity studies, and its importance as an empirical science. The philosophical consequences of classification, as well as its history, will be among the themes explored by this series, including systematic methods, empirical studies of taxonomic groups, the history of homology, and its significance in molecular systematics.

The Evolution of Phylogenetic Systematics

EDITED BY
Andrew Hamilton

UNIVERSITY OF CALIFORNIA PRESS
Berkeley • Los Angeles • London

University of California Press, one of the most distinguished university presses in the United States, enriches lives around the world by advancing scholarship in the humanities, social sciences, and natural sciences. Its activities are supported by the UC Press Foundation and by philanthropic contributions from individuals and institutions. For more information, visit www.ucpress.edu.

Species and Systematics, Vol. 5
For online version, see www.ucpress.edu.

University of California Press
Berkeley and Los Angeles, California

University of California Press, Ltd.
London, England

Library of Congress Cataloging-in-Publication Data

The evolution of phylogenetic systematics / edited by Andrew Hamilton.
pages cm — (Species and systematics ; v. 5)
Includes bibliographical references and index.
ISBN 978-0-520-27658-1 (cloth : alk. paper) —
ISBN 978-0-520-95675-9 (ebook)
1. Cladistic analysis. 2. Biology—Classification—Philosophy. I. Hamilton, Andrew, 1972-
QH83.E96 2014
578.01′2—dc23 2013021973

Manufactured in the United States of America

23 22 21 20 19 18 17 16 15 14
10 9 8 7 6 5 4 3 2 1

The paper used in this publication meets the minimum requirements of ANSI/NISO Z39.48–1992 (R 2002) (*Permanence of Paper*).

Cover illustration: Chart of Evolution (1937) by Borgny Bay, Paleontological Museum, University of Oslo. Photo by Colin Purrington, courtesy of the Gould Library, Carleton College.

Contents

Contributors

Malte C. Ebach
University of New South Wales
mcebach@unsw.edu.au

Andrew Hamilton
University of Houston
ahamilton@uh.edu

Robert E. Kohler
University of Pennsylvania
rkohler@sas.upenn.edu

Manfred D. Laubichler
Arizona State University
manfred.laubichler@asu.edu

Norman MacLeod
Natural History Museum, London
University College London
n.macleod@nhm.ac.uk

Brent D. Mishler
University of California, Berkeley
bmishler@berkeley.edu

Gareth Nelson
University of Melbourne
garethn@unimelb.edu.au

Olivier Rieppel
The Field Museum
orieppel@fieldmuseum.org

Michael Schmitt
Ernst-Moritz-Arndt-Universitaet
michael.schmitt@uni-greifswald.de

Beckett Sterner
University of Chicago
bsterner@uchicago.edu

Quentin Wheeler
Arizona State University
quentin.wheeler@asu.edu

David Williams
Natural History Museum, London
d.m.williams@nhm.ac.uk

Introduction

ANDREW HAMILTON

TRENDS AND TRADITIONS IN PHYLOGENETIC SYSTEMATICS

In spring 2006 I met with the entomologist and systematic theorist Quentin Wheeler to discuss topics of common interest: species concepts, methodology in systematics, the recent history of phylogenetic classification, and biodiversity from a systematics perspective. The conversation turned out to have a surprising twist. Quentin was founding a new research center at Arizona State University—the International Institute for Species Exploration (IISE)—that would have a focus on the history and philosophy of systematics. I jumped at the chance to direct this part of the Institute's program because I saw that there was plenty of work to be done that would best be accomplished through the collaborative efforts of systematists, historians, and philosophers.

With help from Manfred Laubichler, I set about recruiting scholars from several disciplines to participate in a workshop to help define and articulate the agenda for the history and philosophy of systematics at the IISE as well as in my own work. Topics of discussion in the workshop ranged from instrumentalism in contemporary systematics to the relationships between homology and monophyly, Willi Hennig's theoretical commitments, and on to how the history of recent and highly contested biology should be most productively pursued.

No attempt has been made to reproduce the breadth of the workshop discussions here. Rather, I have narrowed the scope considerably by including chapters that address issues of historical or conceptual concern to students and practitioners of phylogenetic systematics. The workshop confirmed that there is a lack of historical work on Hennig and his immediate forerunners and made it clear that for two decades before and after World War II, systematics developed more or less independently in two contexts. Put much too simply, the Anglo-American context in which J.S. Huxley, Ernst Mayr, G.G. Simpson, and other proponents of "evolutionary taxonomy" worked was focused on the modern synthesis and on making systematics an important part of evolutionary biology. As Huxley put it in his 1940 volume, *The New Systematics,* "fundamentally, the problem of systematics, regarded as a general branch of biology, is that of detecting evolution at work" (2). The synthesis and the contributions of its architects have been well studied from both historical and conceptual perspectives.

Phylogenetic systematics—or at least the branch of it that became known in the late 1960s and early 1970s as cladistics—developed in quite a different context from that of evolutionary taxonomy and has been much less studied. A few commentators (Hull 1988; Donoghue and Kadereit 1992; Bowler 1996; Rieppel 2006) have noted that Walter Zimmermann and Willi Hennig, founders of the cladistic approach, both reacted strongly against the resurgence of idealistic morphology in the German-speaking world. The sense in which the new approach they developed came out of this context has not been well explored, though there is an emerging literature. If the new systematics (which included evolutionary taxonomy) was a self-conscious attempt to save taxonomy from being overlooked in favor of an evolutionary theory that included molecular genetics as a preeminent component, to what, exactly, was cladistics a response? What set of conceptual, practical, and empirical problems were Hennig and his collaborators, forebears, and followers trying to solve, and how were they different from those faced by scientists working in the context of the synthesis? Several of the chapters in this volume address these questions directly.

In addition to historical questions, there are closely related conceptual ones: much of what was under discussion in both camps was methodological, epistemological, and ontological. In this last category were concerns about species, the process(es) of speciation, and what kind(s) of groups ought to be regarded as natural and therefore appropriate for study by systematists. In the methodological and epistemological categories were debates about reconstructing phylogenies, homologies and what they mean, the information content of systematic claims, and the best basis for systematics, including the question of whether systematics ought to be phylogenetic at all. It turns out, not surprisingly, that understanding these conceptual debates in detail requires further splitting than that between German-speaking cladists and English-speaking synthesists provides: zoologists and botanists took up these issues in very different ways, even when they were working in the same national context. There were points of shared concern—evolutionary reticulation chief among them—but the differences in the way species were understood and how relationships between taxa were conceived and represented point to a need for a richer understanding of the conceptual structures in use in the decades leading up to the wide-scale embracing of phylogenetic systematics as we now know it.

The meaningful debates, of course, did not end with the publication of Hennig's work in English or with the development of his ideas in his few published interactions with evolutionary taxonomists. In selecting and preparing the chapters for this volume, it became clear that important parts of the more recent history of phylogenetic systematics have not been sufficiently treated. Recent work by systematists and biogeographers who have been exploring their own history have opened new avenues of inquiry (Williams and Knapp 2010). As several authors in this volume point out, many of the systematists who were important in advancing Hennig's ideas in the United States and the United Kingdom first learned of Hennig through the dipterist and biogeographer Lars Brundin. There is a case to be made, however, that though Brundin and Hennig were treated as having the same view by their critics—see, for instance, Darlington's (1970) response to Brundin's (1966) monograph—their views and the views of those that followed diverged in subtle ways that may prove important for understanding and explaining the trajectories cladistics took in the late 1970s and through the 1980s and even today. Several contributions in these pages speak to Brundin's impact and its ramifications for early adopters of the phylogenetic approach to systematics.

Wherever the splitting points between schools of phylogenetic thought are to be located, it is clear that the debates between them are overtly conceptual, even philosophical, in ways that are fairly rare in the sciences. It may be going too far to say, with the microbiologist Susan Perkins (2011, 895), that cladists are "often seen as obsessed with the philosophical underpinnings of science." As the chapters in this volume show, however, it certainly is the case that Hennig paid serious attention to the philosophical literature he took to be relevant to his work and that contemporary disagreements about ranking, species, and the relative importance of pattern and process are rooted in sophisticated, though sometimes conflicting, understandings of the appropriate nature of science. One reason for putting this volume together is that in this case, historical and conceptual foundations are tied together so closely that it is difficult to understand either without the other. And it is difficult to understand recent developments in systematics and how to evaluate them without both.

This volume would have been incomplete without a look at systematics through the lens of contemporary issues in technology, biodiversity, institutions, and practice. In systematics there is much to be said about the changing relationships between field biology, the laboratory, collections, and the use of computers and how all these articulate with concepts about what is collected, curated, and displayed and for what purpose. Here these broader issues are addressed in several ways, including one chapter that contains a nuanced study of the history of systematics in the United States and three chapters that explore the ways computing technology has contributed to, built on, and rendered complex the relationships between systematic theory and practice. As with many other fields of science, technology has made a dramatic impact in systematics. Unlike some other fields, however, technology has changed the way some people think about the basic structure of both the day-to-day work and the theoretical underpinnings of the discipline.

THE STRUCTURE OF THIS VOLUME

This volume is not so much a collection of the papers given at the 2008 IISE meeting as it is an attempt to bring the conceptual, historical, and social foundations of modern phylogenetic systematics into sharper relief and to offer a springboard for further work in the area. As Robert Kohler points out in his contribution, systematics may be the biological science that is least well studied by historians, yet it is of overarching interest because of its connections to biodiversity, collections, the rise of natural history museums, and field studies and the important part it has played in the development of contemporary biology, especially the biodiversity sciences. In addition, there is this: systematics is both the site of a recent genuine revolution in fundamental concepts and an arena of ongoing debate and discussion. What is as stake in the near future is the way we should understand classification, especially ranking, and how the objects of systematics (species and clades) inform our view of the biological world, its study, and its management.

Contributors to this volume come from several disciplines but have not separated historical investigations from conceptual, social, or biological ones. Neither have they taken pains to discuss only the past, present, or future. Most chapters discuss more than one aspect of systematics from more than one disciplinary or temporal perspective. Several threads run through the volume. One of these is the topic of how systematics became phylogenetic. Here the "how" points not only to the necessary conceptual changes—and therefore to Willi Hennig—but also to the necessary social and practical innovations. Another thread concerns what, exactly, is being studied and whether and why phylogenetics is the best or most appropriate approach to understanding and bounding biological objects. A third thread addresses the set of tools and methods for doing the work: from simple nets for collecting insects to the sophisticated computer-based imaging and bioinformatics infrastructure described in the chapters at the end of the volume. A fourth thread takes a phylogenetic approach to phylogenetic systematics itself, trying to locate the points in history at which systematics split into multiple lineages, along with the reasons for these branchings.

I have organized these and other threads in three parts. The first, "Historical Foundations," both locates modern systematics in its larger context and acknowledges the importance of Willi Hennig for systematics in the second half of the twentieth century and into the twenty-first. It also, first implicitly and then explicitly, addresses the role of changes in technology as they bear on conceptions of theory and practice in systematics. Part 2, "Conceptual Foundations," does two jobs at once. In addition to tracing the changes described in part 1 as they are reflected in conceptual debates and discussions about basic notions in systematics, part 2 includes detailed first-person histories that throw light on how and why some systematists both understood and came to embrace the Hennigian approach. The final part, "Technology, Concepts, and Practice," describes technologically enabled approaches to systematics, along with their justifications and limitations.

Because the history of systematics and its conceptual foundations are intertwined, the divisions in this book are largely artificial; they are intended to structure the reader's approach to the volume rather than to insist on an intellectual division of labor between scientists, historians, and philosophers or between past, present, and future. One major motivation for the volume is that the history of systematics has been understudied and that since systematics informs so much of the rest of biology, it will become increasingly important to understand what systematics is, how its concepts developed and what they mean, and where systematics might be going.

Part 1 begins with Rob Kohler's extended look at the history of natural history in the United States in the nineteenth and twentieth centuries. Kohler reminds those of us who spend most of our time working with concepts or archived volumes that systematics begins and ends with natural history collections. He argues compellingly that a comparative history of systematics will be necessary if we are to understand the bigger picture, and he frames some questions that can structure such an inquiry. Kohler's chapter is a challenge to the community as much it is a contribution to knowledge about collections and their use. He points to much more good work that needs doing: what is collected, by whom, to what end, and for what reward varies quite a lot with place, time, and other factors. Understanding collections in detail puts us far along the path to being able to see systematics and its related disciplines in greater detail.

Michael Schmitt, who has emerged in recent years as Hennig's foremost biographer, acquaints readers with Hennig the man, and also with Hennig the founder of modern phylogenetic systematics. Those who are unfamiliar with Hennig's work will find in chapter 2 a concise introduction to several of the most important features of his thinking, including the way he refined and used Ernst Haeckel's notion of monophyly, as well as his own use of the concept of synapomorphy and how his thinking about parent-offspring relations led him to controversial conclusions (including the so-called dichotomy rule) about the relationship between stem species and the entities to which they give rise. Schmitt also points to some interesting omissions in Hennig's work, including the lack of a method for determining character polarity—a crucially important part of a systematics that is based on knowledge of, or at least on hypotheses about, which character states are primitive and which are evolutionarily derived.

In chapter 3 Manfred Laubichler discusses some important conceptual shifts in ideas about homology and phylogeny in the German-speaking context in the early twentieth century. As Laubichler points out, homology is a central concept for all of comparative biology, and it is at the very core of phylogenetic systematics and attempts to understand phenotypic evolution. Not surprisingly, given the kinds of disagreements that characterized biology over the course of the late nineteeth and early twentieth century, homology has been a highly contested notion, and its history provides a window on these disagreements. Laubichler draws out a distinction first made by Hans Spemann in 1915 between a historical notion of homology and a developmental one, arguing that the developmental framing funded some of the conceptual innovations in phylogenetic systematics made by Walter Zimmermann (an early influence on Hennig) and by Hennig, but also reinforced a long-standing and ongoing confusion about the units of phenotypic evolution and phenotypic transformation. Understanding this confusion, Laubichler argues, makes the contours of some of the debates at the intersection between cladistics and developmental evolution clearer.

Part 2 begins with a transitional chapter, chapter 4, which serves to shift the focus slightly, but only slightly, from history to conceptual foundations by examining Hennig's use of what later became known as the individuality thesis in the context of his work. In this chapter I explore the rise of phylogenetic systematics from Darwin to Hennig, detailing some important concerns that were raised in the first half of the twentieth century about phylogenetic approaches to systematics. These challenges came from German-speaking scientists working in the tradition of idealistic morphology, and also from English and American scientists who were trying to reconcile evolution with classification. Both groups had the problem, as Olaf Breidbach (2003, 182) has pointed out, that "at the start of the 20th century there was no definitive argument to secure phylogenetic systematics." I argue that first Zimmermann and then Hennig addressed critics by way of a sophisticated set of ontological arguments that had consequences for how systematists should understand the basis of their science as well as what sorts of things they were studying. On this analysis, the long-standing questions of how to group organisms into species and species into clades have answers that are well informed by theory, but taxonomic ranking remains an important and ongoing problem.

Chapter 5, by Olivier Rieppel, continues a focus on Hennig's ontology but in a wider context. Rieppel provides important details about how Hennig developed some of the more theoretical aspects of phylogenetic systematics, with a special focus on Hennig's philosophically informed thinking about individuals and how he understood their logical and structural relations to each other. Reippel contrasts this analysis of the ontological foundations for Hennig's work with the ontology that came just a little later, and in so doing provides the volume's first real introduction to the pivotal role played by Lars Brundin in the adoption of Hennigian ideas.

Rieppel argues that Hennig's ideas were not just adopted but also adapted, even as early as Brundin's 1966 introduction of Hennig to English-speaking systematists. According to Rieppel, a difference in philosophical backgrounds between the Anglo-American receivers of Hennig's work and its German foundations contributed to a split between theorists regarding how to understand some foundational ideas in phylogenetic systematics in the late 1960s through the 1970s and 1980s, including its relationship to evolution. It is as a part of this conversation that we meet some of the first and historically most important champions of cladistics, among them Colin Patterson and Gareth Nelson.

In a chapter that is partly first-person history and partly an exploration of the written first-person accounts of others, Nelson explores both what happened as he and others discovered Hennig through Brundin and how they understood the conceptual changes that the phylogenetic approach brought with it. Nelson's piece recounts his own discovery of Brundin's monograph in the reading room of the Swedish Museum of Natural History in fall 1966, as well as Patterson's "conversion" (Patterson's term) to Hennig's ideas after reading Brundin.

In addition to offering further documentation of the importance of Brundin to the spread of phylogenetic systematics, Nelson's chapter points to the role paleontologists played in the adoption of some of Hennig's ideas, especially his conception of systematics as the search for sister groups. The extent to which paleontology is a crucial part of the story of the rise of cladistics has not been widely appreciated. Nelson's emphasis on the differences in content between the 1967 Nobel Symposium on Lower Vertebrate Phylogeny and the 1972 Linnean Society of London meeting in honor of the paleontologists Erik Stensiö and Erik Jarvik provides strong evidence of a major shift toward phylogenetics among those who worked on fossil fishes.

Chapter 7, by David M. Williams and Malte C. Ebach, addresses social and scientific aspects of the branching of cladistics through the lens of Colin Patterson's thinking about the relationship between systematics and evolution. In November 1981 Patterson gave an intentionally provocative talk on this topic at the American Museum of Natural History. The incident became famous or notorious, depending on one's perspective, in part because it was surreptitiously recorded and then circulated among creationists, many of whom took it as evidence that even senior biologists doubt the truth, viability, or usefulness of evolutionary theory.

The relationship between systematics and evolution, as several chapters in this volume show, is complex. Through a look at Patterson's work and thought, Williams and Ebach illuminate not only the 1981 talk but also something of the character of biologists' attempts in the early 1980s to understand cladistics in relation to systematic biology generally. These efforts were complicated by the rise of molecular and algorithmic methods for doing phylogenetics in the mid-1960s and unfriendly arguments among cladists about which sets of concepts and methods should be adopted.

Williams and Ebach argue that reading Patterson as a creationist is an understandable tactic for creationists but little more than a bludgeon for his fellow systematists. As they understand Patterson, he thought cladistics had branched into two schools by the early 1980s because of differences in the understanding of key concepts—especially phylogenetic relationships and homology—as well as methods. Patterson was an early critic of molecular approaches to systematics partly because he thought that using algorithms to analyze sequence data does not accomplish the same task as using morphology to discover phylogenetic relationships. He held that morphological and molecular homologies are not the same thing: molecular homologies, he thought, are missing the direct evidence of character transformation that one can find in morphology. As Williams and Ebach show, this view was informed by a particular understanding of homologies and their relation to characters and to ontogeny, rather than by allegiance to creationism.

Williams and Ebach go a step further. Having offered a detailed analysis of Patterson's thinking as a means of making his discussion of creationism clearer, they go on to examine certain rhetorical practices among systematists from the 1980s up to today, especially as they relate to dismissals of Patterson's concerns about molecular systematics on the grounds that he was a creationist. In so doing, they offer insights into the interplay between what they call the "scientific" and "political" content of the debates over concepts and methods that are ongoing in systematics.

In the final chapter of part 2, Brent Mishler offers what is partly first-person history. He takes a historical view as a means of offering suggestions for continued integration of theory and practice in botany and zoology. In so doing, he begins to address an embarrassing lacuna in the literature, pointing out that botany and zoology have importantly different histories, approaches, and conceptual structures for phylogenetic systematics—or that at least they did until fairly recently. As he notes, botany and zoology came to phylogenetics from quite different directions, with most botanists standing outside of the modern evolutionary synthesis partly because of their dislike of the biological species concept and partly because their ways of working were (and are) different from systematists who work on animals.

Mishler argues that the practice of assembling and publishing floras has much to do with these differences, and in particular, that the way botanists have represented relationships among groups in their floras has worked against a phylogenetic understanding of those groups. While the value of Mishler's chapter lies in part in calling attention to the differences between botany and zoology to which historians and philosophers have not paid enough attention, there is also value in his first-person account of the development of phylogenetic systematics in botany since the 1970s; he gives future historians something to push against as they try to make sense of the differences he describes.

Part 3 addresses the changes technology has brought and is bringing to systematics. The first chapter, by Beckett Sterner, is transitional both in the sense that it is historical in nature while the other two chapters in part 3 are forward-looking and in the sense that it shows the importance of computers in the way systematics was conceptualized in the twentieth century.

Drawing out a point of Kohler's—that systematics attempts to get at nature's inner workings and was never about mere cataloging—Sterner explores the ways the use of computers changed some systematists' approach to nature in the mid-twentieth century. Focusing on the extreme and therefore revealing case of numerical taxonomy (introduced in Nelson's chapter), Sterner traces the attempts of Robert Sokal and Peter Sneath to remake systematics according to a method of measurement and analysis that could only be accomplished effectively by machines. Sterner treats this episode as a case study of the reorganization of classificatory work and in so doing offers the kind of information about the relationships between methodologies, concepts, and ways of working that will be necessary if we are to move toward Kohler's bigger picture. As Sterner points out, numerical taxonomy was not an old approach using new tools but a conscious and concerted reinterpretation of the world and how it might best be described and categorized.

This case is important not only for understanding the relationship between new technologies and the work of science but also because many of the numerical taxonomists' innovations have been adapted to new purposes in contemporary systematics, especially molecular systematics. Sterner's chapter also shows just how complex the recent history of systematics is. The main thrust of chapter 4 is that phylogenetic systematics in its modern form was born from conceptual shifts that were driven partly by a rejection of idealistic morphology and partly by shifts in ontology. Numerical taxonomy had none of these concerns. Indeed, its framers wanted it to be theory-free, a stance hardly imaginable without the use of computers for data analysis.

Chapter 10, by Norman MacLeod, marks a turn in this volume toward attention to new tools for systematists. MacLeod's work on computer imaging systems is motivated in part by two studies showing that humans—including taxon experts—are not as good at consistently and correctly identifying biological species as one might have thought: human taxonomists do well to offer identifications that are consistent with other human taxonomists 70 percent of the time. MacLeod and his coworkers are exploring several image- and computer-based methods for identifying species (and other biological objects), all of which performed better than human taxonomists in the set of tests undertaken so far on several species of foraminifers.

This kind of automation goes right to the heart of issues about what twenty-first-century systematics will look like in terms of its social and intellectual organization. And it also has the capacity to reveal—through an analysis of the modeling and computer-learning assumptions built into image-analysis software—the conceptual assumptions systematists make about the edges and boundaries of species as read from their morphology.

The final chapter, chapter 11, continues the discussion of computer-based imaging as a tool for systematic analysis, with a focus on why an emphasis on comparative morphology—digitally captured—makes sense in an age of molecular evolution and bar code–based analyses of biodiversity. Quentin Wheeler and I provide a detailed look at the epistemology of phylogenetic analysis, arguing that the morphology-based description of species is the crucial first step in a rigorous process of understanding new species, their geographies, functions, and places on the phylogenetic tree. Analyses of gene and protein sequences can be helpful as we try to understand new findings and reconceptualize old ones, but Wheeler and Hamilton echo some of the discussion in Williams and Ebach's chapter by arguing that molecular approaches are best understood as morphology writ small. As we move toward a systematics based more extensively in bioinformatics, they argue, we need to find new ways to capture morphology digitally and make it widely available for just the kind of analyses that are presently carried out. This will accelerate the pace of what Wheeler has long called "species exploration" while being faithful to the epistemic standards of phylogenetic systematics. This shift toward cyber-enabled taxonomy, they note, has important implications for how systematists do their work, as well as for how others might consume it.

NEXT STEPS

Much has been said recently in the popular (McClain 2011) and professional (Pearson, Hamilton, and Erwin 2011) literatures about systematics and the challenges it now faces. Many are concerned that systematics is underfunded and undervalued, even as we find ourselves in the midst of an anthropogenic mass extinction. This volume does not address issues of the relationship between taxonomy and biodiversity directly, but it does provide historical and conceptual background that will be necessary as scholars move to understand the place of contemporary systematics among the other biodiversity sciences and their place in science and conservation policy.

It is often claimed, for instance, that understanding species conceptually is a sine qua non for comparative biodiversity studies. Several of the chapters in this volume argue implicitly against this view or at least call for a stronger defense of it than is usually offered. Do we need *a* species concept to establish biodiversity baselines if botanists and zoologists have very different ideas of what species are, and if it is true that tracking lineages rather than species is the best and most useful approach to understanding the evolutionary path of life on Earth? If our categories and units of analysis are importantly informed by professional practices, as they surely are, will it not be profitable to understand in detail what these practices are, how they came to be, and what they reveal about the underlying conceptual and methodological structures we use to understand the biological world?

This volume is a step—joining the efforts of others (Hull 1988; Williams and Ebach 2008; Williams and Knapp 2010)—toward providing the necessary background understanding of what has happened in the relatively recent history of systematics, as well as what will happen and how it is justified logistically, empirically, and conceptually as the discipline moves toward yet another revolution. This revolution—the embracing of bioinformatics as a basic way of working—will likely be one of practice rather than a shift in the basic understanding of what's in the biological world and how to understand it, but as the chapters in this volume show, a great deal of basic work still remains to be done on phenotypic evolution, ranking, the units of homology, the relationships between phylogenetic inference and phylogenetic classification, and how to create and adopt new technologies that move us toward clear goals for science and society.

Acknowledgments

The International Institute for Exploration provided funds and logistical support for meetings that turned into chapters. This volume would not have been possible without a grant from the National Science Foundation (SES-09083935) or without the input of colleagues at Arizona State University, as well as at the Encyclopedia of Life and the Marine Biological Laboratory. Research assistant Erick Peirson went far above the call of duty in performing editorial and logistical tasks, particularly with respect to the figures in the volume. Series editor Malte Ebach was instrumental in pushing for changes in the way this volume is organized and structured. His eye for detail improved this work.

REFERENCES

Breidbach, O. 2003. "Post-Haeckelian Comparative Biology—Adolf Naef's Idealistic Morphology." *Theory in Biosciences,* 122: 174–193.

Brundin, L. 1966. "Transantarctic Relationships and Their Significance, as Evidenced by Chironomid Midges, with a Monograph of the Subfamilies Podonominae and Aphroteniinae and the Austral Heptagyiae." *Kungliga Svenska Vetenskapsakademiens Handlingar,* Fjarde Serien 11, 1: 1–472.

Bowler, P. 1996. *Life's Splendid Drama.* Chicago: University of Chicago Press

Darlington, P.J. 1970. "A Practical Criticism of Hennig-Brundin 'Phylogenetic Systematics' and Antarctic Biogeography." *Systematic Zoology,* 19: 1–18.

Donoghue, M.J., and Kadereit, J.W. 1992. "Walter Zimmerman and the Growth of Phylogenetic Theory." *Systematic Biology,* 41: 74–85.

Hull, D.L. 1988. *Science as a Process: An Evolutionary Account of the Social and Conceptual Development of Science.* Chicago: University of Chicago Press.

Huxley, J.S. 1940. "Towards the New Systematics." In J. Huxley (ed.), *The New Systematics.* Oxford: Clarendon Press, 1–40.

McClain, C. 2011. "The Mass Extinction of Scientists Who Study Species." *Wired Science.* www.wired.com/wiredscience/2011/01/extinction-of-taxonomists.

Pearson, D.L., Hamilton, A.L., and Erwin, T.L. 2011. "Recovery Plan for the Endangered Taxonomy Profession." *BioScience,* 61: 58–63.

Perkins, S.L. 2011. "Beyond Cladistics: The Branching of a Paradigm" (review of Williams and Knapp 2010). *Systematic Biology,* 60: 895–897.
Rieppel, O. 2006. "Willi Hennig on Transformation Series: Metaphysics and Epistemology." *Taxon,* 55: 377–385.
Williams, D.M., and Ebach, M.C. 2008. *Foundations of Systematics and Biodiversity.* New York: Springer.
Williams, D.M., and Knapp, S. 2010. *Beyond Cladistics: The Branching of a Paradigm.* Berkeley: University of California Press.

PART ONE

Historical Foundations

1

Reflections on the History of Systematics

ROBERT E. KOHLER

Of all the life sciences, systematics is probably the one whose history is least studied. Its celebrity founders have been well historified: Linnaeus, whose universal system of binomial nomenclature still endures; Darwin, who gave classification a biological foundation; and a few others. But of the activities of the hundreds of collectors, curators, and classifiers who have found, preserved, named, and ordered the million-plus species whose world we share—of these our knowledge remains scattered and fragmentary. This is paradoxical, because of all the sciences systematics has the deepest living memory, thanks to rules of nomenclature that oblige those who would name a new species to actively engage the literature back to the Linnaean big bang.

This situation is, happily, changing; substantial histories have been quietly accumulating, some by historians with a sustained devotion to the subject. These include Jim Endersby (2005, 2008), Paul Farber (1976, 1985), Jürgen Haffer (1992), Joel Hagen (1984, 1999), David Hull (1998), Gordon McOuat (1996, 2003), Ernst Mayr (1982, chaps. 4–6), Bruce Patterson (2000, 2001), Harriet Ritvo (1997), Peter Stevens (1986, 1994), Keith Vernon (1993), and Mary P. Winsor (1976).

There are also circumstances external to biological systematics that may stimulate greater interest in taxonomy. One is the decided uptick of interest among historians of non–life sciences in classifying, both as practice and as a way of knowing. This is especially marked in the history of chemistry (Ursula Klein, Michael Gordin) and mineralogy

(Matthew Eddy) and also in the history of ecology and generally in sciences that deal with collections of material objects, like archaeology or ethnology (Klein 2003, 2005; Klein and Lefèvre 2007; Gordin 2002, 2004; Eddy 2003; Bensaude-Vincent, García-Belmar, and Bertomeu-Sánchez 2003; Müller-Wille 2003; and on ecology and thing-rich sciences, see Kohler 2007, 2008). It may be, too, that growing public concern about loss of biodiversity and anthropogenic mass extinction could make systematics and its history matters of broad interest. Attitudes are changing, both in the world of scholarship and in political culture, that have for over a century relegated systematics to the low end of the totem pole of prestige in science.

Meanwhile, we can imagine what we would ideally like to know about the subject. Basic facts, for a start: who first gave that great multitude of creatures their scientific names and identities; and when, where, and how. There is the human encounter with nature in the field—my own particular interest: how fieldwork has been organized, who paid for it, how it was experienced; and the encounter in museums and herbaria with the ever-rising flood of specimens and species to be put in order. How have systematists dealt with tens of millions of fragile objects and a million-plus natural kinds? It is a problem of data handling that few other sciences have had to confront.

Systematists themselves are no less interesting subjects of historical inquiry. What sort of people are these energetic finders and sorters, and what are their distinctive folkways—their rules of categorizing and naming; their customs of rewarding discovery and pruning errors and redundancies? How do they maintain (as all communities must) a sense of common identity, and by what social arrangements do they renew that community across generations? Recruitment is a particular problem for systematics, because the science is based in museums rather than in university departments with their ready access to young talent; so training is more like an informal personal apprenticeship than in most sciences.

This comprehensive history of systematics is the larger undertaking to which my own recent book on American systematists and their global activities between 1870 and 1940 is a modest contribution (Kohler 2006).[1] I want to use this chapter, first, to summarize some of the main ideas of my book, which deals mainly with fieldwork; then, to think more broadly how my approach might apply as well to systematics in other times and places.

It turns out that the history of discovery of new species—at least in vertebrate groups—is surprisingly lumpy and episodic. I did not know that at first. Only when I had nearly finished my book did it occur to me that it would not be hard to chart the historical pace of discovery. The data were ready to hand—thanks to systematists' custom, when compiling species lists, of giving the date that each species received its currently accepted name.

I'm not sure what I expected to find when I began to tabulate these data: perhaps a more or less steady pace of discovery, with spurts and pauses reflecting the randomness of individual initiative and serendipity. So I was surprised—riveted, in fact—when the data on vertebrate groups began to show distinct waves. Following the initial Linnaean burst of naming before 1800 there was a strong second wave of activity peaking in the 1830s and 1840s; then a third wave (for most vertebrate groups though not for birds) that topped out in the 1890s and 1900s (my period); and a fourth, much smaller peak (more a bump) of discovery in the 1960s and 1970s (Kohler 2006, 6–7, fig. 1.1).[2] (See fig. 1.1.)

I was not surprised that a heightened pace of discovery occurred in my period of interest. I had found much anecdotal evidence that it did, and knew just how zealously American museums in this period were collecting around the world. What I had not anticipated was that the entire history of species discovery, before and after my period, would be episodic, with decades of intensified activity separated by relative lulls. It was obvious from these data that individual opportunities and initiative were not idiosyncratic and random but structured—by some thing or things, and on a global scale. But by what things? The same ones that I had found operating in the United States circa 1880–1930? Or by others distinctive of their own times and places?

A grand Noachian narrative of species inventory is there to be written—that became clear—but too big for me, a novice in the history of taxonomy, and perhaps too big for any scholar working alone. But seeing a pattern so concretely silhouetted in the data, I had to wonder how the interpretation I had devised for my chapter of the larger narrative might inform the whole.

SURVEY SCIENCE

We aren't used to thinking of the late nineteenth and early twentieth century as a golden age of systematics. The usual story is that after the grand voyages and explorations of the late eighteen and early nineteenth century, natural history was gradually eclipsed by the newly ascendant sciences of the lab. This is true—but only half the truth. Lab culture did

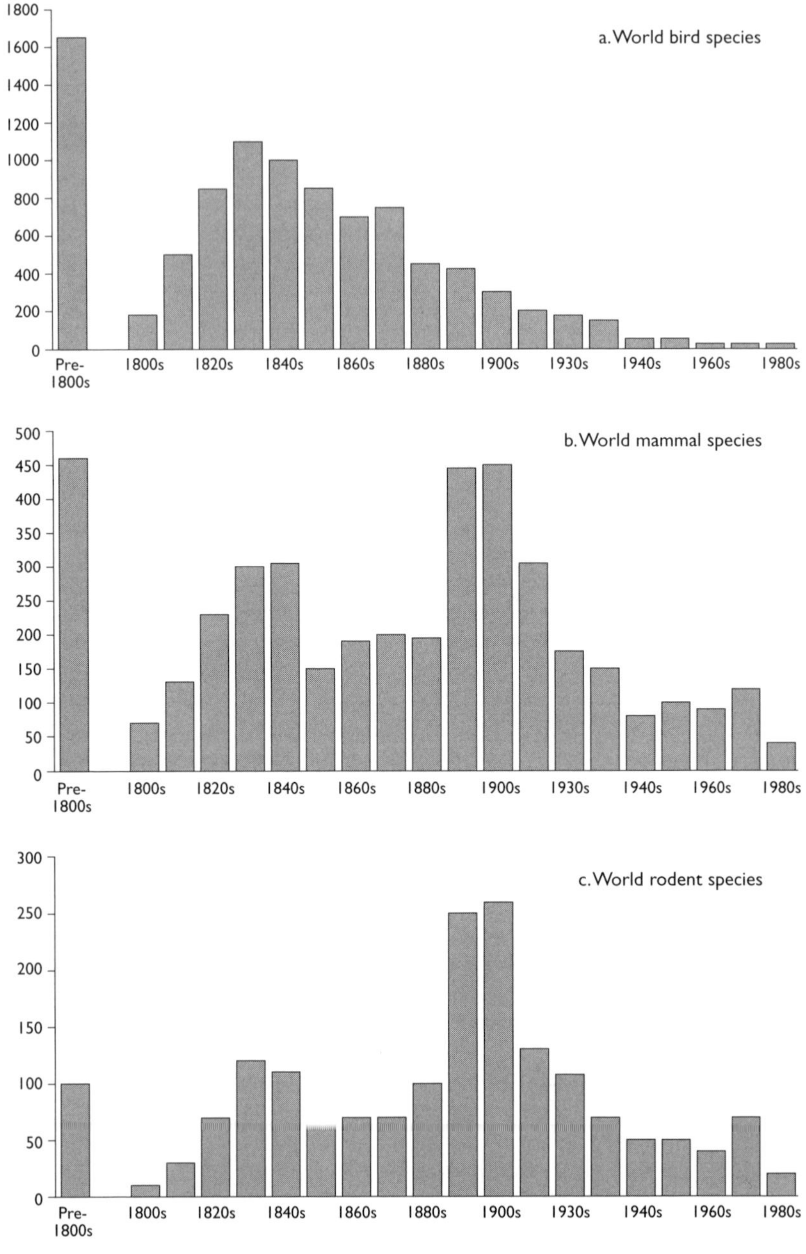

FIGURE 1.1. First descriptions of bird (a), mammal (b), and rodent (c) species, by decade, showing peaks of collecting and discovery. Sources: Sibley and Monroe 1982; Honacki, Kinman, and Koeppl 1982. Graphics courtesy of Jack Kohler, from Kohler 2006, 6–7, fig. 1–1.

not cause systematics to wither away; it continues to flourish. As is often the case, "decline" was only relative and more a matter of perception and attitude than achievement. It was in the early twentieth century that inventories of most vertebrate groups became essentially (say, 90 percent) complete. For birds this point was reached around 1900; for most mammals, circa 1940; for bats and insectivores, by 1950; and for North American snakes and turtles, the 1920s.

The roster of collecting expeditions by American institutions alone is impressive. The U.S. Biological Survey under its founder and chief C. Hart Merriam fielded dozens of parties per year from the late 1880s to the early 1910s, mostly in western North America. A dozen state natural history surveys were also active, most of these small shoestring operations but a few systematic and sustained (e.g., in California, Michigan, and Nebraska). Research museums organized expeditions both in their own regions and abroad: the Museum of Comparative Zoology (Harvard), the Museum of Vertebrate Zoology (University of California), and the University of Michigan Museum. However, it was the larger civic museums that were organizers of expeditions on a truly grand and global scale: most notably the American Museum in New York, the Field Museum in Chicago, and Philadelphia's Academy of Natural Sciences but also midsized museums in cities like Brooklyn, Milwaukee, and San Francisco. The National Museum in Washington was barred from underwriting in-house expeditions, but its curators routinely took part in expeditions organized by federal agencies that had that right.

The scale of museum expeditioning in its heyday is remarkable—all the more so because it is now so little remembered. Between 1887 and 1940 the American Museum dispatched some 206 sponsored expeditions in vertebrate zoology, plus over 200 more in other field sciences (archaeology, paleontology, anthropology). The yearly average in zoology was 3.8 in 1891–1901, 8.5 in 1906–25, and 13.9 in 1926–40. In the peak years 1929 and 1930, 30 and 27 expeditions were dispatched at a cost of $283,000 and $207,000 respectively. In vertebrate zoology, the Field Museum organized 72 expeditions (1894–28), and the Academy of Natural Sciences sent out 57 (1889–1930). In addition to these official, named, and sponsored expeditions there were uncounted unofficial research trips by curators. Overall, thousands of expeditions must have been launched at a cost of millions or tens of millions of dollars (Kohler 2006, 117–123). If there was a big science before the era of Big Science, systematics was it.

Taxonomic field practice in this period was also distinctive: it was a *survey* mode, quite unlike the practices of earlier and later eras. Aiming at a complete inventory of the species of entire regions, survey collecting was both extensive and intensive: extensive in its geographic reach; intensive in the way parties combed every nook and cranny of each locale, crisscrossing and revisiting until every resident species was found and recorded.

Survey collecting was also collecting en masse. It sought, not a single "typical" pair of specimens per species, but large series that represented the full range of intraspecies variability. Expeditions were planned, organized, and (for their time) capital intensive. Whereas collections of tens of thousands of specimens were once thought large, survey collecting produced collections ten or a hundred times as large, and prepared by standardized methods. Survey was rigorous and exacting; and it made systematics if not an exact then an exacting science.

Systematics in the survey mode required an elaborate infrastructure of museums, with their dedicated study collections and elaborate exhibits, and of social networks of well-to-do patrons and local participants. Working in a survey mode also put new demands on its practitioners. To succeed in the business, systematists had of course to be schooled in basic sciences, adept in the arcana of diagnosis and naming, and at home in the vast and sprawling taxonomic literature. But since they were now both field and museum naturalists, they had also to have practical knowledge of how to organize and lead complex field parties and manage sponsors and the media. And there was the curatorial side of tending museum collections, which required skill in managing, fundraising, and museum politics, plus knowledge of the specialized techniques of constructing naturalistic dioramas en masse.

What circumstances in the survey period enabled systematists to create such an elaborate and exacting science and to produce a great wave of species discovery? In my book I set forth a model of three interlocking elements: an environmental element, which gave systematists physical access to relatively unaltered natural areas; a scientific element, which gave them the intellectual incentives to undertake collecting in the demanding survey mode; and a cultural element—wide public interest in nature and natural history—which drew big private money into building museums and collections and into expeditioning on the grand scale.

First the environment. The era of survey collecting coincided almost exactly with a distinctive and fleeting phase in the settlement history of North America that I have called the "inner frontiers." The linear, westward-moving frontier was famously declared closed in the 1890 Census and by Frederic Jackson Turner in his celebrated essay of 1893. There were by then so many pockets of settlement scattered about the West that no distinct line separated settled from unsettled land. However, pockets remained in the interstices of road and railroad nets that were sparsely settled and ecologically lightly altered, yet also relatively easily and cheaply accessible to hunters, tourists, vacationers, and—most significantly for us—naturalists and collectors.

This landscape of inner frontiers was created by Americans' restless mobility and devotion to land speculation, as well as by what William Cronon has called the logic of capital.[3] To preempt rivals, railroad companies built lines into areas in which there were not yet enough residents to produce an actual profit. State and federal giveaways of cheap homestead land likewise encouraged rapid hop-skipping settlement, and speculators would sit tight on large parcels of land and wait for rising prices. The political economy of manifest destiny thus created a landscape of intermingling wild and settled terrain. In the East, abandonment of marginal agricultural land created similar mosaics of fields and second-growth forest.

It was not that naturalists had never before been able to visit the wide West: of course they had. The point is that they had not previously been able to operate there in a survey mode. Transfrontier areas had been accessible only to individual explorers, who lived off the land, or to state-sponsored expeditions on the grand scale. In either case, visitors were limited to rapid transects and opportunistic collecting. Not so in the inner frontiers: there, cheap and relatively comfortable transport enabled naturalists to linger in safety and collect in depth. In inner frontiers expeditions could stay in continual touch by telegraph and rail with home bases, ship out tons of specimens cheaply and securely, and be resupplied en route. They could make repeat visits if necessary, to make species inventories as complete as possible.

The inner frontiers were in hindsight a once-only bonanza of species discovery waiting to happen. Here were large tracts made roughly known by earlier explorer-naturalists who had skimmed the cream of species that were easy to find and catch but who had not lingered—because they could not—to get those that were rare, shy, or elusive. Inner frontiers invited survey. Arguably, survey was possible only in such places. It was a (scientific) form of land use as characteristic of the place and time as bonanza hunting or land speculating. No wonder environment and scientific practice overlapped almost exactly in time.

But physical access to places of opportunity does not explain why scientists would see and take advantage of the opportunities. Naturalists had also to have intellectual incentives to visit inner frontiers, not just casually on working vacations, but systematically, as members of organized surveys. Investment in travel and organized fieldwork had to pay off scientifically—and to be seen in advance to pay off—so that surveys could be planned and patrons secured. Opportunity derived not just from nature but also from changes in taxonomic science.

In fact, systematists could be confident that survey collecting would be worth the trouble, because they already knew that more complete collections made for better and less error-prone taxonomies. This was apparent by the mid-nineteenth century, when the first moves were made toward a more intensive field practice. Naturalists first began to collect systematically and in depth in the 1850s: among them (in the United States) Spencer Baird, who made use of national railroad and boundary surveys in the American West, and Louis Agassiz and Joel Asaph Allen at the Museum of Comparative Zoology. These pioneers in modern collecting were ardent empiricists who took as a matter of faith that denser empirical evidence would make systematics a more exact and secure science. And they were right.

The chief cause of taxonomic error was making designations on the basis of too few specimens or of specimens from a too restricted locale. Without full knowledge of species ranges, for example, it was easy to mistake as new to science species that were in fact already known but simply outside their normal ranges. Likewise, systematists unaware of how variable species can be were prone to designate as new species forms that were only extreme variants of species already named and described. The result was a confusion of false and redundant names.

The conventions of Linnaean taxonomy made such errors not just an individual embarrassment but, in effect, a labor tax on the entire community of classifiers. To keep order in the Ark, systematists had to publicly correct every mistake and to reduce every redundant name to the dustbin category of synonymy. Thus the cost of fixing individuals' mistakes was borne by all. Experimentalists have an easier time of it: their blunders and follies are buried forever in the dark, bottomless bog of uncited and forgotten literature.

But if the first mass collecting made it clear just how error-ridden the taxonomic literature was and why, it also made clear that the problem could be solved by more of the same—by forming deeper and more comprehensive collections. What Baird, Agassiz, Allen, and others had begun as individuals to do became standard practice for all self-respecting and responsible systematists. Darwin's theory of speciation by variation and selection was for some (e.g., Allen) a further incentive to adopt a data-intensive practice. But the chief intellectual incentive for systematists to collect in a survey mode was not theory but improved empirical practices of diagnosis and revision.

The rewards of survey collecting were real and immediate, in terms of quality of work and personal credibility. Those with abundant empirical data got credit not just for the new species they described but also for correcting other taxonomists' mistakes; and they were less at risk of being themselves publicly corrected and discredited. Survey collecting could make less organized modes of gathering seem not just inferior, but irresponsible. If inner frontiers provided the opportunity for systematists to collect more widely and intensively, it was the intellectual and career rewards of improved practice that provided the impetus to exploit those opportunities in an organized way.

Yet the argument is still incomplete. However powerfully drawn by nature's opportunities and the scientific rewards of a data-intensive mode of work, systematists still had to have money and organizational backing to travel and collect en masse—which few individuals possessed. So most were dependent on financing from national or civic museums or well-to-do private patrons. And that was, at first, a problem.

For one thing, museums at the time did not employ curators to collect; curators were expected to stay put and curate. Like amateur naturalists, curators collected during their vacations or in their spare time. The modern role of curator-expeditioner had to be invented for survey to become a general museum practice. Nor did museums at the time engage in active, planned collecting or sponsor in-house expeditions. Collections were assembled from random gifts, purchase, or exchange of duplicates. The rewards of collecting en masse were hardly as immediate or compelling for museum directors and their bourgeois patrons as they were for scientists. Understandably: in their view, museums existed to inform and entertain the public. What benefit was there for museums or the museumgoing public in paying curators to go on junkets to exotic places and fill museums with vast numbers of nearly identical specimens?

Yet in the 1890s and early 1900s museum officials and patrons began to do just that, and the reason they did is the third, cultural, element of our model: popular interest in nature and the culture of outdoor recreation.

It is well known that cultural conceptions of nature were changing dramatically in the late nineteenth century.[4] The prevailing utilitarian view of nature as a warehouse of "resources" to be turned as quickly as possible into money was tempered by alarm over the damage wreaked by all-out, use-it-or-lose-it exploitation, as well as by growing awareness of the recreational values of undisturbed nature. In the United States, preservationists like John Muir and Theodore Roosevelt and institutions like the Sierra Club and the Audubon Society sought to restrain the all-out "war" on nature and to have choice parts of the inner frontiers preserved forever for recreational use.

In literature and the arts, meanwhile, a sentimental and anthropomorphizing view of nature, exemplified by romanticized animal sculpture and nature fables in which animals were endowed with human virtues and vices, gave way to a more naturalistic view of animal life. The nature essay, invented by John Burroughs in the 1870s and widely popularized in the 1890s, was art designed to be scientifically accurate as well as emotionally pleasing. This same ideal of uniting art and science also inspired the invention of the habitat diorama, which occurred in American (and Scandinavian) museums around the turn of the century (Wonders 1993). Diorama builders went to extraordinary lengths to make these objects not just visually beautiful but also true to nature and to the realities of animal behavior and ecology. Conceptions of nature that had been purely cultural—economic, aesthetic—thus became naturalistic: not science exactly, but congruent with a scientific view of nature.

In addition, new and active forms of outdoor recreation evolved in the late nineteenth century that afforded not just new ways of representing nature but also new ways of personally experiencing it that combined pleasure with a naturalistic or scientific interest. Among the more important of these outdoor recreations were sport hunting and fishing, camping and outdoor vacationing, rural perambulating and mountain climbing, summer cottaging in lake and mountain districts, and buying "abandoned farms" as summer vacation homes. These novel cultural practices, I argue, were the soil in which the practices of survey expeditions flourished.

I call these activities "new," but of course they were not quite that. Hunting and fishing, for example, had long been pursued as sport by rural gentry, and by farmers and working folk to put food inexpensively on their tables. Well-to-do families had since the eighteenth century toured rural areas in search of sublime or picturesque sensations. What was new in the late nineteenth century was the meaning these activities acquired when they were taken up by the striving white-collar middle classes. It was these newcomers who made subsistence or leisure pursuits more naturalistic and sciencelike.

Attracted by the pleasures of outdoor recreation but uneasy with the association of leisure with aristocratic idleness and proletarian moral disorder, middle-class nature-goers altered the meaning of these pleasures to square with their values of discipline and self-improvement. This they did by pursuing outdoor vacationing as a kind of work, which they took to be improving and essential to the work of making money and getting ahead. They worked at play, as the historian Cindy Aron (1999) put it, turning leisure into physical and mental recreation (see also Bailey 1978).

The more strenuous forms of outdoor recreation were especially valued by promoters of middle-class vacationing for their physical and moral benefits, as were also instructive pursuits like birding or amateur naturalizing. These afforded an experience of nature-going and learning like those that John Burroughs evoked in prose, carried out in environments like those depicted in museum dioramas. A quasi-scientific interest in nature transformed outdoor recreations into suitably moral and improving activities for middle-class family vacationing.

It is no accident that enthusiasm for these culturally charged modes of nature-going coincided with the period of inner frontiers and with the survey mode in systematics. Inner frontiers, wild yet accessible, were ideal places for the strenuous but not too arduous or risky (or expensive) forms of family vacationing—working at play. This moralized form of outdoor recreation, like natural history survey, was a form of land use specific to inner frontiers.

Nor was it happenstance that middle-class nature-going also sustained survey science. For one thing, it enlarged the pool of sympathizers with—and potential recruits to—survey science. The practices of recreational hunting, camping, and naturalizing were identical in many respects to those of survey science: for example, knowing where to find, recognize, and observe (or catch) animals; and how to travel cross-country and live outdoors in safety and relative comfort. In effect, recreational nature-going constituted a pleasurable and unwitting apprenticeship to the work of scientific collecting.

Of course only a tiny fraction of recreationists ever became career naturalists. However, recreational nature-going also predisposed its devotees to take part indirectly in survey science, by supporting natural history museums or by underwriting surveys and expeditions. Outdoor culture provided wherewithal and created infrastructure: that was the vital connection. Well-to-do families could contribute to survey science by sponsoring expeditions—and, if they so desired, by going along as hunter-collectors—without having to devote their lives to doing science. Recreational nature-going made survey systematics culturally familiar and understandable to the sector of the public that had the inclination and the means to participate in some way large or small. Those who took an active part in expeditions found them a pleasurable and educational blend of science and vacation: it was perhaps the ultimate way to work at play.

Evidence of the connection between expeditioning and nature-going is everywhere. C. Hart Merriam subsidized his own large-scale collecting as an amateur naturalist before he learned to use the politics of government patronage to scale up a personal obsession into a national faunal survey. Annie Alexander, who founded and underwrote the Museum of Vertebrate Zoology and its program of field collecting, was a wealthy sugar heiress and an avid big-game hunter. She needed a public repository for her specimen-trophies (because of legal restrictions on private hunting) but quickly found a higher aim in the scientific program of her director, Joseph Grinnell. Alexander Ruthven ran the University of Michigan Museum and its program of expeditions with the modest financial support of a group of local amateur naturalists and collectors.

Civic museums likewise depended on the patronage of outdoorsy families for their global collecting projects. It was not the prospect of aiding science that first attracted patrons, though; rather it was the desire to assist in building habitat dioramas. It turned out that state-of-the-art dioramas could not be built with materials present in museum storerooms. Their spectacular illusionistic effects required fresh specimens of the highest taxidermic quality, as well as accessory material (dirt, rocks, branches, plants) and paintings or photographs gathered or made in the very places that the dioramas were to depict. And these materials had to be gathered by curators and preparators, because only they knew exactly what was required. If few museum patrons could see the point of vast scientific collections, many were eager to pay for diorama expeditions.

Once in the field, of course, curators and systematists also collected en masse for study collections, and it was not long before expeditions were as much or more for science than for exhibit making. In this way museums were drawn into the business of expeditioning, and curators' professional job descriptions were enlarged to include fieldwork. It was dioramas, initially, that connected museum science to the culture of outdoor recreation. Combining pleasure and science, they embodied the ideal of working at play and afforded museumgoers virtual—and sometimes actual—trips to the inner frontiers where systematists labored to record and order nature's abundance.

The culture of nature-going, conjoined with an exact science of systematics in a landscape of inner frontiers, produced the third wave of species discovery. Inner frontiers gave systematists physical access to places of species abundance, and supported a naturalistic culture of outdoor recreation, which afforded systematists the means to pursue their opportunities and to reap the professional rewards of improved practice. A dynamic of change, which started small and might at any point have stalled, developed into a sustained wave of discovery.

THE BIGGER PICTURE

Because this model was devised for the particular case of North American naturalists, the question will arise, Does it apply as well to others? British, German, French, Scandinavian, and Russian naturalists were active namers of new species, and many were also active field collectors. But did European governments, museums, and botanical gardens organize expeditions on a continental or global scale, as Americans did? Were they drawn by their own or distant inner frontiers? And were they, as Americans were, sustained by a culture of middle-class nature-going? (The science, we may assume, was much the same everywhere; Linnaean systematics has always been a markedly transnational science.)

There are as yet no answers to these questions: the research has not been done. We do know that museums in America and Europe had different operating principles. The British Museum, for example, preferred buying specimens on the open market to in-house collecting. And the Paris and Berlin museums built large collections from specimens gathered by state military or navigational expeditions. Networks of colonial administration were also sources of varying importance for the imperial powers. But civic museums in Germany never developed a system of in-house expeditioning. Nor was there a custom in continental countries of regional natural history surveys (Nyhart and Burkhardt pers. com. [on Germany and France, respectively]; McOuat 2001). But these are bare beginnings of a comparative history. We have no systematic data on collecting and expeditioning by Europeans, and no idea if differences in collecting practices gave the science of systematics distinctive national styles.

Likewise with the cultural element. Some European countries did have a culture of outdoor recreation, especially Britain and Scandinavia (Bailey 1978; Wonders 1993; Allen 1994; Frykman and Löfgren 1987, chap. 2). These (Sweden's in particular) resembled the American mode; but there were also differences. For example, sport hunting was generally not a mass activity in Europe, and feeling for nature seems more strongly linked to the politics of consolidating nation-states. The French were more attached to their domesticated agricultural landscapes than to ones more wild, in contrast to Canadians, Swiss, or Swedes (Nyhart 2009; Applegate 1990; Green 1990; Zeller 1987; Wonders 1993; Zimmer 1998). Whether or not customs of outdoor recreation had the same effects on science in Europe as in the United States remains to be discovered.

On the environmental side as well, little systematic work has been done. But the historical geography of Europe is well developed, so it would not be hard to do. In general, European landscapes in the western core were more intensely settled than those of North America and their flora and fauna more depauperate. But there were accessible frontiers in the far north and east (Scandinavia, Siberia, inner Asia) and in isolated pockets elsewhere. And the westernmost imperial powers—Britain, Spain, France, the Netherlands—had inner frontiers in their transoceanic colonies: they were just not contiguous as in the case of the United States, the Russian empire, and Scandinavia. So national differences may prove to be less stark than we might expect. One thing we can say for sure is that comparative history is in order.

Other lines of comparative inquiry lead from the survey period to earlier and later episodes of species discovery. If the science of one episode was specific to its environmental, scientific, and cultural circumstances, then it is reasonable to suppose that it would be the same story with others. We don't know this, because we don't have the factual evidence; but as a working hypothesis it seems a good way to begin. We would look for a distinctive practice of gathering and sorting, sustained by some combination of geographic access, scientific opportunity, and cultural infrastructure.

There is reason to think that developments in taxonomic science were as vital to earlier and later episodes of heightened activity as they were for the survey era. In the case of the first, eighteenth-century, wave of species "discovery" it was Linnaeus's system of naming that created a scientific opportunity: to refashion a world of competing and incommensurable local systems of classifying into a single, simpler, and universal one. This change in practice gave a second life to older literatures of natural history and materia medica, and data mining became a characteristic practice for a generation of Linnaeans. At a time when travel was costly and arduous, libraries were an accessible and rewarding field for discovery. Intellectual rewards lay in reconceiving what was already "known."

Of course species were also sought in nature. Linnaeans did travel, and those with access to state patronage traveled a lot. And an active international network of local collectors sent specimens to the metropoles of Linnaean science to receive scientific identities (but a project to train a cadre of traveling collectors failed, because unseasoned and inexperienced Europeans too often fell victim to accidents and disease) (Koerner 1999, 147–148).[5] Correspondence and exchange networks constituted the characteristic mode of transnational science in Enlightenment Europe, and natural history collecting was in that mode (Lux and Cook 1998; Harris 1998).

Enlightenment science was also distinguished by its intimate connections with economic production and trade. Lisbet Koerner has shown just how involved Linnaeus and his circle were in mercantilist projects of acclimatizing exotic plant species to northern Europe (Koerner 1999). Here we see both the geographic element—in efforts to remake human and natural biogeography—and the cultural (economic) infrastructure of data gathering. What the culture of outdoor recreation was to the survey period, the political economy of expanding global trade was to the Linnaean. So we see the silhouette of a distinctive Linnaean taxonomic science, shaped by circumstances of science, physical environment, and culture.

Likewise with the second, or "Humboldtian," episode of species discovery. Here again we see a change in taxonomic practice opening up intellectual opportunities: in this case, the invention of natural systems of classification that relied not on single characters arbitrarily selected, as in the Linnaean system, but on whole organisms and their habits and ecology (Stevens 1994). Rewards thus lay in revising the provisional assignments of an artificial classifying system into the more permanent ones of a natural system. This recurrence of data mining, in a new form, may partly account for the bulge of "new"—that is, newly robust—species descriptions in the 1830s and 1840s.

But natural systems of classification depend for their proper working on abundant and comprehensive data, in a way that the simpler artificial systems did not.[6] And getting such data—including knowledge of animals in life—entailed fieldwork. So in the decades following the end of the Napoleonic Wars we find travel becoming a more regular feature of taxonomic practice. Exchange networks remained active; but collecting was increasingly carried out by travelers, both recreational and official, from the European hearth to far-flung parts of the world.

Active on-site collecting was made possible in part by the industrial and economic revolution in global transport—that is obvious. Steamships and railroads, and a denser infrastructure of colonial extraction and administration, made distant outbacks accessible even to individual naturalists. Military doctors and officials in far-flung outposts were major contributors to species inventories. The French and Russians were especially open to combining diplomacy with exploring and natural history collecting. In the 1830s the Russian, Prussian, and French (but, oddly, not the English) consulates in Rio de Janeiro were all staffed by naturalists (Swainson 1839, 391–392; Shearer 2009). Growing imperial rivalries among European states made exploration an affair of state and turned explorers and explorer-naturalists into national celebrities.[7] Systematics was thus enmeshed in the modern political economy of European expansion in its more invasive, coal- and steam-driven, colonizing phase. The science was the incidental beneficiary of these larger changes, as it would be later of the new middle-class culture of nature travel.

In its mobility and long-distance spatial projection, exploration resembled natural history survey; but in most respects, it was distinctly different. Exploration was an activity not of inner frontiers, as survey was, but of distant and sometimes dangerous frontiers that were hard to reach and were occupied by peoples who did not welcome European intrusions. Exploration was meant to be adventure, to afford experience of the exotic, to draw out heroic virtues—which survey decidedly was not. And as noted above, whereas survey was carried out by numerous midsized expeditions, exploration was more typically done either by single collectors traveling fast and light and gathering opportunistically or by large parties in extended, multipurpose expeditions on self-sustaining ships or in overland caravans.

Moreover, whereas survey expeditions were for science, the purpose of exploration was more commonly economic or strategic (commerce, conquest, showing the flag). Itineraries and pacing were dictated by commercial or strategic, not scientific, needs. Collectors for the specimen trade sought what connoisseurs would pay for: namely, rarities.

The market afforded no incentives to gather large series of common species. And in large exploring expeditions collecting was often a sideline, pursued by naturalists as best they were able. Since exploring parties tended to traverse, not linger, naturalists typically could collect widely but not comprehensively or in depth, as survey parties did. Exploratory science was, in short, extensive but not intensive; opportunistic, not planned; not systematic but catch-as-catch-can.

A mode of fieldwork that produced lots of new species relatively quickly but not in large series was in fact well suited to a Linnaean taxonomy and, at first, to natural systems of classification as well. Both were improved by an abundance of species, and for both a deep knowledge of intraspecies variability was a mixed blessing, because it could make categorical boundaries less distinct. Cream skimming thus encouraged a typological conception of species, and a conceptual world of types in turn legitimated an exploratory mode of field practice. Collecting a few representative or "type" specimens for each species was what mobile explorer-naturalists could do, and collections assembled in this way made type species seem natural and real. The only reason to collect in depth was to acquire duplicates for exchange; only later did its advantages for taxonomic science become clear.

This is all somewhat speculative, of course; and it begs the question of when and how exploratory field practice gave way to in-depth survey. One thing is clear, however: it was not a sudden or a simple replacement. The process was more a layering of practices, followed by gradual divergence. Collecting in depth was already fairly common by the 1850s: in the United States, by Spencer Baird and his network of collectors, whom I mentioned earlier; and in Britain, by Joseph Hooker and his network of botanical collectors centered on Kew Gardens and the British Museum (Endersby 2005). And even if traveling naturalists did not collect in depth, firsthand experience of creatures in their natural abundance could open their eyes to the variability of species. Nor did exploration just wither away in the late nineteenth century. Quite the contrary. But its directions and purposes changed: from natural history collecting to physical and human geography; and from places of abundant life to extreme and inhospitable places (deserts, high peaks, the poles) that were poor in flora and fauna but rich in opportunities for heroic display. Natural history collecting thus developed practices distinct from exploration that were then amplified and regularized as intensive survey.

Here again, a combination of environment, science, and culture seems a promising frame for understanding the systematics of the age of steam and empire. It's no more than a model, mind; how well it will stand up to empirical research remains an open question.

A long view of the history of systematics suggests still other lines of inquiry. It is clear, for example, that collecting and discovery are patterned not just in time but in space as well. There is a historical geography of systematics that has been little studied. We know in a general way that there have been hotspots of discovery in various parts of the world, which have come and gone as different regions became accessible or interesting to Europeans. Explorers and collectors often favored areas where their countries were active in commerce or colonizing and where they could count on special access and diplomatic or logistic support. The question is, how often? Science followed flags; but it also followed scientific and biogeographic opportunity. Naturalists were no less drawn to places where biodiversity was rich and scientific problems most enticing, whatever their color on geopolitical maps.[8]

These are hypotheses; but there is evidence that they are sound guides to empirical research. As a test, I charted the 526 rodent species that received their currently accepted identities between 1758 and 1869, tabulating their type locales against the nationalities of the naturalists who described and named them. Geographic patterns of activity stand out clearly in the data (table 1.1). In the discovery of Eurasian species, Germans, Russians (or Germans in Russian employ), and Scandinavians predominated. Americans skimmed most of the cream of North American species, but British and Germans contributed substantially. In South America, Germans described more than twice as many new species as did British and French naturalists. But the British had a virtual monopoly in South and Southeast Asia and Australia, splitting East Indies species with Dutch naturalists. In East and Central Africa, discoveries were pretty evenly divided between British, French, and Germans; while in South Africa, the British predominated over Germans, with the French a distant third.[9] Americans were virtually absent outside the home territory—in sharp contrast to their presence worldwide after 1870.

It is clear from this preliminary mapping that naturalists in any given country were most active where national trade or colonizing gave them an advantage of access and public support but that scientific interest figured as well. German traveling naturalists seem to have been particularly drawn to places of scientific promise (inner Asia, South America, Africa). It stands to reason, since Germany was by then the leading scientific power in most fields but not (yet) a commercial or

TABLE 1.1 WORLD RODENT SPECIES TO 1870 BY NATIONALITY OF NAMERS

	British	French	German	U.S.	Dutch	Russian	Scand.	Italian	Total
N. America	20	8	12	57	—	1	6	1	105
S. America	34	21	63	2	1	1	10	5	137
Eurasia	6	6	19	—	—	27	22	—	80
S. SE Asia	47	6	3	—	—	1	2	—	59
E. Indies	10	3	2	1	9	1	2	—	28
Australia	21	1	—	—	—	—	—	—	21
N. C. Africa	12	10	18	2	5	—	4	—	51
S. Africa	17	3	7	—	2	1	3	—	33
E. Asia	—	6	1	—	4	—	—	—	11
Total	167	64	125	62	21	32	49	6	526

SOURCE: Honacki, Kinman, and Koeppl 1982.

colonial one. Historical research in this period could well begin with these singularly footloose and productive German naturalists (e.g., Haffer 1992, 122–128).

It is also clear that patterns of global activity in systematics have changed over time. For example, Bruce Patterson has shown that in the case of Neotropical (i.e., South and Central American) mammals there have been striking long-term shifts in the centers of scientific activity. Continental systematists predominated between 1758 and 1850; British and (increasingly) Americans, between 1850 and 1950; and Americans and (increasingly) South American naturalists, between 1950 and 2000 (Patterson 2000, 194, fig. 3).[10] Whether these shifts reflect changes in commercial or strategic presence in the region or diverging national trends in scientific interest remains to be established.

How geopolitical and scientific imperatives balanced out is a key question when examining a historical period during which newly modernized taxonomic disciplines were strongly linked to commercial and imperial venturing. The history of systematics is a human biogeography as well as a history of ideas and practices, and we want to discover the principles of systematists' global dispersals. We need to map their taxonomic achievements and explain how that map came to be and why. How deliberate was its creation? Did systematists stake out national spheres of influence and leave other areas to their competitors? Or did all compete in the same strategic regions? To what extent did scientific practices draw upon those of commerce and colonizing?

HOW PERIODS END

Another general question is, What causes episodes of heightened species discovery eventually to wane? The survey model suggests a symmetrical view. If changes in geography, science, and culture create opportunities and incentives for discovery, then further changes in these elements must also cause episodes of discovery to wane. That's the hypothesis.

Some such combination seems to be what in the 1940s brought the period of survey collecting to a close. To be sure, the Great Depression and World War II helped—the one cutting off funding for expeditions, the other making travel of any sort temporarily impossible. But that can't be the whole story. Expeditioning continued through the 1930s, just more discreetly and frugally than before; and when the fighting stopped and prosperity returned (to the United States) after 1945, survey expeditioning did not resume.

What closed the survey era in North America, I have argued, was structural change in the conjunction of environment, science, and culture that had sustained survey science for some fifty years. On the environmental side automobility, interstate highways, and urban sprawl squeezed inner frontiers into enclaves. On the cultural side, outdoor recreation morphed into commercial mass tourism, shedding the intellectual imperatives that turned some recreationists into patrons of survey science. And scientifically, as species inventories neared completion, survey collecting encountered diminishing returns.

It was not a simple matter of natural limits: there were then and still are species left to discover. Limits lie rather in the relation between what nature offers and how naturalists use what is offered to make science and careers. Systematists in the late 1940s could have gone after the last 10 percent of vertebrate species, for example. But their quarry were too few, too scattered, or too evasive to warrant the expense of survey expeditions. It was more cost-effective to rely on individuals and serendipity—which is essentially what happened.

Survey scientists could in theory also have moved on to insects and other invertebrates, many groups of which are little known. But it is hard to see how survey practices could be applied to groups with astronomical numbers of species many of which occupy tiny local ranges.[11] What purpose would be served by total inventories; and how would the public be induced to pay for them? Sampling may be a more appropriate practice in such cases; but would sampling of small locales deflect scientific interest from species inventory to ecosystem dynamics?

Or will the immediate future of systematics be shaped less by what collectors find in nature than by what systematists find in existing collections by applying new methods from the physical sciences—DNA fingerprinting, cladistics, cybernetic data sifting? Perhaps the vast accumulations of specimens and written records from the age of survey will prove to be the next inner frontier (or indoor frontier) of taxonomic science. Perhaps the Linnaean practice of data mining will recur in new ways. As these matters are well beyond my ken, I will leave them as questions—loose ends for someone else to pull on.

What I can say with confidence is that the practice of species survey was specific to its environmental, social, and scientific world. And it seems likely that the practices of Linnaean and Humboldtian systematics were as well to theirs. Doubtless it will be the same with whatever world and practices are coming next.

COMMUNITY DYNAMICS

I have focused thus far on the methods and circumstances of species discovery. However, there are other issues in the history of systematics, no less interesting and significant, having to do with the community of systematists and their customs and with comparisons of systematics as a science with other natural sciences. As this is largely uncharted ground I will just point briefly to some issues that I find especially intriguing.

To start I would like to go back to my earlier observation that systematics is preeminently a science of data management. This will seem obvious for contemporary cyber-systematics—because in our computer age informatics is a familiar category. But systematics has always been a science of data management, from Linnaeus on up. It is, we might say, the type specimen of informatic science. Its communal customs and practices can be understood as tools for managing abundant things (specimens) and categories of things (subspecies, species, genera, families, etc.) that unmanaged would overwhelm the mind.

Many sciences are data-rich, of course. But no other science has anything like the number of categories that systematists must create, tend, prune, arrange, and police. There are millions of organic chemical compounds, for example, but not all that many categories to put them in. They constitute a vast and ever-expanding encyclopedia (Beilstein's *Handbuch der Organischen Chemie*), not a taxonomic classification (Gordin 2005).[12] Likewise, accelerators, geophysical sensing technologies, and gene-sequencing machines produce amounts of data beyond the capacity of human brains to manage. But the game in these sciences is to sift rare signals from discardable noise or to form pictures out of pixels—not to order data into permanent categories. Systematics is special in its superabundance of categories, and what we want to know is how that superabundance has shaped the science and its community of practice.

Systematic biology is also singular in that it has remained preeminently a classifying science. Other thing-y or data-rich sciences have classified at some time in their histories—for example, chemistry, ethnology, ecology, and archaeology. But in these sciences classifying has always proved to be—by necessity or choice—a passing phase (Kohler 2007). Only in systematics have inventorying and arranging remained central and defining activities. Why this was so, and how this fact has affected systematists' culture and community relations are questions historians need to address.

Other features of the culture and customs of systematics are no less distinctive. Of all the sciences systematics has by far the deepest time horizon: that is, the point where working knowledge becomes (mere) history and ceases to be a tool of active use. In experimental sciences that point is typically reached in a few years or even sooner. Systematists, in contrast, operate in a living tradition that goes back 250 years and includes everyone who ever named a species. This is so because of the peculiar conventions of taxonomic nomenclature, especially the rule of priority—which gives precedence, in cases of redundant names, to the first one published—and the rule that rejected and redundant categories, once published, must be preserved forever as "synonyms"—a category of names that are not things but might be mistaken for things if not publicly labeled as phantoms or mistakes. Thus every act of diagnosis, naming, and revision requires systematists to actively engage with the taxonomic literature, and with their scientific predecessors, all the way back to Linnaeus.

Equally distinctive of taxonomic science is the social diversity of its practitioners. In principle anyone can describe and name a species, however expert or inexpert he or she may be. Of course all would-be describers must submit to exacting communal rules of describing and naming. And as the practices of diagnosis have become more exacting and quantitative, systematics has gradually become a game for the credentialed and experienced. Yet systematics remains open in principle to lay participation in a way that few sciences are (observational astronomy is another); and that openness affects the science in ways that have no parallel in the laboratory sciences.

Of course all modern sciences regulate traffic across boundaries with lay publics; but for systematists, because that boundary is unusually open, traffic control is especially tricky. Neither simple exclusion nor free access is an option. Systematists must engage lay naturalists while limiting their ability to disrupt the intricate practices of classifying. Typically they have steered lay participants away from formal taxonomy and toward observing, census, and ecological or life history study (Barrow 1998; Secord 1994; Drouin and Bensaude-Vincent 1996).

Systematics is also distinctive, and perhaps unique, in its relations to other sciences. Practices of naming and classifying are designed to serve the biological ends of systematic biology, obviously; but they also serve other life sciences, whose practitioners may not be interested in species or classifying but need to know (at least) what species they are working with. Systematists are thus stewards of nature's order for all the life sciences. And that fact makes for an ambiguous economy of reward and credit.

All scientists engage in housekeeping, of course, to keep order in the jungle of organic chemical compounds, or in collections of mutant strains of flies or mice or phage, or in databases and DNA and protein archives. But they do this for themselves, and the costs and benefits of housekeeping are assessed by and accrue to those who do it. Housekeeping for oneself is a necessity and brings no dishonor; in contrast, housekeeping for others is seen as low-caste labor and brings little respect. So systematists get no kudos for services to those outside their own ranks. Indeed, their status suffers, because consumers of their taxonomic favors experience the relation as unequal and so perceive those who serve as servants.

Of course, classifying is not mere housekeeping or cataloging (stamp or butterfly collecting is the time-honored gibe): the stereotype is wrong. Systematics is a biological discipline, and classifying gets at the principles of nature's workings just as effectively in its way as comparison and experiment do in theirs. It is an exacting and creative process—as any classifier knows. The puzzle is, why do outsiders persistently mistake what systematists do as routine maintenance and not creative science? The answer, I think, lies again in the social relation of systematists to other biologists, who see only the formalities of naming and ranking (because names are what they use) and not the science. So they take what systematists do as the familiar routine maintenance that they perform for themselves.[13] The galling logic of credit and reward (or lack thereof) is probably inescapable.

At the same time, systematists' deep living past and broad sense of stewardship may make their science a more humane and humanly engaging one than the sciences whose practitioners labor only in the immediate present and only for themselves. We know that laboratory sciences differ greatly in their "moral economies"—that is, in their customs of mutual obligation and reward. Fruit fly geneticists, whose little commensal species is a cornucopia of productive things to do, have for a century sustained a culture of generosity and mutual aid. In contrast, scientists who engage in winner-take-all quests for biopharmaceutical substances tend to develop competitive cultures of every man for himself and the devil take the hindmost (Kohler 1994, 13–14, 98–106, 233).[14] Practices and material cultures shape communal values in varied ways: just how varied, we don't yet know. Systematics, with its exceptionally developed system of communal rules and obligations and its Noachian moral economy, is a premier site for further study.[15]

There is another way in which systematics may be more central to the history of modern science than the stereotypes allow. We believe it is normal for modernizing sciences to abandon description and classification as they acquire theory and experiment (a vestige perhaps of the old belief in universal linear progress). But it seems just as likely that the real bedrock of modern science is neither theory nor experiment but an intensive and exacting empiricism: a capacity to generate, organize, and manipulate ever more abundant factual knowledge. All modern sciences have that in common, however different their sites and modes of practice. Scientists who do not go to nature to record and gather its abundance create artificial environments (labs) and methods (experiment) for making facts even beyond what nature's cornucopia spills forth. Theory and experiment are tools for making and handling facts. And facility in dealing with empirical and categorical abundance is also what science has in common with those other quintessential institutions of modernity: industrial capitalism, the nation-state (Chandler 1977; Scott 1998).[16]

If this is so, then systematics, though it may be low in the prestige hierarchy of sciences, should be seen as the exemplary modern science—because it most fully embodies the defining characteristics of modern society generally. The notion flies in the face of common wisdom, I know—but for just that reason it might be worth pursuing.

The distinctive features of systematists' communal life must derive partly from the fact that systematists cultivate that part of nature that is most abundantly categorical: the part produced by that cunning bit of evolutionary machinery that Darwin was the first to grasp. Nature's endless diversity and abundance keeps them at the work of sorting and arranging. Practitioners of sciences that generate their own facts at will can leave their past behind and give up classifying as unneeded and unmodern.

But natural endowments do not explain everything. In the natural history of the sciences, history matters as much as nature. It makes a difference where and by what historical paths sciences happen to take shape. Systematics was born of an Enlightenment vision of universal and comprehensive, yet thing-y and particular, knowledge. And it evolved into its modern form in a society of economic superabundance that lives by producing, accumulating, ordering, and managing things and facts (e.g., Bowker and Star 1999; Dumont 1980 [1967]). Perhaps humans are by nature collectors and categorizers; culturally we certainly are. We are *Homo categoricus*, and systematics is the science in the doing of which we are most ourselves.

Acknowledgments

I am grateful to Jim Endersby and Peter Stevens for helpful criticisms of a penultimate draft.

NOTES

1. The first section of this chapter is drawn from chapters 1–3 and 7 of this book.

2. Pre-1800 data are lumped because rates per decade would simply reflect the publication of major treatises. It does not follow from these results, of course, that a periodic pattern also obtains for plants and invertebrates.

3. On the logic of capital, see Cronon 1991, esp. chap. 2.

4. The following argument is summarized from Kohler 2006, chap. 3; see also Anker 2007.

5. The Russian Imperial Army operated a successful school for explorer-naturalist diplomats (Shearer 2009).

6. A suggestive account of this change is Swainson 1839, esp. 188–235.

7. The literature on travel and exploration is too vast even to begin to cite. On explorers as celebrities, see Riffenburgh 1994.

8. For example, I have seen little evidence that American museums worked in Central and South America for consciously geopolitical reasons: accessibility and biogeographic interest were what mattered. However, Spencer Baird's activities in the 1850s were clearly part of America's imperial expansion in the far West.

9. Data were taken from Honacki, Kinman, and Koeppl 1982. I chose rodents as a test case because they were not popular with amateurs and show a markedly periodic pattern of discovery. Namers of species were of course not necessarily their collectors; some of the most productive (e.g., Linnaeus and keepers of the British Museum) stayed put and reaped the rewards of what collectors gave or sold to their institutions.

10. Patterson infers nationality from the geographic locations of type specimens and the journals in which descriptions were published.

11. Inventories of groups of insects that are attractive to humans and live in densely humanized areas (e.g., beetles and butterflies in Britain) are nearly complete. And I have it on good authority that arachnologists feel that a complete inventory of world spiders is within reach (Norman Platnick pers. com.). On the other hand, noncelebrity groups in inaccessible places are still barely known.

12. The arcane rules of naming organic compounds, though not officially a classifying system, do in some ways operate as one.

13. For another example of this social logic, see Shapin 1989. The social relations of physicists to chemists, or chemists to biologists, or biologists to behavioral scientists are less fraught, since in these cases it is highly valued theories and reductionist strategies of explanation that are trickled down.

14. The ur-source of the idea of moral economy is Thompson 1991a, 185–258; also Thompson 1991b, 259–351.

15. We might fruitfully compare species inventories and classifications to natural resource commons, which historically have been intricately regulated (and remarkably effective and stable) systems of usufruct rights for diverse communities. The literature on commons is vast, but a good point of entry is Ostrom 1990.

16. For a similar argument about early modern science, see Shapiro 2000; Cook 2007.

REFERENCES

Allen, D.E. 1994 [1976]. *The Naturalist in Britain: A Social History.* London: Allen Lane; reprint, Princeton: Princeton University Press.

Anker, P. 2007. "Science as a Vacation: A History of Ecology in Norway." *History of Science,* 45: 455–479.

Applegate, C. 1990. *A Nation of Provincials: The German Idea of Heimat.* Berkeley: University of California Press.

Aron, C. 1999. *Working at Play: A History of Vacations in the United States.* New York: Oxford University Press.

Bailey, P. 1978. *Leisure and Class in Victorian England: Rational Recreation and the Contest for Control, 1830–1885.* London: Routledge & Kegan Paul.

Barrow, M.V., Jr. 1998. *A Passion for Birds: American Ornithology after Audubon.* Princeton: Princeton University Press.

Bensaude-Vincent, B., García-Belmar, A., and Bertomeu-Sánchez, J.R. 2003. "Natural Classifications in Chemistry: Un impossible rêve?" In U. Klein (ed.), *Spaces of Classification.* Preprint 240. Berlin: Max-Planck-Institut für Wissenschaftsgeschichte, 49–66.

Bowker, G.C., and Star, S.L. 1999. *Sorting Things Out: Classification and Its Consequences.* Cambridge, MA: MIT Press.

Chandler, A.D., Jr. 1977. *The Visible Hand: The Managerial Revolution in American Business.* Cambridge, MA: Harvard University Press.

Cook, H.J. 2007. *Matters of Exchange: Commerce, Medicine, and Science in the Dutch Golden Age.* New Haven: Yale University Press.

Cronon, W. 1991. *Nature's Metropolis: Chicago and the Great West.* New York: Norton.

Drouin, J.M., and Bensaude-Vincent, B. 1996. "Nature for the People." In N. Jardine, J.A. Secord, and E.C. Sparry (eds.), *Cultures of Natural History.* New York: Cambridge University Press.

Dumont, L. 1980 [1967]. *Homo Hierarchicus: The Caste System and Its Implications.* Chicago: University of Chicago Press. Translated from the French.

Eddy, M.D. 2003. "'Reading with Intelligence': The Chemistry of Mineralogical Classification and Geological Composition in Edinburgh's Medical School at the End of the Eighteenth Century." In U. Klein (ed.), *Spaces of*

Classification. Preprint 240. Berlin: Max-Planck-Institut für Wissenschaftsgeschichte, 133–160.

Endersby, J. 2005. "Classifying Sciences: Systematics and Status in Mid-Victorian Natural History." In M. Daunton (ed.), *The Organization of Knowledge in Victorian Britain.* Oxford: Oxford University Press, 61–85.

———. 2008. *Imperial Nature: Joseph Hooker and the Practices of Victorian Science.* Chicago: University of Chicago Press.

Farber, P. 1976. "The Type-Concept in Zoology during the First Half of the Nineteenth Century." *Journal of History of Biology,* 9: 93–119.

———. 1985. "Theories for the Birds: An Inquiry into the Significance of the Theory of Evolution for the History of Systematics." In M. Osler and P.L. Farber (eds.), *Religion, Science, and World View: Essays in Honor of Richard S. Westfall.* New York: Cambridge University Press, 321–339.

Frykman, J, and Löfgren, O. 1987 [1979]. *Culture Builders: A Historical Anthropology of Middle-Class Life.* New Brunswick, NJ: Rutgers University Press. Translated from the Swedish.

Gordin, M.D. 2002. "The Organic Roots of Mendeleev's Periodic Law." *Historical Studies in the Physical and Biological Sciences,* 32: 263–290.

———. 2004. *A Well-Ordered Thing: Dmitrii Mendeleev and the Shadow of the Periodic Table.* New York: Basic Books.

———. 2005. "Beilstein Unbound: The Pedagogical Unraveling of a Man and His *Handbuch.*" In D. Kaiser (ed.), *Pedagogy and the Practice of Science.* Cambridge, MA: MIT Press.

Green, N. 1990. *The Spectacle of Nature: Landscape and Bourgeois Identity in Nineteenth-Century France.* Manchester: Manchester University Press.

Haffer, J. 1992. "The History of Species Concepts and Species Limits in Ornithology." *Bulletin of the British Ornithological Club,* suppl. 112A: 107–158.

Hagen, J.B. 1984. "Experimentalists and Naturalists in Twentieth-Century Botany: Experimental Taxonomy, 1920–1950." *Journal of the History of Biology,* 17: 249–270.

———. 1999. "Naturalists, Molecular Biologists, and the Challenges of Molecular Evolution." *Journal of the History of Biology,* 32: 321–341.

Harris, S.J. 1998. "Long-Distance Corporations, Big Sciences, and the Geography of Knowledge." *Configurations,* 6: 269–304.

Honacki, J.H., Kinman, K.E., and Koeppl, J.W., eds. 1982. *Mammal Species of the World: A Taxonomic and Geographic Reference.* Lawrence, KS: Association of Systematics Collections and Allen Press.

Hull, D.L. 1988. *Science as a Process: An Evolutionary Account of the Social and Conceptual Development of Science.* Chicago: University of Chicago Press.

Klein, U. 2005. "Shifting Ontologies, Changing Classifications: Plant Materials from 1700 to 1830." *Studies in History and Philosophy of Science,* 36: 261–329.

———, ed. 2003. *Spaces of Classification.* Preprint 240. Berlin: Max-Planck-Institut für Wissenschaftsgeschichte.

Klein, U., and Lefèvre, W. 2007. *Materials in Eighteenth-Century Science: A Historical Ontology.* Cambridge, MA: MIT Press.

Koerner, L. 1999. *Linnaeus: Nature and Nation.* Cambridge, MA: Harvard University Press.

Kohler, R. E. 1994. *Lords of the Fly:* Drosophila *Genetics and the Experimental Life.* Chicago: University of Chicago Press.

———. 2006. *All Creatures: Naturalists, Collectors, and Biodiversity, 1850–1950.* Princeton: Princeton University Press.

———. 2007. "Finders, Keepers: Collecting Sciences and Collecting Practice." *History of Science,* 45: 1–27.

———. 2008. "Plants and Pigeonholes: Classification as a Practice in American Ecology." *Historical Studies in the Natural Sciences,* 38: 77–108.

Lux, D. L., and Cook, H. V. 1998. "Closed Circles or Open Networks: Communicating at a Distance during the Scientific Revolution." *History of Science,* 36: 179–211.

Mayr, E. 1982. *The Growth of Biological Thought: Diversity, Evolution, and Inheritance.* Cambridge, MA: Belknap Press.

———. 1995. "Systems of Ordering Data." *Biology and Philosophy,* 10: 419–434.

McOuat, G. 1996. "Species, Rules, and Meanings: The Politics of Language and the Ends of Definitions in Nineteenth-Century Natural History." *Studies in History and Philosophy of Science,* 27: 473–519.

———. 2001. "Cataloguing Power: Delineating 'Competent Naturalists' and the Meaning of Species in the British Museum." *British Journal for the History of Science,* 34: 1–28.

———. 2003. "The Politics of 'Natural Kinds': Practices of Classification in the Age of Reform." In U. Klein (ed.), *Spaces of Classification.* Preprint 240. Berlin: Max-Planck-Institut für Wissenschaftsgeschichte, 97–114.

Müller-Wille, S. 2003. "Eighteenth Century Classifications of Non-Living Nature." In U. Klein (ed.), *Spaces of Classification.* Preprint 240. Berlin: Max-Planck-Institut für Wissenschaftsgeschichte.

Nyhart, L. 2009. *Modern Nature: The Rise of the Biological Perspective in Germany.* Chicago: University of Chicago Press.

Ostrom, E. 1990. *Governing the Commons: The Evolution of Institutions for Collective Action.* Cambridge: Cambridge University Press.

Patterson, B. D. 2000. "Patterns and Trends in the Discovery of new Neotropical Mammals." *Diversity and Distributions,* 6: 145–151.

———. 2001. "Fathoming Tropical Biodiversity: The Continuing Discovery of Neotropical Mammals." *Diversity and Distributions,* 7: 191–196.

Riffenburgh, B. 1994. *The Myth of the Explorer: The Press, Sensationalism, and Geographical Discovery.* New York: Oxford University Press.

Ritvo, H. 1997. *The Platypus and the Mermaid and Other Figments of the Classifying Imagination.* Cambridge, MA: MIT Press.

Scott, J. S. 1998. *Seeing Like a State: How Certain Schemes to Improve the Human Condition Have Failed.* New Haven: Yale University Press.

Secord, A. 1994. "Corresponding Interests: Artisans and Gentlemen in Nineteenth-Century Natural History." *British Journal for the History of Science,* 27: 383–408.

Shapin, S. 1989. "The Invisible Technician." *American Scientist,* 77: 554–563.

Shapiro, B.J. 2000. *A Culture of Fact: England, 1550–1720.* Ithaca: Cornell University Press.

Shearer, D. 2009. "Russian and Soviet Explorers in Central Asia, 1870s to the 1930s." Unpublished lecture, University of Pennsylvania, March.

Stevens, P.F. 1986. "Evolutionary Classification in Botany, 1960–1985." *Journal of the Arnold Arboretum,* 67: 313–339.

———. 1994. *The Development of Biological Systematics: Antoine-Laurent de Jussieu, Nature, and the Natural System.* New York: Columbia University Press.

Swainson, W. 1839. *Natural History: A Preliminary Discourse on the Study of Natural History.* New ed. London: Longman, Orme, Brown, Green, & Longmans.

Thompson, E.P. 1991a. "The Moral Economy of the English Crowd in the Eighteenth Century." In E. P. Thompson, *Customs in Common.* New York: New Press, 185–258.

———. 1991b. "The Moral Economy Reviewed." In E. P. Thompson, *Customs in Common.* New York: New Press, 259–351.

Vernon, K. 1993. "Desperately Seeking Status: Evolutionary Systematics and the Taxonomists' Search for Respectability." *British Journal for the History of Science,* 26: 207–227.

———. 2001. "A Truly Taxonomic Revolution? Numerical Taxonomy, 1857–1970." *Studies in History and Philosophy of the Biological and Biomedical Sciences,* 32: 315–341.

Winsor, M.P. 1976. *Starfish, Jellyfish, and the Order of Life.* New Haven: Yale University Press.

Wonders, K. 1993. *Habitat Dioramas: Illusions of Wilderness in Museums of Natural History.* Stockholm: Almqvist and Wiksell.

Zeller, S. 1987. *Inventing Canada: Early Victorian Science and the Idea of a Transcontinental Nation.* Toronto: University of Toronto Press.

Zimmer, O. 1998. "In Search of National Identity: Alpine Landscape and the Reconstruction of the Swiss Nation." *Comparative Studies in Society and History,* 40: 637–665.

2

Willi Hennig's Part in the History of Systematics

MICHAEL SCHMITT

Willi Hennig's method of assessing phylogenetic relationships is sometimes termed a "revolution" (e.g., Dupuis 1990; Mishler 2000; Wheeler 2008) or at least praised as marking "a milestone in the history of systematic biology" (Richter and Meier 1994, 212) or as a new paradigm (Kühne 1978). It might, therefore, be of interest to ask what is so different in cladistics as compared to traditional systematics. And who was the man who caused that turn in this branch of biology? Here I investigate the development of Hennig's thinking about systematics, tracing the roots of his deviation from traditional methodology and evaluating his contribution to modern cladistics.

WILLI HENNIG, THE PERSON

Emil Hans Willi Hennig (fig. 2.1) was born on April 20, 1913, the first child of Karl Ernst Emil Hennig (August 28, 1873–December 28, 1947), a railroad worker, and Marie Emma née Groß (June 12, 1885–August 3, 1965), the illegitimate child of a maidservant. He had two younger brothers, Fritz Rudolf (March 5, 1915–November 24, 1990) and Karl Herbert (April 24, 1917–January 1943 [missing near Stalingrad]). Rudolf Hennig, who became a Protestant clergyman, bequeathed a handwritten autobiographical sketch that was obviously intended to become the first chapter of a more extensive autobiography (document 1). According to this sketch, Emma Hennig saw to it that her sons

FIGURE 2.1. Willi Hennig, 1973. Courtesy of Willi Hennig Archive, Görlitz

received excellent educations, especially the eldest, Willi, who received private lessons in French while still in primary school. Rudolf describes his parents as Protestants but not especially religious. They were true Prussian patriots but not fanatics. They were not members of any political party, club, or organization.

Although the Hennig children grew up during the last years of World War I and the turbulent postwar years of the early twenties, they were only a little affected by the war and the hunger and political chaos that followed on the defeat and end of the Second German Empire. Emil Hennig did not lose his job, as so many others did, and the family was also able to maintain a small plot it owned. There the Hennigs grew potatoes and other vegetables to contribute to the family's table. Nevertheless, Rudolf recounted regular periods of financial difficulty, and also that his parents took great pains to insulate the children from these problems.

Willi Hennig seems never to have written a diary or an autobiography, or at least no such writings were preserved. Since he was two years older than Rudolf, he likely had even clearer impressions of the war and the period following it. However, it might be that he did not experience the events around him with the same sensitivity as his younger brother.

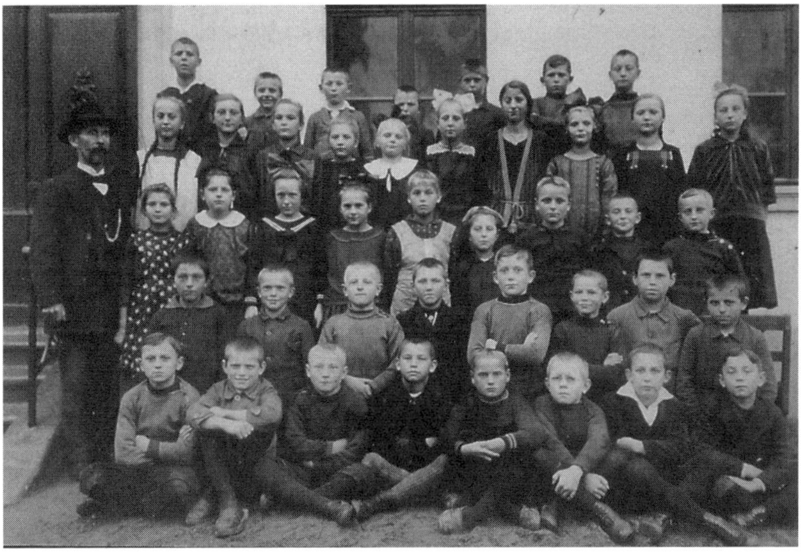

FIGURE 2.2. Willi Hennig (sitting, first row, second from right) as a primary school pupil at Oppach, ca. 1925. Courtesy of Willi Hennig Archive, Görlitz

From early childhood, Willi was described as being introverted and shy (fig. 2.2). He preferred staying at home and reading all sorts of books to playing outside with other children. He also showed an early interest in natural history, stimulated by the retired physician who taught him French. Willi's brother Rudolf mentions in his chronicle (document 1, 41; see also Vogel and Xylander 1998) that Willi was given the nickname "Orang" (after orangutan) because of his introverted nature and his overwhelming interest in natural history.

The Hennig family moved several times between 1913 and 1927, ostensibly due to Emil Hennig's relocations, but Vogel and Xylander (1998) speculate that Emma Hennig's emotional restlessness also may have played a part. Rudolf gives the impression that she was a difficult person who encountered social friction with housemates and neighbors soon after arriving at a new place (document 1).

In 1927 fourteen-year-old Willi Hennig entered a public boarding school in Klotzsche, near Dresden. There he lived in a house ("Abteilung") under the supervision of his science teacher, M. Rost, who brought the young Willi into contact with Wilhelm Meise (November 22, 1901–August 24, 2002) (see Haffer 2003; Hoerschelmann and Neumann 2003) because Willi wanted to learn more about natural history than Rost could teach him. Meise was curator of the non-insect collections at the State Museum of Zoology in Dresden. Even before Willi passed the final examination (*Abitur*) on February 26, 1932, and entered the University of Leipzig for the 1932 summer semester, he worked as a volunteer at the Dresden museum. Meise entrusted him with the task of compiling an index to the list of fishes in the British Museum of Natural History, as a kind of test of his seriousness and perseverance (document 2). Soon Meise recognized how motivated, gifted, and already well trained Hennig was, and he invited him to coauthor two papers on the "flying" snakes of the genera *Chrysopelea* and *Dendrophis* (Meise and Hennig 1932, 1935). Already as a pupil at gymnasium, Hennig had written a composition test of thirty-one pages (including tables, illustrations, and references) on the position of systematics in zoology, dated May 4, 1931 (document 3). This essay gives clear evidence of Hennig's serious interest in and profound knowledge of the subject. He received the highest score for it ("sehr gut"), although the test was on German, not biology.

Soon after meeting Wilhelm Meise, Hennig got into contact with Fritz Isidor van Emden (October 13, 1898–September 2, 1958), then curator of the entomological collections and a respected Coleoptera taxonomist. It is unclear whether Hennig already had an interest in insects before he met van Emden or received the crucial stimuli from him. At any rate, according to Meise (document 2), van Emden guided Hennig to complete a doctoral thesis on the copulatory apparatus of cyclorrhaphous Diptera, under the supervision of Paul Buchner (April 12, 1886–October 19, 1978), founder of the modern science of symbiosis. He received his doctoral degree on April 15, 1936, at the age of twenty-three. By that time he had published eight papers, including a two-hundred-page revision of the fly family Tylidae (Hennig 1934–35, 1935–36) and a sixty-eight-page revision of the agamid genus *Draco* (Hennig 1936b).

Van Emden was expelled from the public service on September 30, 1933, due to the racist Nazi laws (Hennig 1960). His successor was Klaus Günther (October 7, 1907–August 1, 1975), from Berlin, with whom Hennig from then on cultivated a lifelong friendship.

After working several months as a volunteer at the Dresden museum, Hennig received a grant from the German Science Foundation (Deutsche Forschungsgemeinschaft) to work at the Deutsches Entomologisches Institut (DEI, then in Berlin-Dahlem), which was run by the Kaiser-Wihelm-Gesellschaft, the predecessor of the Max Planck Society. There he was appointed junior research assistant (October 1, 1938; document 4) and then regular assistant (January 1, 1939; document 5).

At the University of Leipzig, Hennig had met a biology student, Irma Wehnert (August 29, 1910–April 26, 2000), who switched to the study of the history of arts after five semesters. They married on May 13, 1939. The couple had three sons: Wolfgang (b. 1941, geneticist), Bernd (b. 1943, molecular biologist), and Gerd (b. 1945, teacher of German and history).

Hennig underwent his obligatory military training in spring 1939 and was called for regular army service in September 1939, at the onset of World War II. He was deployed as an infantryman in Poland, France, Denmark, and Russia (Schlee 1978). After he was wounded by shrapnel at the Eastern Front while serving there as a private first class in 1942, he was decorated with the badge for casualties in Black. From then on, he served as a military entomologist at the Institute of Tropical Medicine and Hygiene at the Academy for Military Medicine, Berlin, and was soon sent to join the Tenth Army (Army Group C) in Italy to investigate and control malaria and other epidemics. He held the military rank of "Sonderführer Z," which corresponds roughly to platoon leader or master sergeant (document 4; document 8, 59). On March 6, 1945, he was awarded the military cross of merit, second class with swords. This decoration was bestowed for meritorious service in the homeland or behind the front lines (Klietmann 1996, 37). Until he was taken prisoner by British troops in May 1945 as a member of Malaria Training Troop 1 at Lignano (near the Gulf of Trieste, northern Italy), he was stationed at an ambulance unit in the German military hospital of Abano Terme near Padua (document 7). After a few weeks of captivity, the British troops took him out of the prisoner-of-war camp and put him into the British anti-malaria service at a former German military hospital that the British had taken over. According to short personal notes from that time, he could leave the institution but, at least in the beginning, always accompanied by a British officer. Hennig was released from captivity in October 1945. During his time as a POW, he wrote the manuscript of his *Grundzüge einer Theorie der phylogenetischen Systematik,* which was published in 1950 (see www.cladistics.org/about/hennig.html).

After his return to Germany, Hennig received a temporary appointment at the University of Leipzig as the replacement first of Prof. Dr. Friedrich Hempelmann (January 26, 1878–August 6, 1954), then, from December 1, 1945 to March 31, 1947, of Prof. Paul Buchner, Hennig's doctoral supervisor (who had left the university in 1944). Hennig gave up this appointment voluntarily because he wished to return to the DEI, where he worked from April 1, 1947, and became head of the Department of Systematic Entomology and vice-director on November 1, 1949. On August 1, 1950, he received the *venia legendi* (right to teach) following his habilitation from the Brandenburgische Landeshochschule at Potsdam and the title of professor on October 10, 1951.

The DEI was transferred to the Blücherhof manor in Mecklenburg-Vorpommern, near Waren, from 1943 to 1950. Thereafter it moved to a building in the Waldowstrasse 1 in Berlin-Friedrichshagen in the Soviet sector of the city. The Hennig family lived in West Berlin, on Opitzstrasse 3 in the district of Steglitz. During the days preceding August 13, 1961, Hennig was traveling in France with his youngest son, Gerd. When he learned of the erection of the Berlin wall on that day, he returned immediately to Berlin (on August 14), removed his personal belongings from DEI, and quit his job there (document 9).

In order to continue working at DEI he would have had to move to East Berlin, or to the German Democratic Republic, which would have meant living under difficult political and economic conditions (Peters 1995). His decision to leave the DEI was not an inconsequential one, as he was already envisaged as its future director. Quitting DEI cost him a number of privileges and opportunities, but Hennig was ardently anti-communist and already under suspicion by the East German secret police (Staatssicherheit) (document 10).

Though Hennig opposed communist ideas, he was by no means a follower of National Socialist ("Nazi") politics. Neither was he a member of the NSDAP (Nationalsozialistische Deutsche Arbeiterpartei), as documented in the official dossier on him in the records of the State Museum of Natural History, Stuttgart (SMNS) (document 6). There is also no evidence in the written documents or verbal reports of his contemporaries that he held Nazi views. Neither I nor Hennig's son Bernd have ever seen the Nazi salutation "Heil Hitler" used by him in his private correspondence. Even if a sender had written it, Hennig signed his reply using a conventional complimentary close (document 11). It seems that he added this formula only to official letters to addressees in German institutions (documents 12, 13). To make this point clear: any accusation that Willi Hennig sympathized with Nazi ideology is pure invention. This view is also strongly supported by Rieppel (2011a). Obviously, the singular source of such insinuations is a group of followers of Leon Croizat (July 16, 1894–November 30, 1982) who extensively commented on the question "Was Hennig a Nazi?" on the Internet discussion list Panbiog-L (document 14). The origin of this denunciation is, according to Platnick and Nelson (1988, 415), Croizat's unsubstantiated personal anti-German posture.

In 1961 Hennig was appointed professor at the Technische Universität Berlin, where he gave two lectures and one practical course on invertebrate zoology (Schmitt 2002). In April 1963 he became head of the Department of Phylogenetic Research (established especially for him) at the Staatliches Museum für Naturkunde, Stuttgart (SMNS). This department was housed provisionally in the city of Ludwigsburg, 15 kilometers north of Stuttgart. Here he worked and lived—at Denkendorfer Strasse 16 in Ludwigsburg-Pflugfelden—until the night of November 5, 1976, when he died suddenly from a heart attack (Schlee 1978; Schmitt 2001, 2003, 2013).

WILLI HENNIG, FOUNDER OF PHYLOGENETIC SYSTEMATICS

Hennig's basic ideas about phylogenetic research can easily be summarized. First, the only concept of relationship that can be based on an objective measure (recency of common ancestry) is that of genealogic relationship, as all other attempts to assess relationship are inevitably flawed by arbitrary emphasis on or restriction to certain characteristics of the organisms investigated. Second, focusing on the genealogical relations between species or supraspecific taxa led to the insight that Ernst Haeckel's (1866) notion of "monophyly" had to be refined to allow for nonarbitrary circumscription of taxa. Only defining *monophyletic* as "including a stem species and all of its descendants" avoids the subjectivity of the exclusion of one or some of these descendants from an accepted taxon, that is, forming a paraphyletic taxon. Third, species and supraspecific taxa can be shown to be monophyletic if they share at least one character that can be interpreted as an evolutionary novelty (a synapomorphy) in the character set of the stem species. If these ideas are accepted, the work of phylogenetic systematics is to establish a system containing exclusively species and monophyletic taxa.

Because a stem species is connected to all its immediate descendants through the same type of parent-offspring relations, it is logically unjustified to state that it survives in only one of its daughter species. Consequently, Hennig found it appropriate to state that the stem species always goes extinct when it splits into two or more daughter species. It is, however, important to note that this decision is a convention referring only to the taxonomist's need to draw boundaries within a continuum of generations. Hennig felt obliged to clarify this concept in a publication by Schlee (1971), where he wrote (p. 28) that his conceptual solution to the species boundary problem of "obligatory extinction of the stem species" does not mean that in reality organisms go extinct but simply that it is in his view the most rational way to delimit species (as taxa, not as real entities) in time.

As I have noted elsewhere (Schmitt 2001; 2003; 2013, 166), Hennig did not state or assume that species split obligatorily and exclusively into two descendant species. On the contrary: he discussed at length the possibility of polytomous splitting of species (Hennig 1966, 210ff.). But since only bifurcations can be demonstrated objectively by assessing putative synapomorphies, whereas a polytomy can never be proved positively, his "dichotomy rule" is a methodological principle that aims at resolving phylogenetic relations between the taxa of interest into dichotomies. Even in his 1950 book, where he did not explicitly claim that the "principle of dichotomy" is a methodological one, he only searched for an explanation of his empirical observation that in so many cases taxa, especially the more comprehensive ones, split into two subordinate taxa. He referred to a formal model by which he could demonstrate that even a sequence of multiple splitting can (!) result in a bifurcated tree in the long run (1950, 333). Rieppel (2011b) interprets Hennig's (1950, 332ff.) considerations as evidence that he grounded the principle of dichotomy ontologically in speciation. Although I largely agree with Rieppel's view, I emphasize that Hennig repeatedly claimed that his considerations offered *possible* explanations. I could not find any indication that he assumed bifurcation as the necessary mode of speciation.

Already in 1936, in his paper on the relations between geographic distribution and systematic classification of some Dipteran families, Hennig (1936a, 170) clearly distinguished primitive characters as "independently retained inheritance from the ancestor of all tylids" (unabhängig bewahrtes Erbe von den allen Tyliden gemeinsamen Vorfahren) from the "progressive characters" (die gemeinsamen fortschrittlichen Merkmale) shared by Tylinae and Neriinae. He emphasized that these latter characters "especially suggest" (legen . . . den Gedanken . . . besonders nahe) closer phylogenetic relationships of these two subfamilies. In the same paper, he stated explicitly that a systematic classification must be understood in terms of phylogenetic (which is to say, genealogical) relationship rather than morphological similarity and dissimilarity (see also Schmitt 2001, 328; 2003, 372).

From the general discussions found in his publications between 1936 and 1950 (see Anonymous 1978), it becomes clear without any doubt that the approach initially published in 1950 matured step by step over the years. It might be that Wilhelm Meise, Fritz van Emden, and—above all—Klaus Günther assisted in personal discussions from time to time, but there is definitely no evidence that they contributed more than occasional ideas.

As several authors have recognized (e.g., Hull 1988, 130; Kitching et al. 1998; Richter and Meier 1994; Schmitt 2001, 333), Hennig put remarkably little effort into elaborating a generally applicable and methodologically unobjectionable method for polarizing character states. Even his publications after 1966 remained vague and hardly useful for practical purposes in this respect. Also in personal letters (kept in the archive of SMNS), he denied even the possibility of providing a general tool for assessing character polarity. In the small book that was obviously intended to become the introductory chapter of a planned textbook on phylogenetics, later published by his eldest son, Wolfgang, in 1984, he even wondered that it is "strange to say that sometimes the phylogenetic systematics is blamed for failing to provide or develop methods for the discrimination of apomorph and plesiomorph characters" (merkwürdigerweise wird der phylogenetischen Systematik manchmal zum Vorwurf gemacht, daß sie es versäumt habe, Methoden anzugeben oder zu entwickeln, mit deren Hilfe plesiomorphe von apomorphen Merkmalen . . . unterschieden werden können; 46 f.). Instead of offering a general tool or method, he referred in this booklet to lists of primitive and derived characters for insects to be found in literature (Hennig 1984, 47). Only in a joint paper with Dieter Schlee, posthumously published in 1978, did he write (or agree to) some explicit statements on this topic, among which "distribution of the characters among the taxa" comes closest to what is now known and used as the "outgroup comparison method" (Wiley 1981, 139; Watrous and Wheeler 1981).

THE HENNIGIAN REVOLUTION

If it is not a practical method for revealing genealogical relationships, what, then, is the core of the "Hennigian revolution" of 1950 and 1966? Hennig introduced the need for systematists to make clear statements of the form "A is more closely related to B than either is to C," rather than put a taxon somewhere "in between" others or allegedly solve a taxonomic problem by erecting a separate Linnaean unit for a taxon in question. Moreover, he developed a method that required explicit presentation of supporting evidence rather than statements based purely on intuition or inexplicable experience. For the first time a method was at hand that made phylogenetics a scientific enterprise comparable to the branches of investigation that fall into Popper's concept of science (although there is still an ongoing debate on the question whether this applies to cladistics, i.e., the contemporary version of Hennigian phylogenetic systematics; see, e.g., Rieppel 2007; Kluge 2009) (from Schmitt 2010).

In his own review of the *Grundzüge,* Hennig (1952) elucidates the aspects of his book that he saw as most important. The review emphasizes the relevance of systematics as a biological discipline that provides a general reference system for generalizations. This purpose can only be met, he argues, by a strictly phylogenetic system. He went on to criticize the conventional systems for mingling different concepts of "relationship," especially the type of system "often denoted by the term 'network relationship' [Netzverwandtschaft]." But he does not mention any newly introduced method. It is exclusively the concept (he called it the "theory") of systematics that he felt was noteworthy.

Jahn (1992) mentions as one of Hennig's most outstanding contributions to systematics the new concept of macrotaxonomy—meaning a focus on relations between species and monophyletic groups of species—after about fifty years of prevalence of the relations between species and infraspecific populations. Likewise, Richter and Meier (1994, 212) stated that "Hennig was the person who redirected the interest of systematics to the study of supraspecific taxa after years of focusing on species and infraspecific taxa."

This means that in pre-Hennigian systematics a focus on the phylogenetic relations between supraspecific taxa was lacking and that the ideas then current about how to form hypotheses about such relations did not meet the scientific standards Hennig demanded. That this was indeed the case is highlighted by, for instance, the handbook chapter of Kühnelt (1962), where the possibility of assessing phylogenetic relationships objectively is explicitly denied and most emphasis is put on alpha taxonomy. Richter and Meier (1994) point to a similar picture in the major English-language textbooks of that time.

From a present-day perspective, it may appear surprising how much emphasis Hennig and those who evaluated his achievements put on a "strictly phylogenetic" concept of relationship. The relevance of Hennig's thinking might be judged from a quote from J. S. L. Gilmour (1940, 461 f.): "It is even doubtful whether the real significance of the term 'phylogenetic relationship' is yet fully understood. A resolution of these differences is surely one of the greatest needs of systematic biology." The differences mentioned here are those between "taxonomic trees" reflecting "phylogenetic trees" perfectly and "logical classifications" (based on correlation or coherence of characters) not necessarily being "phylogenetic." This demand mirrors not just the personal opinion of the author, but a general understanding of that time, making clear that the publication of Hennig's *Grundzüge* marks a real innovation in biological systematics.

The actual revolution of systematics was certainly realized by the development of the practical method of polarizing characters on which modern cladistics is based. Hennig would probably have appreciated this course of history. It is, however, highly unlikely that he would have accepted the complete abandonment of the traditional view of systematics as a discipline dealing with real organisms. Consequently, he would probably have criticized Platnick's (1977) view that the sister group hypothesis "A is sister-group to B+C" is isomorphic to, for instance, the "phylogenetic tree" A→B→C (A is the ancestor of B, B is the ancestor of C). Following Hennig, each taxon, including the terminal taxa, has to be monophyletic. If taxa can be shown to be monophyletic, they must possess at least one autapomorphy and can, therefore, under no circumstances be ancestor to any other taxon of the above set (since an ancestor cannot have a character in an apomorph state as compared to its descendant). After all, the debate could only be sensible within a Hennigian framework if A, B, and C were species. There cannot be supraspecific ancestors. As Hull (1988, 137) pointed out, "The only relation represented in cladograms and their isomorphic classifications are sister-group relations."

Although the unsatisfying elaboration of a practical procedure in his fundamental books of 1950 and 1966 leaves some doubt whether it is justified to ascribe to him a new method as the turning point in the history of systematics, Willi Hennig must definitively be credited for laying the foundations for the cladistic method of phylogenetic analysis. Even if his explicit descriptions of his practical method are theoretically insufficient and of only little practical use, he gave in his empirical papers on certain animals—mostly Diptera—valuable examples that could be used as models for further refinement of the method, last but not least, in his general treatments of invertebrates (1957, 1959, 1969). But certainly more important is the fact that he outlined a framework of clear terms and concepts that allows or even compels us to reduce subjectivity and arbitrariness in biological systematics. It is in this way that he made a crucial contribution to the conversion of systematics from an art or a handicraft to a legitimate branch of science.

Willi Hennig can be seen as a rather unassertive (Schmitt 2010) scout of the cladistic revolution to follow him. He prepared the soil for the great change and made the first step but did not show any intention to walk the path to the end.

There is an amusing parallel between Willi Hennig and Carl Linnaeus: the latter is lauded for the introduction or at least the consequent use of binomial names for species, that is, establishing the combination of a genus name and a "trivial" name. Yet he did not mention this achievement when listing his major contributions to science in his autobiography (see Schmitt 2008). Thus we may prize the two of them for achievements that they themselves did not regard as their greatest ones.

Acknowledgments

I cordially thank Prof. Dr. Willi Xylander, Görlitz, Germany, for permission to reproduce photographs from the Willi Hennig Archive; Dr. Bernd Hennig, Freiburg im Breisgau, Germany, for personal communication of additional biographical data on his father; Prof. Dr. Gabriele Uhl, Greifswald, Germany, for carefully reading and improving the manuscript of the present chapter; two anonymous reviewers for helpful comments; and Andrew Hamilton, Tempe, Arizona, for painstakingly and discreetly improving the English of my manuscript.

This chapter is an extended version of a paper presented at the symposium "What's on Your Planet? Species Exploration and Charting Biodiversity," Arizona State University, Tempe, March 3, 2008.

REFERENCES

Unpublished Works

Document 1. Photocopy of a handwritten sketch of an autobiography by Rudolf Hennig. Author's private archive; original at Willi Hennig Archive, Görlitz (Germany).

Document 2. Letter of Wilhelm Meise to M.S., 11 pp., from June 12, 1998.
Document 3. Composition test at school on "Die Stellung der Systematik in der Zoologie," 31 pp., handwritten, May 4, 1931. Photocopy in the author's private archive, original with Dr. B. Hennig, Freiburg im Breisgau (Germany). This essay was posthumously published by Dieter Schlee in *Entomologica Germanica,* 4 (1978): 193–199.
Document 4. "Anmerkungen zu Personalfragen" (comments on personnel affairs), written by Willi Hennig under the date November 26, 1966, on the occasion of an inquiry by the dean of the faculty of mathematics and sciences of the Freie Universität Berlin. Original with the family; photocopy in the author's archive.
Document 5. Index card "Willi Hennig" in the biographical card file of Senckenberg Deutsches Entomologisches Institut, Müncheberg, Germany.
Document 6. "Bericht der Abt. für Phylogenetik 3.1962, Anlage 2," in Willi Hennig's dossier at Staatliches Museum für Naturkunde, Stuttgart (SMNS).
Document 7. Letter from Dr. Bernd Hennig, Freiburg im Breisgau, to M.S., dated January 4, 2001.
Document 8. Sammlung wehrrechtlicher Gutachten und Vorschriften Heft 3, Bundesarchiv, Zentralnachweisstelle Kornelimünster. Printed as a manuscript, 1965.
Document 9. Minutes of a telephone conversation between Mr. Gerd Hennig and M.S., November 28, 1995.
Document 10. Letter from Prof. Wolfgang Hennig to M.S., dated September 1, 1998.
Document 11. Personal communication of Dr. Bernd Hennig to M.S. on May 28, 2010.
Document 12. E-message from Niels Peder Kristensen, Copenhagen (Denmark) of July 17, 2010.
Document 13. Scanned image of two letters of October 12 and 21, 1940, by Willi Hennig to O. Schröder, then curator at the Zoological Museum of Kiel.
Document 14. Digest of the discussion thread "Der Biologe, German neo-Darwinism and Hennig," April 2, 2009, http://groups.google.com/group/panbiog/t/05e0fcb97dcf655e?hl = en.

Published Works

Anonymous [Wolfgang Hennig]. 1978. "In Memoriam: Willi Hennig (20.4.1913–5.11.1976)." *Beiträge zur Entomologie,* 28: 169–177.
Dupuis, C. 1990. "Hennig, Emil Hans Willi." In F.L. Holmes (ed.), *Dictionary of Scientific Biography,* 17, Suppl. 2. New York: Charles Scribner's Sons, 407–410.
Gilmour, J.S.L. 1940. "Taxonomy and Philosophy." In J. Huxley (ed.), *The New Systematics.* London: Oxford University Press, 461–474.
Haeckel, E. 1866. *Generelle Morphologie der Organismen. Allgemeine Grundzüge der organischen Formenwissenschaft, mechanisch begründet durch die von Charles Darwin reformierte Descendenz-Theorie.* Berlin: Georg Reimer.

Haffer, J. 2003. "Wilhelm Meise (1901–2002), ein führender Ornithologe Deutschlands im 20. Jahrhundert." *Verhandlungen des Naturwissenschaftlichen Vereins in Hamburg Neue Folge,* 40: 117–140.

Hennig, W. 1934–35. "Revision der Tyliden (Dipt., Acalypt.). I. Teil: Die Taeniapterinae Amerikas." *Stettiner entomologische Zeitung,* 95: 6–108, 294–330 (1934); 96: 27–67 (1935).

———. 1935–36. "Revision der Tyliden (Dipt., Acalypt.). II. Teil: Die außeramerikanischen Taeniapterinae, die Trepidariinae und Tylinae. Allgemeines über die Tyliden. (Zugleich ein Beitrag zu den Ergebnissen der Sunda-Expedition Rensch, 1927)." *Konowia,* 14: 68–92, 192–216, 289–310 (1935); 15: 129–144, 201–239 (1936).

———. 1936a. "Beziehungen zwischen geographischer Verbreitung und systematischer Gliederung bei einigen Dipterenfamilien: ein Beitrag zum Problem der Gliederung systematischer Kategorien höherer Ordnung." *Zoologischer Anzeiger,* 116: 161–175.

———. 1936b. "Revision der Gattung *Draco* (Agamidae)." *Temminckia,* 1: 153–220.

———. 1950. *Grundzüge einer Theorie der phylogenetischen Systematik.* Berlin: Deutscher Zentralverlag.

———. 1952. "Autorreferat: Grundzüge einer Theorie der phylogenetischen Systematik." *Beiträge zur Entomologie,* 2: 339–331.

———. 1957. *Taschenbuch der Zoologie, Band 2: Wirbellose I (ausgenommen Gliedertiere).* Leipzig: Georg Thieme (and subsequent editions).

———. 1959. *Taschenbuch der Zoologie, Band 2: Wirbellose II (Gliedertiere).* Leipzig: Georg Thieme (and subsequent editions).

———. 1960. "F.I. van Emden †." *Zoologischer Anzeiger, Suppl.* 23 (*Verhandlungen der Deutschen Zoologischen Gesellschaft* 1959): 528–529.

———. 1966. *Phylogenetic Systematics.* Trans. D.D. Davis and R. Zangerl. Urbana: University of Illinois Press.

———. 1969. *Die Stammesgeschichte der Insekten.* Frankfurt am Main: Waldemar Kramer. English ed. *Insect Phylogeny,* translated and edited by A.C. Pont, with revision notes by D. Schlee. Chichester: John Wiley & Sons.

———. 1984. *Aufgaben und Probleme stammesgeschichtlicher Forschung.* Berlin and Hamburg: Paul Parey.

Hennig, W., and Schlee, D. 1978. "Abriß der phylogenetischen Systematik." *Stuttgarter Beiträge zur Naturkunde Serie A (Biol.),* 319: 1–11.

Hoerschelmann, H., and Neumann, J. 2003. "Prof. Dr. Wilhelm Meise 12.9.1901–24.8.2002." *Journal of Ornithology,* 144: 110–111.

Hull, D.L. 1988. *Science as a Process: An Evolutionary Account of the Social and Conceptual Development of Science.* Chicago: University of Chicago Press.

Jahn, I. 1992. "Geschichte der Systematik." In Schmitt, M. (ed.), *Lexikon der Biologie,* vol. 10 (*Biologie im Überblick*). Freiburg im Breisgau: Herder Verlag, 329–333.

Kitching, I.J., Forey, P.L., Humphries, C.J., and Williams, D.M. 1998. *Cladistics: The Theory and Practice of Parsimony Analysis.* 2nd ed. Oxford: Oxford University Press.

Klietmann, K.-G. 1996. *Auszeichnungen des Deutschen Reiches 1936–1945. Eine Dokumentation ziviler und militärischer Verdienst- und Ehrenzeichen.* Stuttgart: Motorbuch-Verlag.

Kluge, A.G. 2009. "Explanation and Falsification in Phylogenetic Inference: Exercise in Popperian Philosophy." *Acta Biotheoretica,* 57: 171–186.

Kühne, W.G. 1978. "Willi Hennig 1913–1976: Die Schaffung einer Wissenschaftstheorie." *Entomologica Germanica,* 4: 374–376.

Kühnelt, W. 1962. "Prinzipien der Systematik." In *Handbuch der Biologie, Bd. 6, Teil 2/1.* Konstanz: Akademische Verlagsgesellschaft Athenaion, 1–16.

Meise, W., and Hennig, W. 1932. "Die Schlangengattung *Dendrophis.*" *Zoologischer Anzeiger,* 99: 273–297.

———. 1935. "Zur Kenntnis von *Dendrophis* und *Chrysopelea.*" *Zoologischer Anzeiger,* 109: 138–150.

Mishler, B.D. 2000. "Deep Phylogenetic Relationships among 'Plants' and Their Implications for Classification. " *Taxon,* 49: 661–683.

Peters, G. 1995. "Über Willi Hennig als Forscherpersönlichkeit." *Sitzungsberichte der Gesellschaft Naturforschender Freunde zu Berlin Neue Folge,* 34: 3–10.

Platnick, N.I. 1977. "Cladograms, Phylogenetic Trees, and Hypothesis Testing." *Systematic Zoology,* 26: 438–442.

Platnick, N.I., and Nelson, G.J. 1988. "Spanning-Tree Biogeography: Shortcut, Detour, or Dead End?" *Systematic Zoology,* 37: 410–419.

Richter, S., and Meier, R. 1994. "The Development of Phylogenetic Concepts in Hennig's Early Theoretical Publications (1947–1966)." *Systematic Biology,* 43: 212–221.

Rieppel, O. 2007. "The Metaphysics of Hennig's Phylogenetic Systematics: Substance, Events, and Laws of Nature." *Systematics and Biodiversity,* 5: 345–360.

———. 2011a. "The Dark Side of the Moon." *Cladistics,* 27: 34–40.

———. 2011b. "Willi Hennig's Dichotomization of Nature." *Cladistics,* 27: 103–112.

Schlee, D. 1971. "Die Rekonstruktion der Phylogenese mit Hennig's Prinzip." In *Aufsätze und Reden der Senckenbergischen Naturforschenden Gesellschaft 20.* Frankfurt am Main: Waldemar Kramer, 66 pp.

———. 1978. "In Memoriam Willi Hennig 1913–1976. Eine biographische Skizze." *Entomologica Germanica,* 4: 377–391.

Schmitt, M. 2001. "Willi Hennig (1913–1976)." In I. Jahn and M. Schmitt (eds.), *Darwin and Co., eine Geschichte der Biologie in Portraits,* vol. 2. Munich: C.H. Beck, 316–343, 541–546.

———. 2002. "Willi Hennig (1913–1976) als akademischer Lehrer." In J. Schulz (ed.), *Fokus Biologiegeschichte.* Berlin: Akadras, 53–64.

———. 2003. "Willi Hennig and the Rise of Cladistics." In A. Legakis, S. Sfenthourakis, R. Polymeni, and M. Thessalou-Legaki (eds.), *The New Panorama of Animal Evolution (Proc. 18th Int. Congr. Zoology).* Sofia-Moscow: Pensoft, 369–379.

———. 2008. "Carl Linnaeus, the Order of Nature, and Binominal Names." *Deutsche entomologische Zeitschrift,* 55: 13–17.

———. 2010. "Willi Hennig, the Cautious Revolutioniser." *Paleodiversity*, 3 suppl.: 3–9.

———. 2013. *From Taxonomy to Phylogenetics: Life and Work of Willi Hennig*. Leiden: Brill.

Vogel, J. and Xylander, W.E.R. 1999. "Willi Hennig—Ein Oberlausitzer Naturforscher mit Weltgeltung." *Berichte der Naturforschenden Gesellschaft der Oberlausitz*, 7–8: 145–155.

Watrous, L.E., and Wheeler, Q.D. 1981. "The Out-Group Comparison Method of Character Analysis." *Systematic Zoology*, 30: 1–11.

Wheeler, Q.D. 2008. "Undisciplined Thinking: Morphology and Hennig's Unfinished Revolution." *Systematic Entomology*, 33: 2–7.

Wiley, E.O. 1981. *Phylogenetics: The Theory and Practice of Phylogenetic Systematics*. New York: Wiley.

3

Homology as a Bridge between Evolutionary Morphology, Developmental Evolution, and Phylogenetic Systematics

MANFRED D. LAUBICHLER

In one of the central conceptual papers on the homology concept, Hans Spemann introduces an important distinction between phylogenetic or historical and developmental or mechanistic conceptions of homology. This distinction continues to shape debates about homology today (Spemann 1915; but see also Wagner 1989; Hall 1992). A central passage is worth quoting:

> We no longer believe that we first can establish the phylogenetic relations between animals in order to subsequently derive developmental laws. Rather we begin to realize, that we first have to determine these laws, before we can understand or even establish the morphological series that we use to classify organisms. (Spemann 1915, 84)

This passage illustrates the conceptual and methodological distinction between historical and mechanistic approaches within biology. Even though the current situation, almost a century after Spemann wrote his essay, is more complex, it still reflects these seemingly incompatible positions. On the one hand, historical biology attempts to reconstruct the history of life based on molecular, morphological, and paleontological evidence, employing an increasingly sophisticated methodological repertoire of phylogenetic reconstruction. The mechanistic tradition, on the other hand, is reflected in approaches within the fields of developmental evolution and evolutionary developmental biology that ultimately aim to explain the developmental origins of phenotypes, of phenotypic variation, and of the transformations of phenotypes throughout evolutionary history.

The relationship between these two approaches represents yet another one of the classic "chicken and egg" problems of evolutionary biology. Trying to overcome these predicaments, many biologists assume that they can use an independently derived phylogeny—generally based on molecular data—that would then serve as the backdrop for further analysis of phenotypic evolution. The literature discussing the pros and cons of this approach is vast and remains ambiguous. In this chapter I take a different approach. Rather than argue about the validity and difficulties of present methods of phylogeny reconstruction, I suggest that by analyzing the history of the homology concept in the twentieth century we can draw some valuable lessons that allow us to see how current approaches in developmental evolution—aided by a much more detailed understanding of "developmental laws" than Spemann had a hundred years ago—make the project of understanding the "morphological series that we use to classify organisms" much more feasible. As a consequence, theoretical discussions in phylogenetic systematics and cladistics would benefit from a better understanding of the principles of developmental evolution.

WHAT CAN WE LEARN FROM THE HISTORY OF THE HOMOLOGY CONCEPT?

The history of the homology concept has been discussed in a number of papers (e.g., Spemann 1915; Panchen 1994, 1999; Rieppel 1994). However, the majority of them focus on nineteenth-century developments (for additional perspectives, see Donoghue 1992; Laubichler 2000; Brigandt 2003). In order to understand the current situation and see how developmental evolution can contribute to a resolution of Spemann's dilemma, we also need to briefly review how the homology concept continued to play a role in twentieth-century debates. As the conceptual discussion in the context of phylogenetic systematics and cladistics has been covered elsewhere, I focus on developmental and mechanistic conceptions of homology that continued and refined Spemann's argument.

Spemann's Critique of the Homology Concept

Hans Spemann was a hard-core experimentalist who was deeply skeptical of theoretical speculation. Only rarely did he diverge from his modus operandi, and when he did, as in the paper on homology, it is an indication of how important he thought this issue was for all of biology. In his 1936 book, *Experimentelle Beiträge zu einer Theorie der Entwicklung,* he compared his theoretical methods to that of an archaeologist. This is more than just a coy cultural reference. Rather it shows a deep methodological understanding of the problems of historical biology and the relationship between mechanisms and history. The archaeologist

> re-creates the image of a god from fragments that only he holds in his hands. He has to believe in the existence of the whole, even though he does not know it; but he also cannot just re-create it according to his own ideas. . . . Foremost he is obliged to honor the fractures. Only then can he hope to fit new findings at their right place. (Spemann 1936, 275)

This was indeed the method that Spemann employed when he traced the origins of the homology concept in the "History and Critique of the Homology Concept" through a period of idealistic morphology (Camper, Goethe, Geoffroy de St. Hilaire, Owen), followed by a historical period of comparative anatomy and phylogeny (Darwin, Haeckel, Gegenbaur, Müller), and finally the causal-analytical period of causal morphology and *Entwicklungsmechanik.* According to Spemann, the basic tenets of the homology concept were already established during the idealistic period of morphology and included a geometric conception of an ideal archetype and the comparison between similar parts of different animals irrespective of their function. This version of the homology concept found its canonical expression in Owen's (1843, 374) definition of homology: "HOMOLOGUE: The same organ in different animals under every variety of form and function." Owen (1848) further distinguished general homology, the similarity between a morphological character and its representation in the archetype, from special homology, the similarity between the same character in two species. This distinction identifying "general" with "ideal" was especially characteristic of the idealistic period of morphology.

In the decades after the publication of the *Origin,* comparative anatomists (especially in Germany) were mainly concerned with the derivation of phylogenetic trees. This endeavor added two temporal dimensions to discussions about homology, one ontogenetic and one phylogenetic. The sameness of morphological structures in different taxa could be explained by common descent—this is the essence of the historical homology concept. In addition, the identification of homologues in different taxa became an important tool for deriving phylogenetic relationships.

To avoid circular reasoning, however, this procedure required independent criteria for the identification of homologies. In the idealistic framework, the ideal archetype served as the reference frame for establishing homologies. In the context of the historical homology concept, embryological evidence replaced reference archetypes. Many leading evolutionary morphologists, including Haeckel and Gegenbaur, argued that true homology can only exist between two parts that have developed from the same anlage, or embryonic stage (Haeckel 1866; Gegenbaur 1878; Laubichler 2003a, 2003b). One type of historical process—ontogeny—thus became crucial in explaining the other—phylogeny. However, ontogeny is intrinsically messy, as Haeckel noticed when he interpreted individual development (ontogeny) as a recapitulation of phylogeny. To account for deviations from a simple linear model, such as shortened or otherwise altered ontogenetic sequences, he further distinguished between palingenesis and caenogenesis (Haeckel 1866). However, attempts to find an explanation for homology between adult characters in the early embryonic stages (anlagen) soon encountered empirical and conceptual difficulties.

Not surprisingly, in his critique of the historical homology concept Spemann focuses on the flexibility of development, and especially the phenomenon of regeneration. These empirical observations provided the main challenges to this preformistic approach to homology. Among the best known of these examples are the phenomena of vertebrate lens induction and regeneration. In the normal course of development the lens is formed out of the head ectoderm at the point of contact with the optic vesicle of the forebrain. After surgically removing the lens, or even larger parts of the eye, the lens is regenerated in certain amphibian species. But it is no longer formed out of the original tissue (the head ectoderm) but rather out of a different source, the dorsal margin of the iris. Furthermore, Spemann himself found that in certain amphibian species transplanted pieces of epidermis that were not part of the anlage of the lens also could be induced to produce a lens.

To address these problems, Spemann reintroduced Lankester's (1870, 39) original distinction between homogeny and homoplasy. In Lankester's definition, *homogeny* refers to those aspects of homology that can be traced directly to the common ancestor. This is a restricted definition insofar as it requires the continuous presence of all features of a particular character from ancestor to descendant species. Therefore only the more general aspects of organismal design will be homogeneous between species, while further differentiations or independent developments would not fall under this category. In Lankester's definition, *homoplasy,* on the other hand, refers to the similarity that is produced "when identical or nearly similar forces, or environments, act on two or more parts of an organism which are exactly or nearly alike." Then "the resulting modifications of the various parts will be exactly or nearly alike" (Lankester 1870, 39). For Spemann this distinction is operational in the sense that it focuses attention on those "forces or environments" that are the mechanistic cause for organic similarity. When seen this way, lenses that originate in different ways and from different materials could still be regarded as homologous, as could the characters of two embryos or adults that develop from cells separated after the first divisions, in which case the reduction of homology to common anlagen reaches its limit. The homology concept of the historical period (homogeny sensu Lankester), with its emphasis on historical continuity (both ontogenetically and phylogenetically), disintegrates because it cannot account for the peculiar features of development. Therefore, Spemann argued, homology has to be approached from within the causal-mechanistic analysis of development as captured by Lankester's original definition of homoplasy.

Gavin de Beer: The Return of Development

In the early 1970s Gavin de Beer (1971) reconnected the question of homology with the old problems of embryology (see also Hall 2000, 2004). In a widely read primer written for students, de Beer discussed several of the still unsolved problems of homology. Not surprisingly, most of these problems were connected with questions of development. De Beer realized that the principle of common descent does not solve all the difficulties that are associated with the homology of morphological characters. While common descent can at least suggest a reason for some apparent oddities in development, it does not always offer a mechanistic explanation. Take, for example, the location of the laryngeal nerve in mammals, which runs backwards, loops around the *ductus arteriosus,* and then runs forward to innervate the larynx. Pointing out the homology of the mammalian *ductus arteriosus* with the sixth arterial arch of the fish and the homology of the laryngal nerve with the fourth branchial branch of the vagus shows conservatism of evolution and therefore something of an explanation, but not a mechanism for bringing it about. For many other features, common descent does not even provide such an explanation.

De Beer, in analyzing the contributions of embryology and genetics to understanding the problem of homology, raised several important issues. For instance, he pointed out that a correspondence between early stages of development is not necessary for adult characters to be homologous, that different organizer-induction processes can lead to homologous adult structures, and that the identity of genes does not guarantee the homology between characters. He essentially restated Spemann's positions of 1915. In doing so he brought development back to the discussions about homology.

Rupert Riedl: The Order of Homology and the Systems Theory of Evolution

A few years after de Beer identified homology as the great unsolved problem in biology, Rupert Riedl presented his "systems theory of evolution" (Riedl 1975, 1978; see also Wagner and Laubichler 2004). Riedl attempted to explain, as the title of his book suggests, the often astonishing manifestations of "order" in the living world. Animal morphologies are clustered, and the morphospace of all possible life-forms is mostly empty (Gould 1977, 1989). These facts cannot easily be reconciled with the idea of gradual evolution that was at the core of models in population and quantitative genetics at that time. In the early seventies, many theorists challenged the canonical neo-Darwinian picture of evolution that was an outgrowth of the Modern Synthesis of the 1930s and 1940s (e.g., Eldredge and Gould 1972; Gould and Vrba 1982; Margulis 1982; Maynard-Smith et al. 1985). Indeed, a newly emerging focus on the role of developmental processes in evolution, the romantic phase of "evo-devo," also blossomed in those years (Wagner, Chiu, and Laubichler 2000). Homology is central to Riedl's theory, probably more so than to most other proposals of that time.

For Riedl, homology is the most visible expression of natural order. In one sense homology is simply a consequence of common descent. However, phylogenetic relationships explain the distribution of homologues but not their mechanistic cause. But neither could the models of quantitative and population genetics account for the remarkable expression of biological order as evidenced by the distribution of homologues. In Riedl's opinion, the answer to the problem of homology could be found only in the systemic conditions of development. These could be seen as an expression of a fourfold order: norm, hierarchy, interdependency, and tradition.

In Riedl's theoretical system, all four expressions of order—norm, hierarchy, interdependency, and tradition—are part of the explanation of homology. Homologues are seen as identical, that is, normative parts, whose identity is maintained by systemic (functional) interdependencies within the developmental processes that produce them, and that form a hierarchical system that is a consequence of tradition (i.e., inheritance and common descent). Another concept that Riedl introduced to describe the mechanistic causes for the identity of homologues is that of burden. Burden is a measure of the degree of systemic integration of specific characters within the developmental process. The more integrated a character is within development, the higher its burden and the more stable the character. The idea of burden is closely related to the notion of developmental constraints that was proposed around the same time. Both concepts acknowledge the intrinsic limitations that the developmental system imposes on the degrees of variation of a specific character. They differ in that developmental constraints focus more on the limitations imposed on variation, whereas burden is defined as a quantitative measurement of the cost of changing a character that is functionally embedded in a complex developmental system of interdependencies. But, insofar as homology can be seen as a statement about the limitations of variation in specific characters (Wagner 1999), the notions of burden and developmental constraint stand at the beginning of the recent interest in a mechanistic explanation of homology that is the core of the biological homology concept (see below). Riedl's approach to the problem of homology contains many elements of earlier conceptions. His distinction between different forms of homology and his reliance on Remane's homology criteria put him in the traditions of comparative anatomy and theoretical biology. However, his emphasis on a causal mechanism for the explanation of homology also makes him an heir to Spemann's causal-analytical approach to homology.

The Last Decades: The Biological Homology Concept, Hierarchical Homology, and Partial Homology

Coinciding with the simultaneous rise of evolutionary developmental biology and cladistics, the problem of homology has received more attention than ever before. Here I cannot discuss all the different contributions in any detail (but see Donoghue 1992; Hall 1994; Bock and Cardew 1999 for excellent reviews of the current state of the discussions). Not surprisingly, in the context of developmental evolution, development plays an important role in conceptualizing homology (Roth 1984, 1988; Wagner 1989, 1994, 1995, 1996; Donoghue 1992). Here I follow Wagner's (1999) outline of the basic principles of the biological homology concept.

The core assumption of the biological homology concept is that homologues are the units of phenotypic evolution. As such they are individuated quasi-autonomous parts of an organism that share certain elements and variational properties. Therefore, if two characters are to be homologous, they can differ only in those aspects of their structure that are not subject to shared developmental constraints. The role of developmental mechanisms is to guarantee the identity of two structures, since they limit the variational properties of quasi-autonomous units. Below I will explore the connections between developmental mechanisms and the problem of homology further.

Here I just want to mention two additional dimensions of homology, which have recently received some attention: hierarchical and partial. Homology can occur between objects at different levels of the biological hierarchy (Riedl 1975; Abouheif 1997). Homology at these different levels is generally recognized by independent criteria of comparison at each level, such as Remane's criteria or sequence comparisons (see also Laubichler 1999). It is an open question to what extent these different forms of homology coincide, that is, whether we can deduce morphological homology based on established genetic homologues (see also below). Developmental processes figure prominently in this context. They mediate between the different levels of homologues (genetic and morphological). It is the goal of the causal-analytical approach to homology to find an explanation for the existence of morphological homologues in the developmental processes that produce them. So far all studies that have explicitly considered this question, rather than just assumed that the developmentally prior objects determine the status of the derived characters, have cautioned against this form of preformism (see, e.g., Wagner 1989; Dickinson 1995; Abouheif 1997).

Related to the question of hierarchical homology is the problem of partial homology (Wake 1999). Partial homology assumes that in the case of certain complex characters not all elements need to be homologous but that it is possible to identify certain parts that are. There are, for instance, cases, such as the paired appendages of gnathostomes, that share certain developmental mechanisms (anteroposterior patterning) but differ in others (skeletogenesis) (Wagner 1999). These cases usually represent a hierarchy of shared derived characters (Hennig 1966) that can be interpreted as an increasingly inclusive set of partial homologues (see also Lankester 1870).

In conclusion, what this brief history of the causal-mechanistic homology concept in the twentieth century tells us is that homology is indeed "the central concept of biology." We have also seen that all contributions to homology fall within the three categories discussed by Spemann in 1915. Even though Spemann already formulated the basic principles of the causal-analytical approach to homology, his program did not become realized until about three decades ago, when new conceptual insights began to challenge the dominance of the Modern Synthesis in evolutionary biology. This challenge coincided with breakthroughs in developmental genetics and has led to the present program of developmental evolution. We have, in a sense, come full circle. Fortunately, we can also draw on the insights gained in the context of other approaches to homology, especially within phylogenetic systematics.

THE HOMOLOGY PROBLEM IN PHYLOGENETIC SYSTEMATICS

Adolf Remane: The Problem of Identifying Homology

In the interwar period theoretical problems of biology, especially those connected to morphology and questions related to developmental physiology and genetics, were widely discussed. These discussions included an intensive debate about homology, although the arguments tended to center on conceptual questions and matters of definition. For instance, Ludwig von Bertalanffy and Adolf Meyer attempted to distinguish different notions of homology, such as typological, ontogenetic-typological, phylogenetic, and developmental physiological homology (Meyer 1926; Bertalanffy 1934). These theoretical debates clarified issues of nomenclature but did little to contribute operational research programs in comparative and historical biology.

The situation changed in 1952 when Adolf Remane published his treatise, *Die Grundlagen des natürlichen Systems, der vergleichenden Anatomie und der Phylogenetik* (Remane 1952). There Remane discusses homology in the context of phylogeny and systematics. He gives the following rationale for his theoretical analysis. At the time he was working on a comprehensive overview of all animal phyla that would later become the basis for his successful textbook in systematic zoology (Remane, Storch, and Welsch 1975). Therefore it was important to clarify the theoretical foundations of both phylogeny and systematics.

For Remane, this was only possible by way of a thorough understanding of the principles of comparative anatomy, which, in turn, entailed an operational account of homology.

As discussed above, natural systems of classification were always based on some notion of similarity or sameness. In the context of phylogeny, a natural system implies that systematic groups are distinguished by the shared characters they inherited from their last common ancestor, in other words, by historical homologues. The problem is how these shared characters can be identified within the practice of comparative anatomy. For this purpose Remane developed a set of criteria that provided the morphologist with a checklist for establishing sameness, that is, homology.

Remane's criteria (three main and three auxiliary ones) lead to a probabilistic argument for homology. If these criteria are fulfilled, then it is more likely that two characters are homologous than that they are completely independent. Homology is likely when there is similarity between relative positions of characters within a common structural plan, when there is similarity in the structural details between these characters, and when transitional forms exist. In those cases where the characters under consideration are too simple and do not have enough structural details to be compared directly according to the three main criteria, they can still be considered homologous if they are present in a large number of related species and if there are other such characters that have a similar distribution. The likelihood of homology is diminished, however, if such characters are also present in nonrelated species.

As might be expected from someone who attempts to produce a phylogenetic system, Remane's approach to homology is historical. The distribution of homologues among different taxa is seen as a consequence of their phylogenetic relationship. But this leaves one with the epistemological problem of how one can identify those homologues that are used to establish phylogenetic relationships between taxa independently of a preexisting phylogeny. Remane's homology criteria are intended to overcome this problem, but he did not stop there. In the second part of his book, which is nowadays mostly ignored, he discusses various "phylogenetic laws," such as the biogenetic law, the principle of conserved earlier stages of development, and various principles of differentiation and specialization that could account for the remarkably ordered transformations observable in phylogeny. All these ideas place him squarely within the historical tradition of comparative anatomy as described by Spemann.

Willi Hennig: The Place of Homology in Phylogenetic Systematics

In the 1950s Willi Hennig also began to develop his method of phylogenetic systematics (cladistics) (Hennig 1950, 1966). Drawing heavily from the rich German tradition of systematics and also from the conceptual discussions in theoretical biology, Hennig developed an operational definition of homology in the context of phylogenetic systematics. Hennig started with the assumption that "evolution is the transformation of organismal form and behavior" (Zimmermann 1953, translation as in Hennig 1966, 88; see also Donoghue and Kadereit 1992). This process of organic transformation includes anagenesis as well as cladogenesis. Hennig assumed that it is possible to reconstruct this process (i.e., phylogeny) by following the sequence of transformations of specific characters. Hennig's main insight was to characterize each new lineage by one or more transformed characters, the so-called synapomorphies. These have to be distinguished from those character states that are shared between different lineages, the symplesiomorphies. Whether a particular character or character state is a synapomorphy or a symplesiomorphy therefore depends on the rank of the taxa one is analyzing. But in any case, both synapomorphies and symplesiomorphies refer to the sameness between characters. They are, however, more inclusive than traditional homology because the absence of a character is also a legitimate character state in phylogenetic analyses. Hennig's methods eventually transformed systematics, especially after the English translation of his book on phylogenetic systematics was published in 1966 (Hennig 1966). But while his methods are operational in the sense that though they allow for a logically consistent reconstruction of phylogeny, they still depend on a prior assessment of the sameness of characters, that is, homology. Henning defined homology to include all the transformed states of a character.

This definition, however, still requires independent criteria for the assessment of homology. Hennig employed a variety of methods to identify homologues, such as paleontological evidence, but he also heavily relied on Remane, whose homology criteria then become auxiliary criteria in the context of Hennig's definition.

I have described both Remane's and Hennig's contributions to homology in some detail, even though they do not make many references to development, because their insights have become the basis for modern phylogenetic systematics. And as Paula Mabee (2000) and others (e.g.,

Wagner 1999) have pointed out, a proper phylogeny is still the basis for all further work in developmental evolution.

THE HOMOLOGY CONCEPT IN DEVELOPMENTAL EVOLUTION

This brief overview of some twentieth-century contributions to the discussion of the homology concept makes it clear that the Spemannian distinction between the historical and causal-mechanistic explanations of sameness is indeed an accurate description of a conceptual divide within the biological sciences. Furthermore, it has become clear that we have to distinguish between historical and causal (or a combination of the two) explanations of organic similarity and various criteria for establishing this similarity (see also Bolker and Raff 1996).

As with most biological concepts, the interpretation of homology depends on a specific reference process (Laubichler 1999; Wagner and Laubichler 2000). In the case of homology, the relevant reference processes are evolution and development. Both are complex, hierarchical processes that are linked in various ways. It is therefore not surprising that depending on which aspect of evolution or development one is studying, different interpretations of homology will be relevant. Attempts to eliminate this inherent multidimensionality of the homology concept and to develop increasingly "sharper" definitions of homology are therefore rather fruitless.

David Wake has suggested that we stop worrying about what homology "is" and that we begin to address the interesting empirical questions that are connected with the notion of homology, such as stasis, modularity, the preservation of design, or latent homology (Wake 1999). But the multidimensionality of the homology concept is also one of the main reasons that it is indeed the central concept of biological classification (Wake 1994).

Homology is no longer tied exclusively to morphological characters; rather it is a notion that applies to everything from genes to behaviors (see Bock and Cardew 1999 for an overview). The ubiquitous presence of the homology concept in different areas of biology makes it all the more pressing that we are clear about the relevant reference process in each case. In the case of sequence homology, the reference process of molecular evolution encompasses the mechanisms of base-pair substitution, the translocation of chromosomal elements, the duplication of genes, and so on. On the other hand, the reference process of phenotypic evolution, which is the basis for the homology of morphological characters, includes a different set of evolutionary mechanisms, such as changes to the regulatory logic in genomic control systems that ultimately control such processes as allometric growth, changes in the timing of developmental events, developmental constraints, canalization, modularity, and other phenomena.

I cannot explore all dimensions of homology here; rather I will limit my analysis to the different connections that exist between the homology of morphological characters and developmental processes. There are at least four different connections between morphological homology and development. Developmental processes can be used to establish the homology between morphological characters, developmental processes can provide an explanation for the homology between morphological characters, developmental pathways can themselves be homologues, and developmental processes are part of the explanation of evolutionary innovations that can also be interpreted as incipient homologues. Below I briefly sketch the assumptions and problems associated with these four dimensions of homology in development. One cautionary note, however, applies to all four cases. None of these cases is clear-cut in the sense that we can establish a priori a general set of rules on how developmental processes are connected to homology. Rather, as new evidence suggests, homology is a systemic property of evolving developmental systems. Only when we understand the properties of these systems can we hope to come any closer to unraveling this "unsolved problem" of biology (de Beer 1971).

The Role of Development in Establishing Homology

Establishing homology between morphological characters is central for understanding evolutionary transformations and innovations. Consequently, in the course of the past 150 years, various criteria for identifying homology have been proposed. However, in recent years many researchers have tended to assume that sequence and gene expression data trump all other forms of evidence when considering homology between morphological characters (e.g., Quiring et al. 1994; Halder, Callaerts, and Gehring 1995). This attitude is reminiscent of earlier attempts to establish homology between morphological characters by comparing their embryological origins (anlagen). But while the earlier focus on the similarity of embryological origins generally limited the number of possible homologues, the present focus on expression patterns is more likely to increase the reach of homology relations, thereby rendering them uninformative. The recent discussion about a possible homology between vertebrate and insect eyes due to a shared master control gene (*Pax-6*) is just one example of this trend (Quiring et al. 1994; Halder, Callaerts, and Gehring 1995; Dahl, Koseki, and Balling 1997).

Both the genetic and the embryological approach tie morphological homology to the presence of specific elements in the ontogenetic sequence that leads to these characters. This is an inherently preformistic notion. The problem with such an approach to homology is that while it is appealing to reduce the problem of establishing homology on the morphological level to the simpler question of the presence or absence of certain identifiable elements, it does not, in most cases, adequately represent the complexities of the developmental processes that create the shared similarities between homologues. The same developmental role of orthologous genes does not guarantee the identity of morphological characters, nor are the same developmental pathways required to create homologous characters. As data on the multiple roles of many transcription factors such as *distalless* demonstrate, the same regulatory module can be employed in different developmental pathways (Panganiban et al. 1997). Similarly, a regenerated lens develops through a different developmental pathway (see above). These problems do not imply that sequence and gene expression data are useless for assessing homology relations between morphological characters. On the contrary, they provide us with additional information that needs to be weighted together with all other forms of evidence, not unlike in systematics, where it is increasingly common to use both molecular and morphological data to resolve questions of phylogeny. We can, however, conclude that morphological identity is a systemic property that needs to be understood in the context of the reference process of morphological and developmental evolution rather than a preformistic concept.

The Role of Developmental Processes in the Explanation of Homology

It is the goal of the causal-analytical (biological) approach to homology to provide a mechanistic explanation for the phenomenon of organic sameness. The relevant reference process for morphological homology is developmental evolution: development, because it is the proximate cause of morphological characters; and evolution, because it deals with organic transformations and stability. Günter Wagner (1999) has outlined a research program to test the biological homology concept empirically. Its basic steps involve identifying putative homologues, determining a proper phylogeny, describing intra- and interspecific patterns of variation, analyzing the modes of development for each putative homologue, and testing whether differences in the modes of development affect differences in variational properties.

The assumption behind this approach is that an explanation for the stability of homologues can be found in the properties of the developmental processes that create them. Stability of morphological characters implies that the potential variation of these characters is limited or at least constrained in particular ways. This can be accomplished by constraints acting on morphogenetic mechanisms as well as by morphostatic mechanisms that maintain or stabilize character identity (Wagner and Misof 1993). The question of homology is therefore connected to the related issues of modularity and canalization (Wagner 1996; Wagner and Altenberg 1996; Wake 1999). We can therefore conclude that in the context of the biological homology concept a mechanistic explanation of the homology of morphological characters has to involve the systemic properties of developmental processes of both morphogenesis and morphostasis.

In addition, recent advances in our understanding of the regulatory genomic control systems for development have revealed more details of the structure of developmental processes that produce phenotypes (for a summary, see Davidson 2006, 2011; Peter and Davidson 2009). These insights have also sharpened the focus of developmental evolution. We now have a much clearer understanding of how the genome, going through a sequence of regulatory states, controls the emergence of phenotypic characters. Furthermore, comparative analysis of gene regulatory networks has revealed both the hierarchical nature and the different degrees of conservatism of network elements (Davidson 2006, 2011; Erwin and Davidson 2006). These findings suggest that there is a correlation between the conserved structure of regulatory elements—to be distinguished from conservation of individual elements and sequences—and the conservation of morphological features.

Another conceptual development in the context of developmental evolution, one that is closely connected to the analysis of the structure of gene regulatory networks, is Wagner's (2007) recent proposal regarding what he calls character identity networks. The argument here is a continuation of his earlier proposal of a biological homology concept. Building on our understanding of gene regulatory networks and their various properties—especially their conserved structures—Wagner develops a conceptual framework that can account for both conservation or homology and innovation or novelty. And building on conceptual insights from phylogenetic systematics, namely, the distinction between characters and character states, Wagner connects these phenotypic observations with the structure of gene regulatory networks that produce them. This leads to a distinction between different types of mutational changes: those that produce a new character state, leaving the basic architecture of the underlying genomic character identity network unchanged, and those that change the character identity network itself, thus establishing a new morphological character or an evolutionary novelty.

The Linked Problems of Homology and Innovation

Developmental processes have also been implicated in the origin of evolutionary novelties or innovations (Müller and Wagner 1991, 1996; Müller and Newman 1999, 2003, 2005; see also *Journal of Experimental Zoology Part B: Molecular and Developmental Evolution,* special issue, *Evolutionary Innovation and Morphological Novelty* [2005]). Evolutionary innovations are incipient homologues. They are apomorphies that are the backbone of phylogenetic systematics (Hennig 1966). In the context of the reference process of developmental evolution, novelties provide us with the biggest challenge, but they are also a window through which we can study the role of developmental processes in shaping morphological transformations.

Müller and Newman (1999, 2003, 2005) have suggested that there might be different phases in the origin and establishment of morphological novelties. They argued that epigenetic processes, such as interactions between cells, tissues, and the environment, as well as the basic biomechanical properties of these parts, play an important role in the generation and integration of new structures. In a second step these incipient homologues would then become integrated, both genetically and developmentally, to function as autonomous organizers of organismic design (Müller and Newman 1999, 2003). This hypothesis suggests an important role for developmental processes in the generation of organic diversity.

Developmental Processes as Homologues

Descriptions of developmental processes now routinely involve characterizations of gene regulatory networks. Recently it has become clear that gene regulatory networks are modular structures (Wagner 1996;

Abouheif 1999; Wake 1999; Davidson 2001, 2006). As modular structures they have the potential to be recognized as a distinct level of homology within the biological hierarchy (Riedl 1975; Abouheif 1997; Davidson 2006; Erwin and Davidson 2006). There are several important questions that are associated with the potential homology of regulatory gene networks. One the one hand, we need to define criteria to assess the homology between different networks. This requires further detailed studies of the interactions between the elements of these networks (the structure of these characters). Complications are prone to arise due to genetic redundancy. Also, these networks acquire additional regulatory linkages and new developmental roles in the course of evolution (Abouheif 1997; Gerhart and Kirschner 1997; Davidson 2006). While this does not change the basic modular structure of networks, it makes it more difficult to delineate the exact boundaries of these characters. It is therefore to be expected that in many cases we will find partial homology between networks (and true homology between certain elements of these networks). Many components are very old, as is evidenced by their remarkable conservation across different phyla (De Robertis 1994) and the fact that they have often recombined with other modules to form new networks (Gerhart and Kirschner 1997; Davidson 2006).

We therefore need a good phylogeny before we can assess the homology between different networks. Focusing on the homology between regulatory networks also raises the question to what extent their homology implies the homology of morphological characters. As we have seen, this is an old problem. But, due to the modular organization of biological systems and the present evidence of multiple functions of many key elements as well as of recombination between the elements of these networks, unambiguous cases will be quite rare.

CONCLUSION

Homology is one of the central concepts in developmental evolution. The expected exponential growth of available gene expression and sequence data only increases the need for an operational approach to homology (Wagner 1999). It also highlights the fact that there is no single concept of homology that would capture all the interesting empirical questions that are associated with biological order (Riedl 1975) or the phenomena of organic sameness (Wake 1999). The causal-analytical approach to homology, that is, "homology in development," has a long tradition that goes back at least to Hans Spemann. A mechanistic, or biological, explanation of the causes for homology will be the key to understanding the transformations of organic forms, that is, of phenotypic evolution (e.g., Müller and Wagner 1996; Raff 1996).

The success of the causal-analytical approach to homology depends on the availability of reliable phylogenies (Wagner 1999; Mabee 2000). A historical approach to the problem of homology has been at the core of the development of modern phylogenetic systematics (cladistics). Therefore, different approaches to the problem of homology converge in the context of developmental evolution. We might thus conclude with Spemann (1915, 63), "There are concepts of such centrality, that their origin, change and disintegration, in short, their history captures the development of the science they are part of. Homology is such a concept for comparative anatomy." (And, we might add, for developmental evolution.)

REFERENCES

Abouheif, E. 1997. "Developmental Genetics and Homology: A Hierarchical Approach." *Trends in Ecology and Evolution,* 12: 405–408.

———. 1999. "Establishing Homology Criteria for Regulatory Gene Networks: Prospects and Challenges." In G.R. Bock and G. Cardew (eds.), *Homology. Novartis Foundation Symposium* 222. Chichester: Wiley, 207–225.

Abouheif, E., Akam, M., Dickinson, W.J., Holland, P.W.H., Meyer, A., Patel, N.H., Raff, R.A., Roth, V.L., and Wray, G.A. 1997. "Homology and Developmental Genes." *Trends in Genetics,* 13: 432–433.

Bertalanffy, L. v. 1934. "Wesen und Geschichte des Homologiebegriffes." *Unsere Welt,* 28.

Bock, G.R., and G. Cardew, eds. 1999. *Homology. Novartis Foundation Symposium* 222. Chichester: Wiley.

Bolker, J.A., and Raff, R.A. 1966. "Developmental Genetics and Traditional Homology." *BioEssays,* 16: 489–494.

Brigandt, I. 2002. "Homology and the Origin of Correspondence." *Biology and Philosophy,* 17: 389–407.

———. 2003. "Homology in Comparative, Molecular, and Evolutionary Developmental Biology: The Radiation of a Concept." *Journal of Experimental Zoology, Part B: Molecular and Developmental Evolution,* 299B: 9–17.

Butler, A.B., and Saidel, W.M. 2000. "Defining Sameness: Historical, Biological, and Generative Homology." *BioEssays,* 22: 846–853.

Dahl, E., Koseki, H., and Balling, R. 1997. "Pax Genes and Organogenesis." *BioEssays,* 19: 755–765.

Davidson, E. 2001. *Genomic Regulatory Systems: Development and Evolution.* San Diego, CA: Academic Press/Elsevier.

———. 2006. *The Regulatory Genome: Gene Regulatory Networks in Development and Evolution.* San Diego, CA: Academic Press/Elsevier.

Davidson, E., and Erwin, D. 2006. "Gene Regulatory Networks and the Evolution of Animal Body Plans." *Science,* 311: 796–800.

de Beer, G. 1971. *Homology, an Unsolved Problem.* Oxford: Oxford University Press.

De Robertis, E.M. 1994. "The Homeobox in Cell Differentiation and Evolution." In D. Duboule (ed.), *Guidebook to the Homeobox Genes.* Oxford: Oxford University Press.

Dickinson, W.J. 1995. "Molecules and Morphology: Where's the Homology?" *Trends in Genetics,* 11: 119–121.

Donoghue, M.J. 1992. "Homology." In E.F. Keller and E.A. Lloyd (eds.), *Keywords in Evolutionary Biology.* Cambridge, MA: Harvard University Press.

Doyle, J.J., and Davis, J.I. 1998. "Homology in Molecular Phylogenetics: A Parsimony Perspective." In D.E. Soltis, P.S. Soltis, and J.J. Doyle (eds.), *Molecular Systematics of Plants.* Norwell: Kluwer Academic.

Eldredge, N., and Gould, S.J. 1972. "Punctuated Equilibria: An Alternative to Phyletic Gradualism." In T.J.M. Schopf (ed.), *Models in Paleobiology.* San Francisco: Freeman, 82–115.

Farris, J.S., and Kluge, A.G. 1986. "Synapomorphy, Parsimony, and Evidence." *Taxon,* 35: 298–306.

Gegenbaur, K. 1878. *Grundriss der vergleichenden Anatomie.* Leipzig: Engelmann.

Gerhart, J., and Kirschner, M. 1997. *Cells, Embryos, and Evolution.* Malden, MA: Blackwell.

Ghiselin, M.T. 1976. "The Nomenclature of Correspondence: A New Look at 'Homology' and 'Analogy.'" In R.B. Masterson, W. Hodos, and H. Jerison (eds.), *Evolution, Brain and Behavior: Persistent Problems.* Hillsdale, NJ: Lawrence Erlbaum, 129–132.

Gilbert, S.E., Opitz, J.M., Raff, R.A. 1996. "Resynthesizing Evolutionary and Developmental Biology." *Developmental Biology,* 173: 357–372.

Gould, S.J. 1977. *Ontogeny and Phylogeny.* Cambridge, MA: Harvard University Press.

———. 1989. *Wonderful Life.* New York: Norton.

Gould, S.J., and Vrba, E. 1982. "Exaptation: A Missing Term in the Science of Form." *Paleobiology,* 8: 4–15.

Haas, O., and Simpson, G.G. 1946. "Analysis of Some Phylogenetic Terms, with Attempts at Redefinition." *Proceedings of the American Philosophical Society,* 90: 319–349.

Haeckel, E. 1866. *Generelle Morphologie der Organismen.* Berlin: Reimer.

Halder, G., Callaerts, P., and Gehring, W.J. 1995. "Induction of Ectopic Eyes by Targeted Expression of the *Eyeless* Gene in *Drosophila.*" *Science,* 267: 1788–1792.

Hall, B.K. 1992. *Evolutionary Developmental Biology.* London: Chapman & Hall.

———. 2000. *Homology: The Hierarchical Basis of Comparative Biology.* San Diego, CA: Academic Press.

———. 2004. *Homology: The Hierarchical Basis of Comparative Biology*. San Diego, CA: Academic Press.

———. 2007. "Homology and Homoplasy." In M. Matthen and C. Stephens (eds.), *Handbook of the Philosophy of Science: Philosophy of Biology*. San Diego, CA: Academic Press.

Hennig, W. 1950. *Grundzige einer Theorie der Phylogenetischen Systematik*. Berlin: Deutscher Zentralverlag.

———. 1966. *Phylogenetic Systematics*. Trans. D.D. Davis and R. Zangerl. Urbana: University of Illinois Press.

Holmes, E.B. 1980. "Reconsideration of Some Systematic Concepts and Terms." *Evolutionary Theory*, 5: 35–87.

Inglis, W.G. 1966. "The Observational Basis of Homology." *Systematic Biology*, 15: 219–228.

———. 1988. "Cladogenesis and Anagenesis: A Confusion of Synapomorphies." *Journal of Zoological Systematics and Evolutionary Research*, 26: 1–11.

Jardine, N. 1967. "The Concept of Homology in Biology." British Journal for the Philosophy of Science, 18: 125–139.

Kleisner, K. 2007. "The Formation of the Theory of Homology in Biological Sciences." *Acta Biotheoretica*, 55: 317–340.

Lankester, R. 1870. "On the Use of the Term Homology." *Annals and Magazine of Natural History, Zoology, Botany and Geology*, 6: 34–43.

Laubichler, M.D. 1999. "A Semiotic Perspective on Biological Objects and Biological Functions." *Semiotica*, 127: 415–431.

———. 2000. "Homology in Development and the Development of the Homology Concept." *American Zoologist*, 40: 777–788.

———. 2003a. "Carl Gegenbaur (1832–1903): Integrating Comparative Anatomy and Embryology." *Journal of Experimental Zoology: Part B Molecular and Developmental Evolution*, 300B: 23–31.

———. 2003b. "Units and Levels of Selection in Developing Systems." In B. Hall and W. Olson (eds.), *Keywords and Concepts in Evolutionary Developmental Biology*. Cambridge, MA: Harvard University Press, 332–341.

Lauder, G.V. 1994. "Homology, Form, and Function." In B.K. Hall (ed.), *Homology: The Hierarchical Basis of Comparative Biology*. San Diego, CA: Academic Press, 151–196.

Love, A.C. 2007. "Functional Homology and Homology of Function: Biological Concepts and Philosophical Consequences." *Biology and Philosophy*, 22: 691–708.

Mabee, P.M. 2000. "Developmental Data and Phylogenetic Systematic Evolution of the Vertebrate Limb." *American Zoologist*, 40: 789–800.

Margulis, L. 1982. *Symbiosis in Cell Evolution*. San Francisco: Freeman.

Maynard-Smith, J., Burian, R., Kauffman, S., Alberch, P., Campbell, J., Goodwin, B., Lande, R., Raup, D., and Wolpert, L. 1985. "Developmental Constraints and Evolution." *Quarterly Review of Biology*, 60: 265–287.

Meyer, A. 1926. *Logik der Morphologie im Rahmen einer Logik der gesamten Biologie*. Berlin: Julius Springer.

Mooi, R.D., and Gill, A.C. 2010. "Phylogenies without Synapomorphies—A Crisis in Fish Systematics: Time to Show Some Character." *Zootaxa*, 2450: 26–40.

Müller, G.B., and Newman, S.A. 1999. "Generation, Integration, and Autonomy: Three Steps in the Evolution of Homology." In G.R. Bock and G. Cardew (eds.), *Homology. Novartis Foundation Symposium* 222. Chichester: Wiley.

———. 2003. "Origination of Organismal Form: The Forgotten Cause in Evolutionary Theory." In G.B. Müller and S.A. Newman (eds.), *Origination of Organismal Form*. Boston, MA: MIT Press, 3–10.

———. 2005. "The Innovation Triad: An EvoDevo Agenda." *Journal of Experimental Zoology Part B: Molecular and Developmental Evolution*, 304: 487–503.

Müller, G.B. , and Wagner, G.F. 1991. "Novelty in Evolution: Restructuring the Concept." *Annual Review of Ecology, Evolution, and Systematics*, 22: 229–256.

———. 1996. "Homology, *Hox* Genes, and Developmental Integration." *American Zoologist*, 36: 4–13.

Owen, R. 1843. *Lectures on the Comparative Anatorny and Physiology of the Invertebrate Animals, Delivered at the Royal College of Surgeons, in 1843*. London: Longman, Brown, Green, and Longmans.

———. 1848. *On the Archetype and Homologies of the Vertebrate Skeleton*. London: John van Voorst.

Panchen, A.L. 1994. "Richard Owen and the Concept of Homology." In B.K. Hall (ed.), *Homology: The Hierarchical Basis of Comparative Biology*. San Diego, CA: Academic Press, 21–62.

———. 1999. "Homology: History of a Concept." In G.R. Bock and G. Cardew (eds.), *Homology. Novartis Foundation Symposium* 222. Chichester: Wiley, 5–23.

Panganiban, G., Irvine, S.M., Lowe, C., Roehl, H., Corley, L.S., Sherbon, B., Grenier, J.K., Kimble, J., Walker, M., Wray, G.A., Swalla, B.J., Martindale, M.Q., and Carrol, S.B. 1997. "The Origin and Evolution of Animal Appendages." *Proceedings of the National Academy of Science U.S.A.*, 94: 5162–5166.

Patterson, C. 1982. "Morphological Characters and Homology." In K.A. Joysey and A.E. Friday (eds.), *Problems of Phylogenetic Reconstruction*. Systematics Association Special Vol. 21. London: Academic Press.

———. 1988. "Homology in Classical and Molecular Biology." *Molecular Biology and Evolution*, 5: 603–625.

Peter, I. and Davison, E. 2009. "Modularity and Design Principles in the Sea Urchin Embryo Gene Regulatory Network." *FEBS Letters*, 583: 3948–3958.

Pinna, M.C.C. de. 1991. "Concepts and Tests of Homology in the Cladistic Paradigm." *Cladistics*, 7: 367–394.

Quiring, R., Walldorf, U., Kloter, U., and Gehring, W.J. 1994. "Homology of the Eyeless Gene of *Drosophila* to the Small Eye Gene in Mice and Aniridia in Humans." *Science*, 265: 785–789.

Raff, R.A. 1996. *The Shape of Life: Genes, Development, and the Evolution of Animal Form*. Chicago: University of Chicago Press.

Remane, A. 1952. *Die Grundlagen des natürlichen Systems, der vergleichenden Anatomie und der Phylogenetik*. Leipzig: Akademische Verlagsgesellschaft.

Remane, A., Storch, V., and Welsch, U. 1975. *Systematische Zoologie*. Stuttgart: Gustav Fischer.

Riedl, R. 1975. *Die Ordnung des Lebendigen*. Berlin and Hamburg: Parey.

———. 1978. *Order in Living Organisms*. New York: Wiley.

Rieppel, O. 1980. "Homology, a Deductive Concept?" *Journal of Zoological Systematics and Evolutionary Research*, 18: 315–319.

———. 1994. "Homology, Topology, and Typology: The History of Modern Debates." In B.K. Hall (ed.), *Homology: The Hierarchical Basis of Comparative Biology*. San Diego: Academic Press, 63–100.

Roth, V.L. 1984. "On Homology." *Biological Journal of the Linnaean Society*, 22: 13–29.

———. 1988. "The Biological Basis of Homology." In C.J. Humpries (ed.), *Ontogeny and Systematics*. New York: Columbia University Press.

Schmitt, M. 1995. "The Homology Concept—Still Alive." In O. Breidbach and W. Kutsch (eds.), *The Nervous System of Invertebrates*. Basel: Birkauser Verlag, 425–438.

Sluys, R. 1996. "The Notion of Homology in Current Comparative Biology." *Journal of Zoological Systematics and Evolutionary Research*, 34: 145–152.

Smith, H.M. 1962. "Classification of Structural and Functional Similarities in Biology." *Systematic Biology*, 11: 45–47.

Spemann, H. 1915. "Zur Geschichte und Kritik des Begriffs der Homologie." In C. Chun and W. Johannsen (eds.), *Allgemeine Biologie*. Leipzig and Berlin: B.G. Teubner.

———. 1936. *Experimentelle Beitrage zu einerTheorie der Entwicklung*. Berlin: Springer.

Szarski, H. 1949. "The Concept of Homology in the Light of the Comparative Anatomy of Vertebrates." *Quarterly Review of Biology*, 24: 124–131.

Szucsich, N.K. 2007. "Homology: A Synthetic Concept of Evolutionary Robustness of Patterns." *Zoologica Scripta*, 36: 281–289.

Wägele, J.-W. 1996. "Identification of Apomorphies and the Role of Groundpatterns in Molecular Systematics." *Journal of Zoological Systematics and Evolutionary Research*, 34: 31–39.

Wagner, G.P. 1989. "The Biological Homology Concept." *Annual Review of Ecology, Evolution, and Systematics*, 20: 51–69.

———. 1994. "Homology and the Mechanisms of Development." In B.K. Hall (ed.), *Homology: The Hierarchical Basis of Comparative Biology*. San Diego, CA: Academic Press, 273–299.

———. 1995. "The Biological Role of Homologues: A Building Block Hypothesis." *Neues Jahrbuch für Geologie und Paläontologie, Abhandlungen*, 195: 279–288.

———. 1996. "Homologues, Natural Kinds, and the Evolution of Modularity." *American Zoologist*, 36: 36–43.

———. 1999. "A Research Programme for Testing the Biological Homology Concept." In G. Bock and G. Cardew (eds.), *Homology. Novartis Foundations Symposium* 222. Chichester: Wiley, 125–134.

Wagner, G.P., and Altenberg, L. 1996. "Complex Adaptations and the Evolution of Evolvability." *Evolution,* 50: 967–976.

Wagner, G.P., Chiu, C., and Laubichler, M. 2000. "Developmental Evolution as a Mechanistic Science: The Inference from Developmental Mechanisms to Evolutionary Processes." *American Zoologist,* 40: 819–831.

Wagner, G.P., and Laubichler, M.D. 2000. "Character Identification in Evolutionary Biology: The Role of the Organism." *Theory in Biosciences,* 119: 20–40.

———. 2004. "Rupert Riedl and the Re-Synthesis of Evolutionary and Developmental Biology." *Journal of Experimental Zoology, Part B,* 302B: 92–102.

Wagner, G.P., and Misof, B.Y. 1993. "How Can a Character Be Developmentally Constrained Despite Variation in Developmental Pathways?" *Journal of Evolutionary Biology,* 6: 449–455.

Wake, D.B. 1994. "Comparative Terminology." *Science,* 265: 268–269.

———. 1999. "Homoplasy, Homology and the Problem of 'Sameness' in Biology." In G. Bock and G. Cardew (eds.), *Homology. Novartis Foundations Symposium* 222. Chichester: Wiley, 24–33.

Williams, D.M. 2004. "Homologues and Homology, Phenetics and Cladistics: 150 Years of Progress." In D.M. Williams and P.L. Forey (eds.), *Milestones in Systematics.* Boca Raton, FL: CRC Press, 191–224.

Van Valen, Leigh M. 1982. "Homology and Causes." *Journal of Morphology,* 173: 305–312.

Zimmermann, W. 1953. *Evolution. Geschichte ihrer Probleme und Erkenntnisse.* Freiburg and Munich: Alber.

HALF PRICE BOOKS®

Half Price Books
1835 Forms Drive
Carrollton, TX 75006
OFS OrderID 37558336

Thank you for your order, Douglas Warren!

Thank you for shopping with Half Price Books! Please contact service74@hpb.com. if you
any questions, comments or concerns about your order (114-1928724-9006654)

Visit our stores to sell your books, music, movies games for cas

SKU	ISBN/UPC	Title & Author/Artist	Shelf ID	Qty	OrderSKU
S462384031	9780520276581	The Evolution of Phylogenetic Systematics ...	SCI6.3	1	

SHIPPED STANDARD TO:
Douglas Warren
21 Nw 1st Ave PO BOX # 1802
01_26471UWB
Dania FL 33004
h2vyzg6m1r82nmr@marketplace.amazon.com

ORDER# **114-1928724-9006654**
AmazonMarketplaceUS

have

PART TWO

Conceptual Foundations

4

Historical and Conceptual Perspectives on Modern Systematics

Groups, Ranks, and the Phylogenetic Turn

ANDREW HAMILTON

DOUBTFUL SPECIES, DUBIOUS RANKING

Charles Darwin is sometimes taken to have been confused about species (Mayr 1982), or to have thought that there isn't really any such thing (Ereshefsky 2010), despite the fact that his best-known book was putatively about them. These readings of Darwin are not unfounded. Mayr's (1982) discussion focuses partly on Darwin's lament that it is difficult to distinguish between species and varieties:

> Practically, when a naturalist can unite two forms together by others having intermediate characters, he treats the one as a variety of the other, ranking the most common, but sometimes the one first described, as the species, and the other as the variety. But cases of great difficulty, which I will not here enumerate, sometimes occur in deciding whether or not to rank one form as a variety of another, even when they are closely connected by intermediate links; nor will the commonly-assumed hybrid nature of the intermediate links always remove the difficulty. . . . Hence, in determining whether a form should be ranked as a species or a variety, the opinion of naturalists having sound judgment and wide experience seems the only guide to follow. (1859, 47)

Some biologists and some philosophers have been led by the kinds of concerns Darwin expressed to try to define the species taxon carefully, which has led to a proliferation of species concepts as well as to a large and growing literature and meta-literature on species concepts. The debate shows no signs of being exhausted, and there are few indications of progress.

Even if it is the case, as Mayr argued, that he confused the true nature of species, Darwin was on to something. Long debates about species concepts aside, Darwin was at least as much concerned with *ranking* as he was with grouping. Darwin, after all, thought he had solved the grouping problem: "our classifications will come to be, as far as they can so made, genealogies" (1859, 486). This, Darwin supposed, would lead to a situation in which "we shall at least be freed from the vain search for the undiscovered and undiscoverable essence of the term species" (485), because the species rank, like the genus rank, isn't real. What's real, of course, is the group.

Having been freed by the knowledge that descent with modification will mean that we should not expect to find clean lines between species or between species and varieties, and therefore that the search for the true meaning of "species" is a fool's errand, we have looked high and low for the one true species concept ever since. We have also struggled mightily with ranking, though this pursuit has not explicitly been given the attention it deserves. The history of systematics since Darwin might just as well be viewed as a series of attempts to understand and label taxonomic ranks as an attempt to understand and label species. Making classifications genealogies or even phenologies, after all, does nothing to fix ranks, and despite what Darwin says in the *Origin,* very few taxonomists have seen fit to assign superspecific ranks in ways that are "merely artificial" and "made for convenience" (1859, 485). The quest to make principled sense of ranks had an important impact on systematics before Darwin, and has had no less an impact since. Nowhere is this clearer than in the case of the early history and development of phylogenetic systematics.

This chapter traces ideas and counter-ideas about ranking that follow on a connected suite of related arguments about grouping, the reality of species and higher taxa, and the practical aspects of classifying. In the next section I offer a look at some of the objections to phylogenetic systematics before Hennig, drawing out the importance of concerns about ranking. Following is a survey of objections to pre-Hennigian phylogenetic systematics that have to do with basic considerations about what is real, knowable, and how notions about reality and what we can know about it bear on the basis for systematics. The next sections contain discussions of the ontological commitments of Walter Zimmermann and Willi Hennig, two important architects of modern phylogenetic systematics. In particular these sections deal with Zimmermann's claims that lineages are real and with Hennig's arguments leading to the conclusion that species and groups of species are both real and individuals. These ideas formed an important basis for Hennig's theory of systematics. The discussion then comes back to ranking and addresses the ways in which Hennig's mature theory does and does not offer a solution to the ranking problem. I conclude with a brief look at the legacy of ranking problems for present-day systematics.

GROUPING, RANKING, AND PHYLOGENETIC SYSTEMATICS

Phylogenetic systematics in the cladistic form we now have was born in Germany in the late 1930s and reached early adulthood in 1950 with the publication of Willi Hennig's *Grundzüge einer Theorie der Phylogenetischen Systematik*. The theory reached conceptual maturity in the mid-1960s, signaled by the publication of Hennig's *Phylogenetic Systematics* in 1966. This book is sometimes taken to be an English translation of the *Grundzüge,* but it was significantly revised and expanded by Hennig and thoroughly edited by D. Dwight Davis and Rainer Zangerl.

Phylogenetic systematics in a broader sense predates the twentieth century, of course. As we have seen, it, or something like it, was suggested by Darwin in 1859. Ernst Haeckel, who coined the terms *phylogeny* and *phylogenesis* in 1866, went on to draw famous phylogenetic trees but offered no real advice about how to make the science of ordering nature reflect the new theory of descent with modification. Indeed, the paleontologist and malacologist Francis Arthur Bather (whose work is discussed below) wrote that Haeckel proceeded in his tree-drawing endeavor "with the enthusiasm and imagination of a poet, and with a poet's license" (Bather 1929, 96). In the German-speaking world, many theorists, especially those whose work was informed by paleontology, were not enthusiastic about the possibility of a phylogenetic systematics well before the systematics wars of the latter part of the twentieth century. While many saw the value of phylogenetics for understanding evolutionary history, there were widespread concerns on both sides of the Atlantic about making systematics phylogenetic. One important concern was about the relationship between ancestral groups and their lack of an appropriate taxonomic rank.

The Austrian paleontologist Othenio Abel, a founder of paleobiology, argued in 1914 that systematics and phylogenetics are incommensurable on the grounds that the classificatory system is capable only of expressing a relatively recent and synchronic view of the biological world:

> Originally, the "system" had been merely the presentation of the upper cross section laid in the present by the genealogy of the animal kingdom. . . . At the moment, though, that we managed to trace back larger groups to their root and to detect that two, three, or more families, two, three, or more suborders, etc., merge into one stem group, it had to become clearly obvious that the "system" is never able to give a clear overview of the phylogenetic processes in the course of the history of a stem group. (33; my trans.)

Part of what underlies Abel's thinking here is his understanding of "stem group" (*stammgruppe*). He seems to have thought that tracing phylogenies leads to the identification of stem groups of species that gave rise to families. These he sometimes also called "root genera" (1914, 3). According to Abel, these groups of species do not themselves form a genus or family, and in fact they do not correspond to any taxonomic rank, so there is no way to identify them in the classificatory system developed by Linnaeus (cf. Willmann 2003). For him, the system seems to have been built for taking a synchronic "snapshot" of what is in recent, horizontal layers of rock strata. "It really seems," he wrote, "as if the system and phylogeny cannot be matched, either in the form of a 'phylogenetic systematics' or in the form of a 'systematic phylogenetics'" (Abel 1914, 34).

Abel's thinking echoes somewhat similar worries voiced earlier by the German zoologist and developmental biologist Alexander Goette and by the Swiss zoologist and paleontologist Adolf Naef. Goette (1898) was concerned that the units of analysis and description in phylogenetics and in systematics cannot map onto one another because all proper taxa are discrete, but lineages are continuous; Naef (1919, 48) argued that "a natural (phylogenetic) system" that includes stem forms is "impossible" because "when it was extant, the ancestral kind of a descendant kind was certainly itself a kind, as are the family, order, and class." The implication is that the Linnaean system does not have a ranking for distantly extinct ancestral kinds that marks them as both ancestral and of higher rank than the species in their lineage, as well as at their appropriate lower rank for their time. The system, then, in Naef's words, cannot "easily accommodate" stem forms.

Some, perhaps most, of these concerns about ranking were driven by the ascendance of idealistic morphology in the German-speaking world in the early twentieth century. After having been mostly abandoned in the nineteenth century in favor of evolutionary thinking, several versions of idealistic morphology, associated with Abel, Naef, Otto Shindewolf, and others, enjoyed a resurgence. The idealistic morphologists, of course, were not of the view that the history of lineages could or should become a fundamental consideration for classification, because they held versions of the notion that, as Naef (1922, 296) put it, the archetype is the appropriate basis for reconstructing fossil organisms and their lineages. To reconstruct ancestors based on fossils would be to abandon the search for the true natural system that one could best (or only) detect by understanding similarities of types. For many German-speaking paleontologists in the 1920s through the 1940s, a lot was at stake with respect to how the relations between ancestors, stem groups, *urformen,* and their ranks were to be understood. As we have seen, some prominent thinkers were more than a little resistant to phylogenetic systematics at least partly because of these relationships.

Adherence to idealistic morphology is not, of course, the whole explanation for skepticism among German-speaking researchers who rejected phylogenetic systematics. As Peter Bowler (1996) has pointed out at length and in wonderfully rich detail, paleontologists working at the end of the nineteenth century and in the first half of the twentieth had very different ideas about the mechanisms for change in the biological world from those that became orthodox during the Modern Synthesis. Influential thinkers whose names are not well known to contemporary students of systematics—Karl von Zittel, Franz Hilgendorf, and Melchior Neumayr—raised important questions about the very possibility of responsible reconstruction of lineages (von Zittel) and about the relationship between speciation and environmental change (Hilgendorf, Neumayr) in ways that complicated the phylogeneticist's job enormously. While both Hilgendorf and Neumayr were committed Darwinists, many biologists in Europe and America, including Abel, subscribed to some version or another of Lamarckian evolution, while others, including Schindewolf and Karl Beurlen, were orthogenicists. Most of these views were driven, or at least supported, on the basis of large and growing amounts of fossil evidence, so it is no surprise that the modern view of evolution as a more or less parsimonious process that can be retrieved by studying characters that diagnose lineage relations seemed at odds with the data and not directly in line with the best theory at the time.

Returning now to concerns about rank that bear directly on the question of whether a phylogenetic systematics was possible or even desirable, it should be noted that it is not the case that concerns about rank that were voiced in German dissolve when one considers phylogenetic systematics outside of the context of twentieth-century versions of idealistic morphology. On the contrary, objections in the twentieth-century English-speaking world—again centered in paleontology—in the twenties and thirties and even into the forties and fifties had something of the same flavor, even if the motivations were somewhat different. This should not be surprising, given that Haeckel took only very preliminary steps toward a method for phylogenetic systematics (Ghiselin 1997; Breidbach 2003) and there were no other prominent methodological or conceptual innovators to speak of until Walter Zimmermann in the late 1930s.

In England, as well as in various other Commonwealth countries, there were concerns about what forming and naming groups based on phylogeny might mean for ranking and indeed for the integrity of the Linnaean system as a whole. In his 1927 presidential address to the Geological Society of London, for instance, F. A. Bather expressed concern about the proliferation of genera and other problems of phylogenetic systematics that he took to result from phylogenetic imposition on "a classification based originally on different principles" (ciii). Bather, who was a keeper at the Natural History Museum, London, the winner of the 1911 Lyell Medal of the Geological Society of London, and a Fellow of the Royal Society, had an early interest in phylogenetics, having delivered, as he reported in his 1927 paper, a lecture to the London Amateur Scientific Society in approximately 1890 in which he entertained the notion that systematists might "base all our categories on genetic affinities."

In his 1927 address and in a separate but related paper that was delivered in the same year at the International Congress of Zoology in Budapest (Bather 1929), he argued that "the whole of our System, from the great Phyla to the very unit cells, is riddled through and through with polyphyly and convergence" (1929, ci) to the extent that phylogeny "is not necessarily the most suitable basis of classification" (ciii). He was specifically concerned that a phylogenetic systematics would lead to an undue and unwieldy proliferation of taxa, especially genera, because the Linnaean system was not based on phylogenetic principles, and the more sophisticated and detailed phylogenies become, the harder it would be to represent this knowledge without breaking up many of the existing groups into smaller groups that contain fewer subordinate groups. A system with nearly as many genera as species, of course, will not be very useful.

The main concern that drove Bather's worries about the proliferation of genera in the 1927 paper was the relationship between phylogenetic groups and how they are to be ranked. "The real difficulty," he wrote, "arises when we proceed to diagnose and name the groupings" (lxxxix). Notice that it is ranking, not grouping, that is problematic: the groups are phylogenetic. In the 1929 paper, he argued along the same lines, writing that "phylogenetic analysis will continue but our systematic nomenclature will be unable to follow it" (100). This is partly because by the standards of the time, according to Bather, it was possible to see polyphyly everywhere, particularly for paleontologists, "because their knowledge of time-relations is more complete than that of neontologists" (1927, lxxxix). This polyphyly, Bather lamented, led phylogeneticists of the time to break apart those genera that could be shown to encompass more than one lineage.

In addition to this concern, there is another, related one. Bather, who clearly wanted to maintain the Linnaean system, was convinced that it is not a natural one and that while lineages are real, no taxa are, on the grounds first that "species are abstract concepts" (xcvi) given by definition rather than by nature and second that

> to delimit a species a horizontal line must be drawn across the lineage at a level arbitrarily fixed. In practice, the levels are generally decided for us by gaps in the series of strata; but theoretically, when gradual transition is postulated, the horizontal divisions must be arbitrary. (lxxxvii)

Bather's language here echoes that of Abel and others in the German literature. Abel and Bather both conceived of systematics as faced with a problem of drawing horizontal lines through rock strata and finding names (ranks) for the groups delineated thereby. And both thought that the system could accommodate this practice well enough but that attempts to include "vertical" information caused the relation between the Linnaean system and phylogenetics to break down.

In addition to concerns voiced by paleontologists properly so called, the community of British evolutionary biologists who were concerned with systematics during the time of the neo-Darwinian synthesis saw problems similar to those pointed to by Bather and by the German-speaking critics of phylogenetic systematics. In his introduction to *The New Systematics* (1940), J. S. Huxley pointed to a dilemma that arises when species are considered synchronically, "as palaeontologists perforce must do":

> When, as often happens, more or less parallel evolution occurs in undoubtedly distinct lineages, should generic names be given to the horizontal stages—in which case the genus is not monophyletic—or to the lineages—when extreme practical inconvenience will result? (18)

Huxley's adoption of the distinction between horizontal and vertical approaches is not coincidental. He was noticing a conversation in the literature that traces to the German paleontologist Otto Shindewolf's work on ammonites in the middle and late 1920s, which was commented on by W. J. Arkell and J. A. Moy-Thomas in *The New Systematics*. Schindewolf, who was perhaps the leading theorist of idealistic morphology, asked in a 1928 paper what should be done about parallel lineages of Devonian ammonites whose transitions could be traced upward through rock strata.

Classifying these groups "horizontally" by assigning a single genus name to all the ammonites in particular strata yields a classification that does not respect the fact that the lineages had been shown to be independent of one another but is still informative and relatively stable. Classifying each lineage as a genus "vertically" and then marking species changes within each genus (lineage) respects phylogeny but has the important problem that without consensus on a method for arriving at and testing phylogenetic hypotheses, phylogenies and the species names that go with them are liable to be reorganized by other workers at will (as happened to some extent with Schindewolf's ammonites) (Brinkmann 1929; George 1933; Beurlen 1937; see also Arkell and Moy-Thomas 1940). In this case, respecting phylogeny would sacrifice stability. Shindewolf opted to work horizontally, giving generic names to the forms in the strata.

Haeckel's focus on phylogeny in 1866 yielded famous trees and optimism in some quarters about building a new systematics that was informed by the theory of descent. By seventy-five years later, when the neo-Darwinian synthesis was in full swing in the English-speaking world and idealistic morphology had made a strong resurgence in the German-speaking world, many of the influential theorists and practitioners who were in a position to speak for or against a phylogenetic basis for systematics offered arguments against it. Often they were at least as concerned about the impact of considerations of phylogeny on ranking as they were about the principles by which organisms might be grouped.

OTHER OBJECTIONS: REALITY, EPISTEMOLOGY, AND MESSY LINEAGES

Not all of the objections to phylogenetic systematics in the first half of the twentieth century had to do with ranking, of course. Several commentators raised other kinds of issues, including the problem of polyphyly (and paraphyly) for existing groups and questions about the appropriate basis for systematics. These are worth pausing to consider because they are fundamental and show something of the full set of challenges faced by modern phylogenetic systematics in its infancy and, as discussed below, bear importantly on Hennig's theory.

There were early concerns about whether anyone ought even to develop phylogenetic systematics as such, because of differing ideas about what a natural system is and does. In *The New Systematics,* Huxley argued that there are cases where "taxonomy does *not* have a phylogenetic basis," including those in which evolution has been convergent or reticulate and those in which the "almost impossible ideal" of knowing where evolution has been "parallel" and where it has been "divergent" has not been achieved (1940, 19; original emphasis). While Huxley was sympathetic to the goals of phylogenetic systematics, he also offered a tentative argument that phylogeny might not be the best basis for classification on the grounds that it is not "natural" in the right way:

> A natural system is then one which enables us to make the maximum number of prophecies and deductions. It also in the majority of cases follows the lines of phylogenetic descent, though these are not always discernible; but there are a certain number of exceptions where a phylogenetic interpretation is meaningless, and others where taxonomy and phylogeny cannot be made to square with one another. (20)

Concerns with what is natural and therefore the appropriate basis for the system are also addressed at length in the Huxley volume by the British botanist John Scott Lennox Gilmour, who would later become Director of the Cambridge University Botanic Garden. Gilmour argued against a phylogenetic basis for systematics on the grounds that the correlation of a large number of attributes gives the most natural set of groups and that phylogenetic classification, which focuses on "the relationship between genealogy and other attributes," is just one among many special-purpose classificatory schemes (1940, 473). This line of thinking was influential for decades, partly because it was taken up in the early 1960s by the pheneticists, particularly Sokal and Sneath (1963) (Hamilton and Wheeler 2008).

While Gilmour had an important impact on systematics in the English-speaking world in the second half of the twentieth century, it should be pointed out that his arguments were anticipated by the Dutch botanist Cornelis Eliza Bertus Bremekamp in the early 1930s. Bremekamp was the author of a well-read general introduction to biology (Bremekamp 1963) and was very active and productive in his work on the Acanthaceae of southern Africa, the Malay Archipelago, and Java. He rejected the possibility of a phylogenetic systematics entirely, arguing that it is not the best or even the only "explanation of the 'natural system' of affinities" (Bremekamp 1931, 8). On his view, concern with lineages is built on circular reasoning, and the fossils are better left alone because "the construction of genealogical trees by the aid of evidence derived from the study of the fossil records appears impossible" (3). Such a study is also unnecessary: the task of the taxonomist, Bremekamp wrote, "is still the same as in pre-Darwinian days: the perfection of the 'natural' system" (7). By this he meant the arrangement of organisms into groups, those groups into larger groups, and so on. Names for taxon categories, he thought, indicated relative group inclusiveness and did not correspond to anything objective in the world. Linnaeus's arguments to the contrary, he wrote, were "based on a quaint mysticism" and are of "historical value only" (9).

Objections like these were raised again and again in the literature well into the 1950s, after which they took on a less general character and were more specifically aimed at Hennig. In a pair of papers published in *Systematic Biology* in 1956 and 1958, for instance, the entomologist Robert Sidney Bigelow argued against "monophyletic classification" on the grounds that "overall similarity and recency of common ancestry are two separate and distinct phenomena, which must be considered separately (1956, 145). Bigelow's target in these two pieces, which he identifies as "a 'phylogenetic school' of thought" (1958, 49) is the tradition most closely associated with Ernst Mayr and G. G. Simpson that later came to be known as evolutionary taxonomy. Despite its later rejection of Hennig's work, evolutionary taxonomy was after all partly phylogenetic in nature, as both Mayr (1942) and Simpson (1945) advocated a version of monophyletic grouping. Contra Mayr and Simpson, Bigelow pointed out that phylogenetic classification would lead us to group crocodiles with birds. Hennig had already argued for precisely this arrangement in his 1950 book, presumably unbeknownst to Bigelow. Bigelow (1956, 146) argued that grouping in ways that does not respect overall similarity "tends to confuse the philosophical basis on which the science of taxonomy rests."

Bigelow was right, of course: without a rethinking of the conceptual foundations of systematics there was much about a new phylogenetic systematics to object to. He raised, as did others before him, central problems about the nonuniform pace of evolution, about the quality of the assumptions underlying conclusions about recency of common ancestors, and about confounding evolutionary processes. While he was most concerned with Mayr and Simpson, whom he criticized directly, Bigelow's arguments apply just as well to what was already happening in Germany, where Zimmermann and Hennig were developing a new approach to systematics. As we shall see, Hennig (1966, 88) would reject as "impossible" the "equation that 'community of similarity = community of descent'," while at the same time arguing explicitly against Bigelow's assertion that what matters is overall similarity. To these developments we now turn.

PHYLOGENETIC SYSTEMATICS: ZIMMERMANN ON GROUPING

In addition to answering fundamental questions about the relationship between grouping and ranking in a phylogenetic system—both in and out of the context of idealistic morphology—proponents of phylogenetic systematics in the second half of the twentieth century would have to deal with long-standing and important issues. These concerned the proper task of systematics, the most natural basis for the system, whether a natural or artificial system should be preferred, and whether there could be a method for detangling the relevant sorts of phylogenetic relationships in a world that is characterized by messy evolutionary histories.

Contemporary phylogenetic systematics, with its focus on monophyly as the only allowable grouping relationship, on sister groups as a centrally important unit of study, and on shared, derived characters (synapomorphies) as diagnostic of phylogenetic relationships, can be traced to the early 1930s. This approach addressed some of the issues described in the last two sections above directly, made others less important, and, while being very clear about grouping, has been criticized by some (Mayr 1969; Darlington 1970) for not including a well-worked-out theory of how groups are to be ranked.

Phylogenetic systematics as we now know it, began, at least in part, as a reaction against idealistic morphology. In an important but long-neglected paper of 1931 titled "Arbeitsweise der botanischen Phylogenetik und anderer Gruppierungswissenschaften" (Methods of Botanical Phylogenetics and other Grouping Sciences), the botanist and systematic theorist Walter Zimmermann made several innovations that would become important to Hennig's understanding of the relationship between systematics and phylogenetics, including an insistence on the objective reality of stem groups and on phylogenetic groups as naturally given, a careful explanation of what phylogenetic relationships are, and a new sense of what Zimmermann called the "task" of systematics.

For Zimmermann, the "*urform,*" whatever its rank, is objective in the sense that it does not depend on a subject for its existence: it is a matter of logic, he argued, that descendants of all sorts have ancestors (1931, 949). This is a move against both the idealists, whose grouping schemes Zimmermann and the idealists themselves regarded as subjective, and various others who offered purpose-specific groupings. Part of the force of Zimmermann's thinking lies in a reorientation of systematic and phylogenetic studies. For Zimmermann, "the task of historical phylogenetics is to find out 'how it was'" (1931, 981; in Donoghue and Kadereit 1992), and the implication is that "it" was just one way—that there is a unique fact of the matter about the history of life on Earth.

This understanding stands in sharp contrast to the set of views, prevalent in the German literature at the time, that systematics should be chiefly concerned with logical, rather than historical, relationships and with making lists of species (Hertwig 1914; Plate 1914; Naef 1917; Horn 1929). Because Zimmermann's approach was historical, it was also a move away from the centrality of capturing what is found in "horizontal" layers and from the widespread view in the German-speaking world that grouping ought to be primarily morphological and typological, with phylogenetic reconstruction a secondary task (Steiner 1937; Danser 1950). Given this approach, it is not surprising that Zimmermann (1931, 949) argued that systematists should "group phylogenetically, that is, following naturally given relationships." On his view, lineages are natural and objective as opposed to constructed and artificial.

Grouping phylogenetically, of course, requires an understanding of what phylogenetic relationships are. Zimmermann's definition reads as a very contemporary one, because Hennig incorporated it directly into his own thinking by the late 1940s. When understood phylogenetically, a dendrogram like that in figure 4.1 means, Zimmermann (1931, 989) wrote, that "B and C are more closely related to one another than to A." As Donoghue and Kadereit (1992, 78) have pointed out, this way of thinking represents a fundamental advance in phylogenetics because it defines the systematists' relation of study as the relation between sister descendants of a common ancestor, and because it points to hierarchy as the appropriate general structure for a phylogenetic system. Species B and C combine to form higher group B + C; Species A, B, and C combine to form a still more inclusive higher group A + B + C, and so

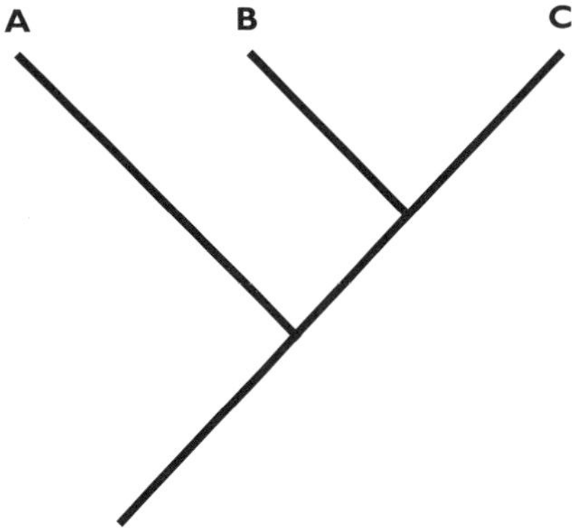

FIGURE 4.1. Simple dendrogram showing relations between three taxa.

on for groups in which there are more taxa (see fig. 4.3 below). Zimmermann argued that whatever relationships could not be expressed using such a dendrogram are not phylogenetic (1931, 990).

It is worth pausing to note, as we will again in the next section below, that this approach to phylogenetic systematics leads to principles for grouping that are well-supported theoretically. That is, no matter what one thinks of the idea that phylogeny is the best or most appropriate basis for systematics, Zimmermann pointed toward a clear theory of phylogenetics that once coupled with a method would constrain and inform the diagnosing of phylogenetic groups. In providing arguments about what is objective and real, he also set the conditions for Hennig to argue for the conclusion that phylogeny is the best general reference system for biology. Zimmermann did not, however, provide much in the way of well-grounded advice about ranking. He argued that "the relative age relationship of ancestors . . . is the only direct measure of phylogenetic relationship" (1931, 990). The locution "*P* is more closely related to *Q* than to any other entity" will hold for monophyletic groups at any rank, and one can understand historical (phylogenetic) relationships between lineages thereby, but there is nothing here that points to a principled, theoretically constrained way to rank more and less inclusive monophyletic groups within the Linnaean system.

THE DEVELOPMENT OF MODERN PHYLOGENETIC SYSTEMATICS: HENNIG'S THEORETICAL WORK

Hennig adopted and adapted Zimmermann's arguments about the reality of phylogenetic groups and used them to develop a justification for phylogeny as the basis of systematics. He also used them to develop a theory, as well as a method, for phylogenetic systematics. While Hennig's methodological innovations are now well known, the nature and importance of his ontological commitments have been widely neglected (but see Rieppel's chapter in this volume and his 2007 paper). Indeed, inattention in the literature to Hennig's concerns about what is real and the ways in which his arguments for the reality of species and of higher (monophyletic) groups support his work dates to the same year that *Phylogenetic Systematics* was published. Many, perhaps most, English-speaking systematists first learned of Hennig's thinking by way of the forty-page methodological and theoretical introduction to the Swiss entomologist and biogeographer Lars Brundin's 1966 monograph-length treatment of the "transantartic relationships" (phylogeny and biogeography) of chironomid midges (Williams and Ebach 2007; Nelson this volume; Rieppel this volume). There, however, Brundin did not describe, explain, or even mention Hennig's arguments about the reality and individuality of species and higher taxa. Though there is a passing reference of this part of Hennig's thinking in a response by Brundin (1972) to a set criticisms offered by the entomologist P. J. Darlington (1970), most commentators have had little to say on this topic.

This lacuna is surprising given that Hennig's attention to ontology dates to his earliest work on the conceptual foundations of systematics. In his first theoretical paper, "Probleme der Biologischen Systematik" (Problems of Biological Systematics), published in 1947, Hennig argued for the reality of superspecific groups on the grounds that if species are real and in speciating give rise to a new group of species, the new group is no less real, so long as all the species in it stand in the sort of relation that Zimmermann called "phylogenetic" (279). Hennig's *Grundzüge* contains arguments along the same lines (1950, 115).

By the time his 1966 book was published, Hennig had developed sophisticated views about biological individuality as a way of dealing with higher taxonomic categories and as a way of providing a justification for phylogenetics as the appropriate basis for systematics. In a section of the first chapter titled "The Phylogenetic System and Its Position among the Possible and Necessary Systems in Biology," Hennig concluded "that the choice of a general reference system for biological systematics is not at all free, but for intrinsic reasons must be the phylogenetic system" (23). This claim comes after a set of ontological arguments to the conclusion that genetic and genealogical relationships between individual organisms form a "complex of individuals" that have a reticular structure. These complexes, he argued, are species, and gaps in this structure are species boundaries (18).

According to Hennig, who drew heavily on the logician John Gregg's (1954) reading of the philosopher J.H. Woodger (1952), phylogenetic relationships *necessarily* take the form of a nested hierarchy like that represented in Zimmermann's dendrogram of 1931. Henning notes, following Gregg, that descent with modification also "necessarily" leads to a hierarchical structure, and goes on to point out that when seeking a general reference system for all of biology "the hierarchic system is the adequate form of representation for the phylogenetic relationships between species" (20). This is to argue that if one takes descent with modification (understood as phylogenetics) as the basis for systematics, then one is obliged to represent phylogenetic relationships using hierarchical structures, but it is not, by itself, quite to argue in favor of taking phylogenetics to be the best basis for systematics, though a tight relationship between the process of evolution and its pattern of outcomes is strongly implied.

Hennig offered four numbered arguments for a phylogenetic basis for systematics (and for phylogenetic systematics as the general reference system for all of biology), but he took the first of them to be "decisive":

> Making the phylogenetic system the general reference system for special systematics has the inestimable advantage that the relations to all other conceivable biological systems can be most easily represented through it. This is because the historical development of organisms must necessarily be reflected in some way in all relationships between organisms. Consequently, direct relations extend from the phylogenetic system to all other possible systems, whereas there are often no such direct relations between these other systems. (22)

While the meaning of this passage is not obvious with the first or second look, Hennig's thinking is somewhat clearer in context. His view of the ontological structure of the objects studied by phylogenetics, captured in figure 4.2, was that phylogenetics studies nested groups: clades are groups of species, species are groups of organisms, and organisms are groups of semaphoronts. Semaphoronts—organisms at particular periods in their development—are, Hennig argued, "the element of all biological systematics" (6), because the semaphoront is all the properties (morphological, physiological, and psychological) of an organism at a particular developmental stage. Arrows in the circle at the bottom right of the diagram indicate that the "individual" is a group of semaphoronts.

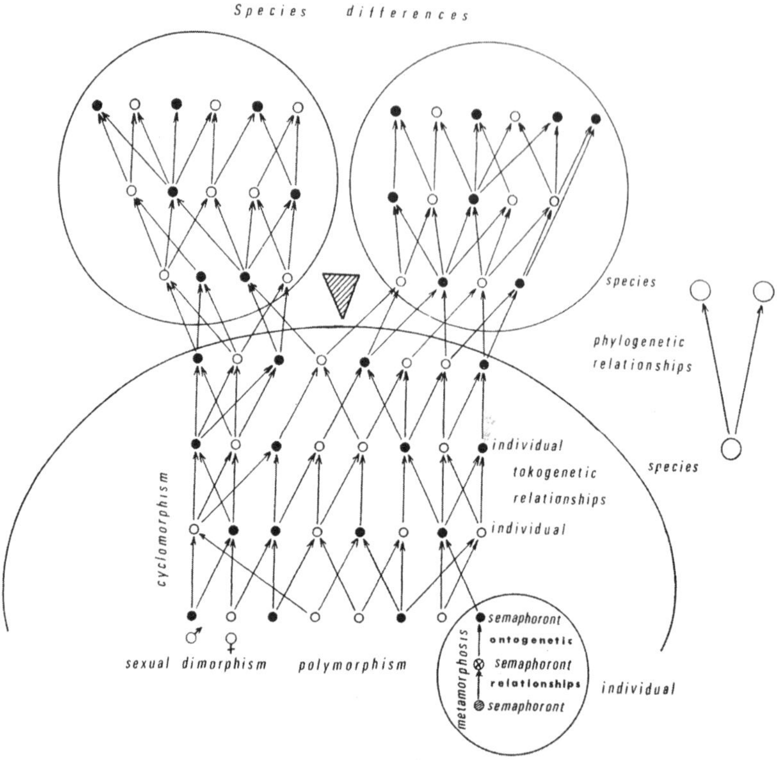

FIGURE 4.2. Hennig's (1966) figure 6, showing the relationship between temporal and developmental parts of organisms, organisms, species, and groups of species. See text for further explanation.

Arrows in the main body of the diagram point from organism to organism (open and closed circles, which represent females and males, respectively) and weave a "fabric of tokogenetic relationships." These are relationships that hold between organisms such that they form species (Hennig 1966, 30). On Hennig's view, new species arise when tokogenetic relationships between groups of organisms break down and new ones are formed. This process is represented by the wedge at the fork of the main diagram. Notice that at the terminal ends Hennig has circled two daughter species. The relation between them is phylogenetic, not tokogenetic, though tokogenetic relations are the "glue" that holds each species together. The phylogenetic relationship between these two species is represented by a branching shape on the right of the diagram. It is clear from Hennig's treatment of higher groups in his

1947 paper, his 1950 book, and in the 1966 book that he thinks organisms, species, and clades are all individuals.

The argument for the primacy of phylogenetics as the basis of the system, then, is ultimately that in representing phylogenies (relations between groups of species), phylogenetics also captures relations between organisms and between developmental stages of organisms in ways that other bases for the system do not. This is a virtue, Hennig argued, because alternative ways of approaching the system can be expressed through a phylogenetic system but not vice versa. The ideological and morphological approaches leave one or another level of analysis out. The other three arguments in favor of phylogenetic systematics are largely practical, and it is telling that Hennig thought that the argument about what phylogenetic systematics can represent, as well as how organismal, tokogenetic, and phylogenetic relationships should be represented, turned importantly on his understanding of what these relationships and entities are.

Hennig did not explicitly discuss his ideas of individuality in chapter 1, waiting instead to address the issue in the context of a subsection of chapter 2 on the origin and reality of higher taxa. As Hennig (1966, 78) understood the situation, the old and new conventional wisdom was that organisms and species could be taken to be real but that authorities including Plate (1914), Kinsey (1936, 1937), Simpson (1951), and Claus, Grobben, and Kühn in their widely circulated textbook, *Lehrbuch der Zoologie* (1932), were of the opinion that nothing at the genus level or above corresponds to any real object. There were others, of course, who were equally willing to deny the reality of species (Martini 1929; Thompson 1952; Blackwelder 1959). The Canadian entomologist and later philosopher William Robin Thompson (1952, 10ff.), for instance, argued against the reality of species on the grounds that the species taxon is an "abstraction" that is pulled together from the features of many individuals but that species themselves are not "the individual or the collectivity of individuals."

This is precisely what Hennig would deny. In contrast to most of his peers, Hennig was quite willing to accept what Michael Ghiselin (1974) later approvingly called the "radical" view that anything—including species and groups of species—that is held together by the appropriate causal, spatial, and temporal connections is appropriately recognized as an individual. Hennig's arguments, which are independent of Ghiselin's, draw heavily on Gregg's (1954) reading of Woodger (1952), as well as on the conceptual work of the zoologist and theoretical biologist

Max Hartmann (1947), the philosopher Nicolai Hartmann (1942), and the psychologist and philosopher Theodor Ziehen (1934). In arguing for the reality of higher groups, Hennig points out, echoing his earlier chapter, that groups of individuals can themselves be individuals. Hennig's examples of higher-level individuals include metazoans that are individuals made up of cells that are also individuals, bee colonies that are made up of organisms (bees) that are also individuals, and groups of species: "there can be no doubt that all the supra-individual categories from the species to the highest category rank, have individuality and reality" (1966, 81). It is not quite correct, of course, to say that the categories are real in either the morphological or the phylogenetic system; they are abstract. Hennig's point is that lineages are real and can be divided into segments that correspond to ranks in the system. The categories, he wrote, "are all segments of the temporal stream of successive interbreeding populations" (1966, 81).

Hennig's view of biological individuality was essentially the contemporary one: individuals are concrete entities located in space and time, and they have a beginning and an end (Hamilton, Smith, and Haber 2009). He cautioned that organisms are not the only kind of individuals and that "the space-matter (bodily) relationship of the parts is not decisive for the concept of individuality and reality" (Hennig 1966, 81). That is, there can be scattered individuals the spatial boundaries for which are provided by something other than skin or membrane. For Hennig, cohesion for species comes in the form of tokogenetic relationships, and for groups of species it comes in the form of phylogenetic relationships.

This view is now fairly intuitive at the species level at least, because of close ties to the biological species concept of Ernst Mayr. Hennig offered a reinterpretation of Mayr's understanding of species in terms of individuality: species begin when the appropriate interbreeding relationships obtain and end when they cease to hold. Species-hood, for Hennig, was a relationship between organisms and not directly about the persistence of member organisms through time and certainly not about the presence or absence of essential traits. Similarly, Hennig individuated groups of related species—clades—by their beginning and ending points: a clade begins when a species branches and ends when there is a further branching in one of the daughter species. If one of the daughter species in figure 4.2 were to split, the new diagram would show the beginning both of a phylogenetic relationship and of a clade.

In championing phylogenetics as the basis for systematics and phylogenetic systematics as the general reference system for all of biology by way of arguing that species and groups of species are both real and individuals, Hennig was arguing not only against the idealistic morphologists in the German-speaking world but also against many others who were publishing in English and had very different ideas both from Hennig and from each other about what is appropriately natural, what is real, and what systematics could and should accomplish. By the middle of the twentieth century, systematics was 250 years old, and its foundations were being renegotiated as ideas about kinds, biological processes, the natural world, and the goals of systematics changed both over time and across disciplinary, cultural, and national boundaries.

PHYLOGENETICS, INDIVIDUALS, AND RANKING

As Olivier Rieppel (2007) has pointed out, Hennig was very much offering a *theory,* and he responded to the challenges discussed above not by dismissing them as unimportant but by offering a new theoretical justification for his ideas about what systematics is and does. Much of this new foundation was shaped by his ontological commitments. These bear on the early development of twentieth-century systematics in important ways. First, they allowed Hennig to address the concerns about reticulation that Bather and Huxley voiced, because in Hennig's system tokogenetic relations can be reticulate, but phylogenetic relations cannot; they are necessarily hierarchical. Hennig's thinking also gives a principled distinction between the species taxon and other taxa: species are the smallest group among which phylogenetic relations obtain.

Reticulation, parallel evolution, and convergence, of course, are not small problems, and it is not generally claimed that Hennig solved them once and for all. Hennig's system has the advantage, however, that it offers a clear method for analyzing characters, testing for homology, and diagnosing lineage relationships in a way that does not lead inexorably to the problems with stability that Schindewolf and Huxley were worried about. Hennig (1966, 140–141) clearly thought that classifying "vertically" in paleontological cases like Schindewolf's ammonite example causes no special difficulties for reliability and stability. The method is the same there as it is in dealing with extant species. The difference lies in access to characters and in being limited in the paleontological case to semaphoronts as opposed to organisms. Hennig thought that his approach provided a way to diagnose lineage splitting at the species level and therefore to map phylogenies successfully, or at least as later terminology would have it, to formulate and test phylogenetic hypotheses. This is the very job that Schindewolf had given up on when he described the problem of understanding his ammonites diachronically and claimed that "another methodology than that which is built up from the principles of idealistic systematics is impossible" (1928, 145). Naef (1919) had gone still further, claiming, as Reif (1986, 176) puts it, that there was and could be "no acceptable methodology for phylogenetic systematics." Hennig surely did not have the final word on this topic, but he did provide a method, and one built on a reconceptualization of what is real and what biological individuals are.

Notice, however, that little in Hennig's theoretical framework throws light on the problem of how to rank groups above the species level. Despite all the innovations in Zimmermann's and Hennig's development of phylogenetic systematics, it would seem that the species rank was the only one in the hierarchy on which progress had been made between Darwin's suggestion in 1859 that classification should be made genealogical and Hennig's offering of a mature theory and method for doing so. Even if one grants that Hennig minimized Huxley's concerns about polyphyly, reticulation, convergence, stability, and so on, while responding to Gilmour and the idealistic morphologists through his arguments about what is natural, objective, and real, there was still the problem in the late 1960s that it was not clear how far Hennig had gone toward assuaging the concerns of sympathetic critics like Bather and unsympathetic critics like Bremekamp, who really wanted classification to be at the fore in systematics.

Hennig knew full well that understanding phylogenetic groups as individuals and superindividuals organized hierarchically as more and less inclusive groups gives a sequence of phylogenetic splitting and uniquely points to the species-level ancestor of any group but that doing so only yields what he called "the relative ranking of categories" and not "absolute rank" (1966, 83). He was concerned about ranking, writing that "many systematists do not recognize that determining the absolute rank of systematic categories is a very serious and important problem" (154). The ranking problem as it applied to Hennig is shown in figure 4.3, a cladogram in which the phylogenetic relations between the terminal taxa are clear, but no information is given about what "size" clade corresponds to a given taxonomic rank. Which, if any, of the ellipses captures the group that is appropriately recognized as a genus or a family under the assumption that the terminals are species?

Hennig's answer to this question rejects the idea, widely accepted at the time (Cain and Harrison 1958; Mayr and Ashlock 1991 [1969]),

FIGURE 4.3. Cladogram with many terminal taxa. The ellipses enclose monophyletic groups at three levels of inclusiveness.

that the degree of divergence of form should play a role in ranking. The outcome of any divergence-of-form argument, he thought, depends on what one measures and how one analyzes the data (1966, 156). Echoing Zimmermann, he argued instead that in a hierarchical system "coordination and subordination of groups is by definition set by their relative age of origin" (160): older groups will be of higher ranks, new breakdowns in tokogenetic relationships will create species, and sister groups will always be at the same level in the hierarchy. Essentially, Hennig tried to leverage relative ranking into a ranking scheme properly so called by suggesting that absolute ranking should be assigned based on age of origin for the monophyletic group in question. For ranking, all one has to do is determine the absolute age of origin for each group and then assign ranks based on conventions for dividing the history of life on Earth into stages. This plan has the advantage, according to Hennig, that it makes ranks in different groups directly comparable to one another. On pain of being unscientific, a class of beetles and a class of primates ought to be the same kind of thing (Hennig 1966, 154).

This proposal for ranking, Hennig noted, was not nearly as well constrained theoretically as his work on grouping:

> The requirement that rank designations must express the comparability of categories—however remotely related the groups—is not a fundamental principle of phylogenetic systematics to the same degree as the requirement that the system must contain only monophyletic groups and that sister groups must be coordinate and given the same rank. (191)

This acknowledgment does not quite go far enough. While Hennig did give a reason for continuing to have ranks—ranks make certain monophyletic groups of different lineages directly comparable with one another—he gave no arguments at all for continuing the *Linnaean* system of ranking or for having seven ranks and attendant subranks rather than six or sixty, and his suggestion that ranking should be based on absolute age does not make direct contact with his arguments about the reality of species and higher taxa or with his methods for diagnosing monophyly.

In 1966, then, systematists had the set of documents that laid out a new theory and method for understanding phylogenetics and its relationship to systematics, as well as a monograph that would begin what has justifiably been called a revolution in systematics (Mishler 2009; Nelson 2004). As Ernst Mayr and Peter Ashlock pointed out, however, the new approach was "without any clear-cut theory of ranking." Systematics was very much in the same position as it had been in 1859 with respect to what to do about classification. For the proponents of phylogenetic systematics, the question of grouping had been solved in principle by Darwin and then solved in practice by Hennig and the innovators that followed him, but ranking remained comparatively undertheorized. Darwin's remark about the freedom associated with the recognition that ranks are "artificial" or "merely convenient" has proven to be well off the mark. Mayr and Hennig, along with many others, saw the lack of a well-articulated *theory* of ranking as an important challenge.[1]

PERSPECTIVES ON THE PRESENT: CONCEPTUAL FOUNDATIONS FOR RANKING?

Systematics in the twenty-first century has inherited an old problem. Attempts at solutions have come in three forms. One is to offer criteria for ranking, as Hennig and others have done. The problem with this strategy is that multiple criteria have been offered, and there has been no clear winner. There are several reasons for this, but an important one is that decisions about ranking are not theoretically constrained in the same way as those about grouping, even in those cases where grouping has been well worked out. A second strategy is to take a pluralist stance and allow multiple ranking schemes that are each purpose-relative. This allows the systematist to use the ranking scheme that best respects her classificatory goals or the peculiarities of her study group but has the obvious problem that it can lead to difficulties in translating across ranking schemes, study systems, or schools of thought. A third strategy is to give up on ranking altogether, or rather to read classifications directly off phylogenies in a rank-free way. This approach has the advantage that it obviates the need for further consideration of how ranks should be discovered or assigned, but it has the disadvantages that it upends several hundred years of convention, introduces a host of practical difficulties, and is not clearly a classificatory scheme in the way many systematists and users of systematic information—policy makers, conservation biologists, ecologists, and others—understand that notion.

It should be noted that the debates about which of these approaches to adopt (as well as which considerations are most salient within each of them) are largely independent of the set of concerns that motivated the systematics wars of the last third of the twentieth century (Hull 1988). Those arguments about theory, practice, and the best basis for systematics have been largely settled in favor of phylogenetics. It is not the case, however, that questions about ranking are a special problem for phylogenetic systematics or for Hennig. Phenetics has precisely the same problem—much to say about grouping but little to say about ranking—as does the evolutionary taxonomy of Mayr, Simpson, and Dobzhansky.

What all of this suggests is that it is time to ask after a theoretical basis for ranking, a ranking *concept* to go along with our attention to species concepts. At present we have no way of choosing between competing criteria for ranking, though there are plenty of arguments for and against particular schemes. Age of origin has advantages and disadvantages just as degree of similarity does, and so on with the other proposed criteria. Which of these sets of advantages and disadvantages is more important or meaningful depends on what ranking is supposed to do, whether and how it is supposed to reflect nature, and how much emphasis should be placed on communicating about the biological world where we have to do so at the expense of "natural" groupings.

If history is any guide, outstanding questions about ranking will not be resolved by the kind of pluralism Marc Ereshefsky (2001) has advocated or by consensus about what is practical or expedient. Consider the debate over the PhyloCode, which is very much about ranking and not about grouping. It stands as another indication that Darwin was wrong about what systematists will do once the ranking problem is solved. Without a *theoretically* constrained scheme for ranking, it is likely that these debates will be with us to stay.

What, then, should we do? It may not be too much to hope that a theoretical understanding of ranks can be developed from an ontologically informed understanding of what the objects of ranking are. It may also not be too much to hope that recent work on individuality, especially as Hennig used and understood the notion, might be helpful in this endeavor. As Rieppel (this volume) has argued, however, getting to this point will likely require attention to the prior problem of how the ontology of phylogenetic systematics should be understood now. Very few systematists pay much attention to the ontological details of Hennig's arguments, even though these details are—or at least were—foundational. Given this and Rieppel's arguments, there is also reason to think that these foundations have shifted or been reshaped over the past five decades. If Rieppel is right that the cut between cladists and transformed or pattern cladists ultimately lies in different ways of conceptualizing the objects and processes under discussion, then there remains much to be done at the theoretical foundations of phylogenetic systematics before we can expect to make progress on the ranking problem.

Acknowledgments

This material is based on work supported by the National Science Foundation under grant number SES-09083935, for which I am thankful. This chapter is a substantially expanded, refocused, and differently framed version of some ideas previously presented at the 2011 ISHPSSB meeting in Salt Lake City and in *Cladistics* (Hamilton 2011).

NOTES

1. It is worth mentioning that Mayr and the evolutionary school of taxonomy to which he belonged also did not have a theory of ranking, despite their frequent criticisms of others on this point. They had an established set of practices on which "hierarchical rank is determined by degree of difference" (Mayr and Ashlock 1991 [1969], 141). This idea is an old one, and was advocated by Darwin, who wrote, "the amount of difference between the several groups—that is the amount of modification each has undergone—is expressed by such terms as genera, families, orders, and classes" (1888, 181). But Darwin, as we have seen, did not think of these ranks as given by his theory, and he did not think of degree of difference as something that could be read off the world.

REFERENCES

Abel, O. 1914. *Die vorzeitlichen Säugetiere.* Jena: G. Fischer Verlag.

Arkell, W. J., and Moy-Thomas, J. A. 1940. "Palaeontology and the Taxonomic Problem." In J. Huxley (ed.), *The New Systematics.* Oxford: Clarendon Press, 395–410.

Bather, F.A. 1927. "Biological Classification: Past and Future." *Quarterly Journal of the Geological Society of London*, 83: lxii–civ.
———. 1929. "Quo Vadis? A Question from the Palaeontologist to the Systematist." In *Comptes-Rendus Xe Congrès International de Zoologie*. Budapest: Imprimerie Stephaneum, 95–101.
Beurlen, K. 1937. *Die Stammungsgeschichtlichen Grundlagen der Abstammungslehre*. Jena: G. Fischer Verlag.
Bigelow, R.S. 1956. "Monophyletic Classification and Evolution." *Systematic Zoology*, 5: 145–146.
———. 1958. "Classification and Phylogeny." *Systematic Zoology*, 8: 49–59.
Blackwelder, R.E. 1959. "The Functions and Limits of Classification." *Systematic Zoology*, 8: 202–211.
Bowler, P. 1996. *Life's Splendid Drama*. Chicago: University of Chicago Press.
Breidback, O. 2003. "From Haeckel to Hennig: The Early Development of Phylogenetics in German-Speaking Europe." *Cladistics*, 19: 449–479.
Bremekamp, C.E.B. 1931. "The Principles of Taxonomy and the Theory of Evolution." *Pamphlets of the South African Biological Society*, 4: 1–8.
———. 1963. *The Various Aspects of Biology: Essays by a Botanist on Classification and Main Contents of the Principal Branches of Biology*. Amsterdam: Noord Hollandsche Uitg. Mij.
Brinkmann, R. 1929. "Statistisch-biostratigraphische untersuchungen an mitteljurrassischen Ammoniten über Artbegriff und Stammesentwicklung." *Abhandlungen der Gesellschaft der Wissenschafter zu Gottingen, Mathematisch-Physikalische Klasses, Neue Folge*, 13, 3.
Brundin, L. 1966. "Transantarctic Relationships and Their Significance, as Evidenced by Chironomid Midges, with a Monograph of the Subfamilies Podonominae and Aphroteniinae and the Austral Heptagyiae." *Kungliga Svenska Vetenskapsakademiens Handlingar*, Fjarde Serien, 11 (1): 1–472.
———. 1972. "Phylogenetics and Biogeography." *Systematic Zoology*, 21 (1): 69–79.
Cain, A.J., and Harrison, G.A. 1958. "An Analysis of the Taxonomist's Judgment of Affinity." *Proceedings of the Zoological Society of London*, 131: 85–98.
Claus, C., Grobben, K., and Kühn, A. 1932. *Lehrbuch der Zoologie*. Jena: G. Fischer Verlag.
Danser, B.H. 1950. "A Theory of Systematics." *Bibliotheca. Biotheoretica*, 4: 1–20.
Darlington, P.J. 1970. "A Practical Criticism of Hennig-Brundin 'Phylogenetic Systematics' and Antarctic Biogeography." *Systematic Zoology*, 19: 1–18.
Darwin, C.R. 1859. *On the Origin of Species*. London: John Murray.
Donoghue, M.J., and Kadereit, J.W. 1992. "Walter Zimmerman and the Growth of Phylogenetic Theory." *Systematic Biology*, 41: 74–85.
Ereshefsky, M. 2001. *The Poverty of the Linnaean Hierarchy: A Philosophical Study of Biological Taxonomy*. Cambridge: Cambridge University Press.
———. 2010. "Darwin's Solution to the Species Problem." *Synthese*, 175: 405–425.

George, T.N. 1933. "Palingenesis and Palaeontology." *Biological Reviews,* 8: 107–135.

Ghiselin, M.T. 1974. "A Radical Solution to the Species Problem." *Systematic Zoology,* 23: 536–544.

———. 1997. *Metaphysics and the Origin of Species.* Albany: SUNY Press.

Gilmour, J.S.L. 1940. "Taxonomy and Philosophy." In J. Huxley (ed.), *The New Systematics.* Oxford: Clarendon Press, 461–474.

Gregg, J.R. 1954. *The Language of Taxonomy: An Application of Symbolic Logic to the Study of Classificatory Systems.* New York: Columbia University Press.

Goette, A. 1898. "Einiges über die Entwickelung der Scyphopolypen." *Zeitschrift für Wissenschaftliche Zoologie,* 63: 292–378.

Haber, M.H., and Hamilton, A. 2010. "Clade Selection and Levels of Lineage: A Reply to Rieppel." *Biological Theory,* 4: 214–218.

Hamilton, A. 2011. "From Types to Individuals: Hennig's Ontology and the Development of Phylogenetic Systematics." *Cladistics,* 27: 1–11.

Hamilton, A., and Haber, M. 2006. "Are Clades Reproducers?" *Biological Theory,* 1: 381–391.

Hamilton, A., Smith, N. and Haber, M. 2009. "Social Insects and the Individuality Thesis: Cohesion and the Colony as a Selectable Individual." In J. Gadau and J. Fewell (eds.), *Organization of Insect Societies: From Genome to Sociocomplexity.* Cambridge, MA: Harvard University Press, 572–589.

Hamilton, A., and Wheeler, Q.W. 2008. "Taxonomy and Why History of Science Matters for Science: A Case Study." *Isis,* 99: 331–340.

Haeckel, E. 1866. *Generelle Morphologie der Organismen. Allgemeine Grundzüge der organischen Formenwissenschaft, mechanisch begründet durch die von Charles Darwin reformierte Descendenz-Theorie.* Berlin: G. Reimer.

Hartmann, M. 1947. *Allgemeine Biologie.* Jena: G. Fischer Verlag.

Hartmann, N. 1942. *Systematische Philosophie.* Stuttgart: Kohlhammer.

Hennig, W. 1947. "Probleme der Biologischen Systematic." *Forschungen und Fortschritte,* 21–23: 276–279.

———. 1950. *Grundzüge einer Theorie der Phylogenetischen Systematik.* Berlin: Deutscher Zentralverlag.

———. 1966. *Phylogenetic Systematics.* Trans. D.D. Davis and R. Zangerl. Urbana: University of Illinois Press.

Hertwig, R. 1914. "Die Abstammungslehre." In P. Hinneberg (ed.), *Die Kultur der Gegenwart.* Teil 3, Abt. 4, 4. Abstammungslehre—Systematik—Paläontologie—Biogeographie. Berlin: Teubner, 1–91.

Horn, W. 1929. "Heteropod Zoology and Entomological Complexes." *Entomological News,* 39: 172–178.

Hull, D.L. 1988. *Science as a Process: An Evolutionary Account of the Social and Conceptual Development of Science.* Chicago: University of Chicago Press.

Huxley, J.S. 1940. "Towards the New Systematics." In J. Huxley (ed.), *The New Systematics.* Oxford: Clarendon Press, 1–46.

Kinsey, A.C. 1936. "The Origin of Higher Categories in *Cynips.*" Indiana University Publications, Science Series, no. 4. Contribution from the Depart-

ment of Zoölogy, Indiana University, no. 242 (Entomological Series, no. 10).
———. 1937. "Superspecific Variation in Nature and in Classification." *American Naturalist,* 71: 206–222.
Martini, E. 1929. "Diskussionbemerkungen zu den Tagesthemata 'Artbegriff' und 'Phylogenie.'" *3. Wand. Versamml. dtsch. Ent., Giessen,* 94–98.
Mayr, E. 1942. *Systematics and the Origin of Species: From the Viewpoint of a Zoologist.* Cambridge, MA: Harvard University Press.
———. 1982. *The Growth of Biological Thought.* Cambridge, MA: Belknap Press.
———. 1969. *Principles of Systematic Zoology.* New York: McGraw-Hill.
Mayr, E., and Ashlock, P. 1991. *Principles of Systematic Zoology.* 2nd ed. New York: McGraw-Hill.
Mishler, B.D. 2009. "Three Centuries of Paradigm Changes in Biological Classification: Is the End in Sight?" *Taxon,* 58: 61–67.
Naef, A. 1917. *Die individuelle Entwicklung organischer Formen als Urkunde ihrer Stammesgeschichte.* Jena: G. Fischer Verlag.
———. 1919. *Idealistische Morphologie und Phylogenetik.* Jena: G. Fischer Verlag.
———. 1922. *Die fossilen Tintenfische. Eine paläozoologische Monographie.* Jena: G. Fischer Verlag.
Nelson, G. 2004. "Cladistics: Its Arrested Development." In D.M. Williams and P.L. Forey (eds.), *Milestones in Systematics.* Systematics Association Special Volume 67. Boca Raton, FL: CRC Press, 127–147.
Plate, L. 1914. "Prinzipien der Systematik mit besonderer Berücksichtigung des Systems der Tiere." In P. Hinneberg (ed.), *Die Kultur der Gegenwart.* Teil 3, Abt. 4, 4. Abstammungslehre—Systematik—Paläontologie—Biogeographie. Berlin: Teubner, 92–164.
Reif, W.-E. 1986. "Evolutionary Theory in German Paleontology." In M. Grene (ed.), *Dimensions of Darwinism: Themes and Counterthemes in Twentieth-Century Evolutionary Theory.* Cambridge: Cambridge University Press, 173–204.
Rieppel, O. 2007. "The Metaphysics of Hennig's Phylogenetic Systematics: Substance, Events and Laws of Nature." *Systematics and Biodiversity,* 5: 345–360.
———. 2009. "Do Clades Cladogenerate?" *Biological Theory,* 3: 375–379.
Schindewolf, O. 1928. "Prinzipienfragen der biologischen Systematik." *Paläontologische Zeitschrift,* 9: 122–169.
Simpson, G.G. 1945. "The Principles of Classification and a Classification of the Mammals." *Bulletin of the American Museum of Natural History* 85: 1–350.
———. 1951. "The Species Concept." *Evolution,* 5: 285–298.
Sokal, R., and Sneath, H.A. 1963. *Principles of Numerical Taxonomy.* San Francisco: Freeman.
Steiner, B. 1937. *Stilgesetzliche Morphologie.* Leipzig: Geest & Portig.
Thompson, W.R. 1952. "The Philosophical Foundations of Systematics." *Canadian Entomologist,* 84: 1–16.

Williams, D., and Ebach, M. 2007. *Foundations of Systematics and Biogeography.* New York: Springer.
Willmann, R. 2003. "From Haeckel to Hennig: The Early Development of Phylogenetics in German-Speaking Europe." *Cladistics,* 19: 449–479.
Woodger, J.H. 1952. "From Biology to Mathematics." *British Journal for the Philosophy of Science,* 3: 1–21.
Ziehen, T. 1934. *Erkenntnistheorie. Zweite Auflage. Erster Teil. Allgemeine Grundlegung der Erkennnistheorie. Spezielle Erkenntnistheorie der Empfindungstatsachen einschliesslich Raumtheorie.* Jena: G. Fischer Verlag.
Zimmermann, W. 1931. "Arbeitsweise der Botanischen Phylogenetik." In E. Abderhalden (ed.), *Handbuch der Biologischen Arbeitsmethoden,* Abt. IX, Teil 3/II. Berlin: Urban and Swartzenburg, 941–1053.

5

The Early Cladogenesis of Cladistics

OLIVIER RIEPPEL

Looking back on the formation of the Willi Hennig Society in 1980, and its journal *Cladistics,* which started to appear in winter 1985, Lorenzen—in a paper titled "Phylogenetic Systematics Yesterday, Today and Tomorrow"—raised the question, "How could it happen that Hennig's phylogenetic systematics, which was explicitly developed on the basis of evolutionary theory, was robbed of this very foundation, and even met broad approval in such an a-phylogenetic rendition?" (1994, 201). Dupuis (1984) similarly concluded that Willi Hennig's (1913–76) insights form at best the foundation only of modern systematics, while Hennig's biographer, Michael Schmitt, found modern systematics to reflect "an enhanced version [of Hennig's phylogenetic systematics] now called 'cladistics'" (2001, 341). The principal culprits in this transformation of Hennig's phylogenetic systematics identified by Lorenzen (1994) were Gareth Nelson and Norman Platnick from the American Museum of Natural History in New York, as well as Colin Patterson from the Natural History Museum London. The philosopher John Beatty (1982) provided the first analysis of this transformation, asserting that the practitioners of cladistics were wrong in thinking that systematics could be rendered independent of evolutionary theory. Instead, Beatty (1982) concluded that transformed or "pattern cladism" is incompatible with evolutionary theory. The goal of this chapter is to trace the origins of Hennig's (1950, 1966) phylogenetic systematics, its subsequent transformation into pattern cladism (see also Hull

1988), and the consequences of these developments for present-day systematics.

When Hennig (1950) developed his theory and method of phylogenetic systematics, he was looking back on earlier developments in German comparative biology, which pitted self-proclaimed phylogenetic systematists against idealistic morphologists (see Zimmermann 1930, 1937–38, 1943; Zündorf 1940, 1943; Donoghue and Kadereit 1992; Hamilton this volume). In Hennig's (1950) view, it was phylogeny, and hence time, not similarity, which was to provide the backbone for a truly phylogenetic system. Hennig (1949, 1950, 1953) chose time as the standard of measure (*tertium comparationis*) for the phylogenetic system.[1] This requires time to be unidirectional and irreversible. Whereas character evolution may be reversed (e.g., the Apterygota among insects), phylogenetic time is irreversible. The process of phylogeny naturally stretches through time; the phylogenetic system consequently had to be structured according to the temporal sequence of dichotomous species lineage splitting events (Rieppel 2011d). It is this emphasis on cladogenesis (a term first introduced by Rensch [1947]) that led Ernst Mayr to call Hennig's (1950, 1966) approach "cladism," or the "cladistic approach," because phylogenetic relationships are analyzed in terms of "recency of common descent" (Mayr 1968, 547) or "nearest common branching point" (Mayr 1965, 78). Mayr's term quickly caught on: "In Hennig's writing, phylogenetic means cladistic" (Byers 1969, 105). David Hull claimed to have been the first to introduce the term *transformed cladists* during the banquet address he delivered at the first meeting of the Willi Hennig Society in Lawrence, Kansas, in 1980: "Just as a character can become modified successively through time in biological evolution, scientists can also gradually change their minds" (Hull 1988, 191). Adopting Beatty's (1982) term *pattern cladism* for this new brand of cladistics, Patterson (1988, 72) located the beginning of "the 'transformation' of Hennigian phylogenetics" in Gareth Nelson's insight (in a 1976 manuscript, now part of Nelson and Platnick 1981) that "cladograms and phylogenetic trees had wrongly or confusingly been treated as equivalent." What had happened?

THE BRUNDIN AFFAIR

Hennig's theorizing became available to "most English-speaking systematists" (Byers 1969, 105) only after a somewhat abbreviated and revised version of the original treatise (Hennig 1950) was published in

English translation (Hennig 1966). But this was not Gareth Nelson's original introduction to cladistics. Instead, Nelson discovered cladistics in the Swedish entomologist Lars Brundin's (1966) exegesis of Hennig's work (1950, 1953, 1957, 1965) (Hull 1988; Williams and Ebach 2008; Nelson this volume), which appeared in the same year as the English translation of Hennig's own writing (Hennig 1966). In Nelson's (2004, 133) recollection, Brundin's "critique (1966: 11–64) began the [cladistic] revolution." In his address delivered at the Second Annual Willi Hennig Society Meeting in 1981, Colin Patterson recalled how he in turn discovered cladistics: "Then, one day early in 1967, Gary Nelson, who was spending six months in the BM [British Museum (Natural History)], told me that something had just appeared in the library that I might find interesting. . . . [I]t was Brundin's monograph on chironomids just arrived. I was bowled over by it—it was like discovering logic for the first time."[2] Brundin was invited to present at the 1967 Nobel Symposium on the topic "Current Problems of Lower Vertebrate Phylogeny" the insights on systematics and its relationship to paleontology that he had gained through his reading of Hennig (Brundin 1968), a presentation that both Nelson and Patterson attended (Hull 1988, 144). In Patterson's (1989, 472) recollection, "The heart of Brundin's paper was one message: 'phylogenetics is the search for the sister group.' That message eventually got through" (see also Nelson 2004, 132). Is it possible that Brundin's (1966) exegesis of Hennig's work was responsible for an early cladogenesis in the phylogenetic systematics/cladistics research program? Or was it the way Nelson and Patterson understood Brundin's interpretation of Hennig that turned systematics away from the temporal tree of life of phylogeneticists to the logically structured cladogram of pattern cladists? What if Hennig himself had planted the seeds from which germinated different interpretations of his work?

Opponents of cladistics saw not much difference in Hennig's (1950, 1966) and Brundin's (1966) writing, lumping them together in their campaign (Byers 1969; Darlington 1970; see also Brundin 1972), which later became known as the "Brundin affair" (Hull 1988, 149). The term was coined by Gareth Nelson, member of the editorial board, in a letter (dated February 28, 1972) to Albert J. Rowell, editor of the journal *Systematic Zoology*, relating to the controversy surrounding the publication of Brundin's (1972) reply to Darlington's (1970) critique. Nelson closed this letter in a statement that again bundled Hennig and Brundin together, claiming that both had been badly misunderstood in the United States, a situation the journal could help to improve.[3] As noted above, Patterson (1988, 72) anchored the beginning of the transformation of cladistics in Nelson's distinction of a cladogram from a phylogenetic tree (manuscript 1976; see also Platnick 1977; Nelson and Platnick 1981). While this is unquestionably an important starting point, another related one is the different concepts of "hierarchy" employed by phylogenetic systematists (Hennig 1950, 1966) and pattern cladists (e.g., Patterson 1982, 1988). Guided by Hull, Williams (1992, 136–137) recognized the pattern cladist hierarchy as one of boxes within boxes, whereas she identified Hennig's hierarchy as a system of species lineages splitting and splitting again.

THE DIVISION HIERARCHY

Because the publication of his 1950 book was delayed due to scarcity of paper in the postwar era, Hennig published two short papers summarizing his ideas (1947, 1949). Hennig stated that species, that is, reproductive communities, "decay by fission," that such dichotomous speciation results in "phylogenetic relations," and that therefore the "general *relational* system" that is the object of systematic research must assume a hierarchical structure (1947, 278; emphasis added). Hennig (1949, 136) continued to stress that "the general *structural* system that portrays the phylogenetic relationships that exist between species results from what we know about the origin of new species," that is, dichotomous speciation. Commenting further on the phylogenetic system, Hennig (1947, 279) called it a "division hierarchy [*Teilungshierarchie*] with the species as its basic divisible entity" (see also Hennig 1966, 72). In this context he explicitly drew an analogy between species and the histosystems invoked by the Tübingen anatomist Martin Heidenhain (1923, 1937) that form the building blocks of the enkaptic hierarchy that is instantiated by an organism.[4] Histosystems, such as cells, are characterized by *the* fundamental property of life, which is divisibility. According to Heidenhain (1923, 1937), every cell is an individual, a complex whole, composed of parts, its organelles. As cells divide, they form organs, which are individuals or complex wholes of a higher degree of complexity, characterized by emergent properties that show the complex whole to be more than the mere sum of its parts. Together, the organs form an organism, an individual or complex whole of yet greater complexity, again instantiating emergent properties. That way, the organism forms an enkaptic hierarchy, a nested system of individuals or complex wholes of increasing degrees of complexity (see also

Hueck 1926; Rieppel 2009). In drawing an analogy between the splitting of cell lineages (or of histosystems more generally) and of species lineages, Hennig (1947) drew an analogy between the enkaptic hierarchy that is instantiated by an organism and the phylogenetic system that likewise forms an enkaptic hierarchy of complex wholes of increasing degrees of complexity. The first publication in which Hennig traces the general *relational,* or *structural,* system that captures the phylogenetic relations between species to the work of Woodger (1952) and Gregg (1954) was his 1957 contribution to the centenary of the German Entomological Society, which indicates that his initial adoption of the concept of an enkaptic hierarchy (*Teilungshierarchie*) most likely was based on Bertalanffy's (1932) discussion of Heidenhain's and Woodger's work (Rieppel 2007a, 2009). Indeed, Hennig (1947, 1949, 1950) would weave together Heidenhain's concept of enkapsis with Woodger's treatment of the division hierarchy as presented by Bertalanfy (1932).

Hull (1988, 105) recounts how in 1928 Joseph Woodger (1894–1981) discovered the second edition of Alfred N. Whitehead and Bertrand Russell's *Principia Mathematica* (1925) in the library of University College London and, starting to read it on the train on his way home that evening, was excited to find in it the formal language he had been looking for that could be used in an attempt to disambiguate biology (see also Cain 2000). It was the new logic (Carus 2007), a calculus of relations, that set Woodger to work on—among other things—the logical structure of biological hierarchies as they are manifest in the organism and its ontogeny. This research would culminate in Woodger's 1952 paper, which contains a formal definition of the concept of a hierarchy—in the formulation of which Willard Quine had lent a helping hand (Woodger 1952, 11 n. 1)—that became the object of Gregg's (1954) analysis. It so happened that Ludwig von Bertalanffy (1901–72), who developed his philosophy of biology under the auspices of members of the Vienna Circle (Stadler 1997), had independently conceived of similar ideas—only reluctantly acknowledging Woodger's priority (Bertalanffy 1932, 29). Bertalanffy (1932, 261) went on to outline not only Heidenhain's concept of the enkaptic hierarchy but also Woodger's "logistical analysis of animal organization" (263), which is where Hennig (1950, 20) found a discussion of Woodger's concept of a "division hierarchy" (*Teilungshierarchie:* Bertalanffy 1932, 265).

Bertalanffy's *Theoretical Biology* (1932) exerted a major influence on Hennig (Rieppel 2007a), and Hennig's (1947, 1949, 1950) understanding of the phylogenetic system as a relational or structural system is indeed deeply rooted in Bertalanffy's philosophy of biology. Hennig (1950) adopted a dynamic (i.e., processual) conception of nature (Rieppel 2007a), citing in its support a paper by Bertalanffy in which the latter identified himself as the "father of organicism" (Bertalanffy 1941, 247; but see Phillips 1970 for deeper roots of organicism). During the early twentieth century, organicism constituted a popular trend in ecology (Jax 1998), with the limnologist August Thienemann (1882–1960) and the entomologist Karl Friederichs (1879–1969) its most prominent exponents in Germany (Potthast 2003; Rieppel 2011b); both authors were cited by Hennig (1950). It consisted in a holistic approach to nature, conceptualizing ecosystems in terms of an enkaptic hierarchy: "More than fifty years ago (1887), an American (S. A. Forbes) already designated the lake as a 'microcosm'" (Thienemann 1941, 4; for an account of organicism in Anglo-American ecology, see McIntosh 2011). In his analysis of ecosystems, Thienemann (1925) referred back to Haeckel (1866), who had distinguished a hierarchy of individuals of different levels of complexity in morphology. Thienemann (1935, 338) correspondingly went on to characterize the individual organism as an "organism [individual] of first order," the biocoenosis as an "organism of second order," and the integrated unit of biotope plus biocoenosis as an "organism of third order"—a nested hierarchy he would later explicitly identify as an enkaptic one. The anatomist Alfred Benninghoff (1890–1953) explicitly praised Friederichs for having expanded the concept of enkapsis from the individual organism to ecosystems and indeed the whole of life (Benninghoff 1935–36, 102). An enkaptic hierarchy, whether instantiated by an individual organism (Heidenhain), an ecosystem (Thienemann, Friederichs), or the phylogenetic system (Hennig), is characterized by its dynamic structure, as it results from the divisibility of its constituent parts. The subdivision of a cell, or of a species, is a spatiotemporally located process, and in that sense—unlike a spatiotemporally unrestricted set or class—is tied to individuality and causality, ultimately to reality. Through their successive subdivision, therefore, the constituent parts build up an inclusive (nested) hierarchy that is not subject to the membership relation (as would be a hierarchy of sets or classes) but to the part-whole relation. The different levels of inclusiveness form complex wholes, or individuals, of increasing degrees of complexity, with the added characteristic that each level of inclusiveness instantiates emergent properties and hence is more than the mere sum of its parts.

Hennig's (1950) phylogenetic system, which is a "general relational system" (Hennig 1947) or a "general structural system" (Hennig 1949), is thus revealed as an enkaptic hierarchy (Rieppel 2009). It is a relational, or structural (which in the language of the new logic says the same), system, as its parts are tied together through causal relations, that is, through the process of species lineages splitting and splitting again. Its processual nature ties that system to space and time, hence to causality and individuality, and ultimately to extramental reality. Hennig (1950, 118) specifically cited Friederichs (1927) when he characterized species as "causally efficacious entities." Species, as well as all higher (monophyletic) taxa, are therefore individuals, complex wholes of increasing degrees of complexity nested within one another. Species names, as well as taxon names, consequently are proper names: "In the phylogenetic system, the name of groups of animals at all levels of inclusiveness . . . are *proper names*" (Hennig 1953, 3, original emphasis; for a history of the species-as-individuals thesis in German biology, see Rieppel 2011c).

TRANSFORMING CLADISM: THE REPLACEMENT OF HISTORICAL BY LOGICAL RELATIONS

Brundin's (1966) explication of Hennig's principles is less sophisticated but much more concise than Hennig's (1950, 1966), which may explain his success in motivating the cladistic revolution (Hull 1988). In some respects, this conciseness is coupled with some degree of hardening of the doctrine, as pertains for example to Brundin's (1966) mostly misguided appeals to logic, or to his account of Hennig's "principle of dichotomy" (Rieppel 2011d; see also comments in Byers 1969; Darlington 1970). There is, however, no apparent schism between Brundin's (1966) and Hennig's (1949, 1950, 1953, 1957, 1965) theorizing. Brundin (1966, 14) followed Hennig in calling the cladogram a "structural picture" that is "in full agreement with the definition of hierarchy, as delivered by Woodger in accordance with the methods of symbolic logic." Brundin again rooted the dichotomous structure of the cladogram in allopatric speciation, a "speciation process" that "presupposes progression in space and development of spatial isolates" (17). In his response to Darlington (1970), Brundin (1972, 69) asserted, "The units of this truly phylogenetic system are the biological species and the strictly monophyletic species groups all of which have individuality and reality." However, Brundin (1966) also adopted from Hennig the convention of expressing the phylogenetic hierarchy in terms of Venn diagrams or a graphic variation thereof (e.g., Hennig 1950, fig. 37; 1957, figs. 4, 6; 1965, figs. 3, 4), which introduced conceptual ambiguity with respect to the nature of the phylogenetic hierarchy. This is, indeed, the juncture where Hennig himself prepared the ground for alternative interpretations of his work, especially by Anglo-American authors thoroughly unfamiliar with Heidenhain's concept of enkapsis. Venn diagrams represent logical relations prevailing between sets, not historical (processual) relations tying together hierarchically structured complex wholes. "The way in which a multidimensional multiplicity is to be treated classificatorily (i.e., 'systematically') is taught by set theory, which is a subdiscipline of mathematics, or of mathematical logic respectively," Hennig (1953, 6) proclaimed, without differentiating between the logic of Venn diagrams and the new logic of the relational calculus employed by Bertalanffy and Woodger. Given such conceptual ambiguity, it was easy to misapprehend Hennig's historical (relational, structural) system of complex wholes as an ahistorical hierarchy of nested sets, as indeed happened in the transformation of cladism.

Commenting on his reaction to Brundin's (1966) exposition of Hennig's principles, Patterson (1981, 195; emphasis added) again recalled "the excitement with which [he] realized that there is a *logical* basis to evolutionary relationships." Based on this insight, Patterson (1982) formalized a set-theoretical approach to the cladistic reconstruction of phylogenetic relationships with his "test of congruence." Confronting conflicting character distribution, organisms are sorted into sets based on shared derived characteristics, and—according to the principle of parsimony—the preferred phylogenetic hierarchy will be the one that, among all the sets under consideration, maximizes the relations of inclusion and exclusion while minimizing the relation of overlap. The result is not a spatiotemporally located enkaptic hierarchy of complex wholes, but a spatiotemporally unconstrained hierarchy of sets within sets (or classes). The logic of set theory, or more precisely the logic of Venn diagrams that was also implicitly appealed to by Hennig (1953, 1957) and Brundin (1966), thus divorced cladistic epistemology from the historicity of its objects of investigation. Whereas for Patterson (1981, 195; emphasis added) there existed a "logical basis to *evolutionary* relationships," he drew an even stronger conclusion in his talk delivered to the Systematics Discussion Group at the American Museum of Natural History in November 1981: "Evolution may well be true, but basing one's systematics on that belief will give bad systematics" (Patterson 2002, 31). In essence, then, the logical basis of cladistic analysis resulted in a divorce of systematics from evolutionary theory. Systematics accordingly deals with characters and their distribution; evolution deals with speciation and its consequences (Patterson 1988, 72 [also in Nelson MS. 1976]; Platnick 1977; Nelson and Platnick 1981).

Both systematic classification sensu Patterson (1982) and the phylogenetic system sensu Hennig (1950) represent nested hierarchies, but whereas the first is a spatiotemporally unconstrained hierarchy of sets (classes) subject to the membership relation, the second is a spatiotemporally constrained hierarchy of complex wholes subject to the part-whole relation. Hull writes, "Although the [move from one to the other] is metaphysically quite drastic, it does not alter any traditional inferences" (1988, 399). What Hull means to say here is that systematic inferences about the organisms under consideration, their characteristics and dispositions, remain unaltered whether one considers them as members of a hierarchical classification or as parts of the tree of life; what changes is the underlying ontology. Whether Mammalia is considered an atemporal set (class) or a spatiotemporally located (historical) entity, it still holds on both accounts that every mammal has a furry coat, a single lower jaw bone, three middle ear ossicles, and so on. But the metaphysical account of what a taxon such as Mammalia *is*, is what changed drastically with the move from Hennig's (1950, 1966) phylogenetic systematics to Patterson's (1982) pattern cladism. And in the wake of this metaphysical change follows a change in the understanding of the biology of Mammalia: diagnostic features and dispositions are no longer necessary (i.e., defining) but instead historically contingent attributes. There are, after all, mammals without furry coats (or tetrapods without limbs, such as snakes).

SYSTEMATIC PATTERNS AND EVOLUTIONARY PROCESS

Patterson (1988) confronted the metaphysical contrast between the "two hierarchies" (Williams 1992)— one a hierarchy of boxes within boxes, the other a system of species lineages splitting and splitting again—by first addressing the "species problem." Granting species individuality in the context of evolutionary theory, he found this a problematic ontology for systematic theory. The reason is that species, continuously subject to anagenetic (Rensch 1947) change as they are according to Darwin's (1859) theory, cannot have any defining properties; or conversely, if species were classes marked out by defining properties, they could not evolve (Ghiselin 1974, 1997; Hull 1976, 1999). But Patterson (1988, 67) thought that for systematists species do "have defining characters and so may not be individuals" (see also Nelson 1985). "So species concepts are not necessarily the same for systematics and neo-Darwinian theory" (Patterson 1988, 67). The same conclusion pertains to monophyletic taxa (groups of species of unique common descent) (Rieppel 2010). Although Patterson (1978) had earlier distinguished monophyletic taxa as individuals from para- or polyphyletic groups as classes, he now harbored second thoughts about that distinction:

> If Ghiselin and others are right that genealogical taxa, as individuals, may not have defining characters, we again encounter paradox. It follows that monophyletic taxa may not have defining characters, whereas non-monophyletic taxa, as classes, can. The experience of systematists is reverse, and many have found the class/individual distinction to be problematic in considering evolutionary groups. (Patterson 1988, 69f.)

For Ghiselin (1974, 1997) and Hull (1976, 1999), the nature of species, and higher taxa, follows from evolutionary theory, which says that species and higher taxa are historical entities, as Hennig (1950, 1966) likewise recognized. For Patterson (1988), the nature of species, and higher taxa, follows from the practice of systematics, where species and higher taxa appear to be marked out by diagnostic = defining properties. For Patterson (1988), the consequence was a theoretical divide in comparative biology. Systematics is about epistemology; it is about the classification of organisms into a hierarchy of sets within sets on the basis of shared derived characters according to some optimality criterion such as parsimony. Evolutionary theory is about ontology; it is about the real world, its furniture, and how it came to be in the course of Earth history. "It follows that evolutionary theories have no necessary impact on systematics, and systematic theories have no necessary impact on evolution" (Patterson 1988, 86). And yet the results of systematic investigations are supposed to provide a link between the two realms, epistemology and ontology: the patterns discovered by systematics are to be causally explained by evolutionary theory (86). What this means is that evolutionary theory, for example, the theory of allopatric speciation, is called upon to causally (historically) explain the dichotomously structured hierarchy of sets within sets that the systematist has worked out according to the tools of his trade. But if, on the systematist's account, species and higher taxa are abstract, atemporal sets; and if, on the evolutionary account, species and higher taxa are historical entities; then the cost of this theoretical divide is the reification of sets through the hypothetical derivation of their members from a common evolutionary origin.

SYSTEMATICS AS AN AUTONOMOUS SCIENCE

The decoupling of systematics from evolutionary theory had another major motivation, which was the historical, logical, as well as epistemological priority of pattern analysis over process explanation (Brady 1985, 1994; see also Williams and Ebach 2008), the major one of the few uniting themes in German neo-idealistic morphology of the early twentieth century (Weber 1958; Rieppel 2006). Ronald H. Brady formulated what at the time was considered by many to constitute the historical and philosophical foundation of pattern cladism and its independence from evolutionary theory (Brady 1982b, 1994). For this reason he became one of several authors—along with Lars Brundin (1907–93), Colin Patterson (1933–98), and the idealistic morphologists Agnes Arber (1879–1960, botanist), Adolf Naef (1883–1949, zoologist) and Rainer Zangerl (1912–2004, paleontologist)—to whom Williams and Ebach dedicated their book, *Foundations of Systematics and Biogeography* (2008). In his comments on the third annual meeting of the Willi Hennig Society, held in 1982 at the University of Maryland, Stevens (1983, 287) assessed the consequences of pattern cladism, and Brady's defense thereof, as follows: "Earlier, Hennig (1966) had made sharp criticisms of idealistic morphology, so the wheel has turned full circle." The focus of Hennig's (1950, 1966) critique was, of course, the independence of systematics from evolutionary theory that was proclaimed by idealistic morphologists and its metaphysical consequences outlined in the preceding section. Commenting on the same issue, Hull (1988, 375) noted that whereas "idealistic morphologists . . . picked Goethe as their patron saint," this icon of German Romanticism (Richards 2002) had not been invoked in the transformation of cladistics, except that "one of its earliest participants[,] . . . Ron Brady, is partial to Goethe." Indeed, Brady had gone to the Goetheanum in Dornach near Basel, Switzerland, in 1971 to write his Ph.D. dissertation, "Towards a Common Morphology for Aesthetics and Natural Science: A Study of Goethe's Empiricism" (Maier, Brady, and Edelglass 2006, 162), and also unexpectedly died there on March 27, 2003, "on the way to the auditorium where he was to give a talk" (9). The Goetheanum comprises the headquarters and cultural and research center (the School of Spiritual Science)[5] of the Anthroposophical Society, founded in 1923–24 by the former secretary of the German branch of the Theosophical Society, the philosopher Rudolf Steiner (1861–1925; an earlier similar initiative of his dates to 1912). Ron Brady was a frequent guest at the

Goetheanum,[6] and also published in the journal edited by its Natural Sciences Section (Brady 2001). During Brady's visits, the physicist Georg Maier acted as a conduit between Brady and the German-speaking Jochen Bockemühl, who from 1971 through 1996 headed the Natural Sciences Section. Bockemühl, an expert in biodynamic agriculture, worked on the metamorphosis of plants in a Goethean spirit and on the phenomenology of the spiritual world in the spirit of Rudolf Steiner, two topics that Brady was enthusiastically inquiring about.[7] As any perusal of Goetheanum publications on Darwinism and evolutionary theory shows,[8] the anthroposophical discussion of these issues moves in contrasts that were prominent in early-twentieth-century German biology (Forman 1971; Bäumer 1990; Ash 1995; Harwood 1996), especially in debates between idealistic morphologists and phylogeneticists (Rieppel 2011b, 2012): organicism/holism versus reductionism, vitalism versus mechanism, the internalism of neo-Lamarckism versus the externalism of Darwinian selectionism, and so on. Collectively, these contrasts represented a clash between the Goethean versus the Newtonian paradigm of natural science as famously sketched by Oswald Spengler in his *The Decline of the West,* with quantum mechanics as a complicating factor hiding in the wings (Rieppel 2012). Steiner's philosophy, as well as that of his followers, is firmly rooted in the Goethean paradigm, and so is Brady's (1986). At least in Brady's case,[9] then, there might have been an underlying agenda in his campaign to divorce systematics from Darwinian evolutionary theory based on (random) mutation and (externalist) natural selection, values that are critically reflected on in anthroposophical philosophy of nature (see Brady 1982a).

The link to Rudolf Steiner and anthroposophy is interesting, as Steiner's philosophy had its roots in the same neo-Romanticism that is manifest in the writings of the leading German idealistic morphologist Wilhelm Troll (1897–1978; Weber 1958; Rieppel 2012), who sought a resurrection of morphology in a German, that is, Goethean, spirit (e.g., Troll 1928, 1935–36). Troll was, indeed, the main target of Hennig's predecessors, mentioned in the introduction, who called themselves phylogenetic systematists (Zimmermann 1930, 1937; Zündorf, 1939, 1940, 1943). Their opposition to idealistic morphology was anchored in their defense of "Darwinism," which—in the German tradition of the time and following Haeckel (1866)—was equated with the theory of natural selection. Zündorf (1939, 1943), for one, characterized the battle that phylogenetic systematists were called on to wage against idealistic morphologists as a fight aiming to propagate Darwinian selection theory against residual Lamarckism in German comparative biology. A second line of criticism against idealistic morphology pursued by phylogenetic systematists was anchored in logical positivism (Zimmermann 1937–38), a philosophical critique of the nature mysticism (Carus 2007) that is manifest in Troll's (1928, 1935–36) rendition of idealistic morphology (see also Weber 1958; Nickel 1996).

No trace of putative (neo-)Lamarckism, let alone nature mysticism, is apparent in the writings of other non-German (i.e., Swiss) idealistic morphologists (Kälin 1931, 1944; Zangerl 1948; see also Rieppel 2006), among whom Adolf Naef (1883–1949) was the most articulate and theoretically best grounded (e.g., Naef 1917, 1919, 1931). Naef argued along the lines previously sketched by Tschulok (1908, 1922), whose class on the theory of descent he had attended when he was at the University of Zürich (Naef 1913, 359 n. 2; Tschulok 1922 is a later publication of his lecture notes). Like Tschulok (1922) before him and Brady (1985, 1994) after him, Naef argued that only an autonomous and independent systematics free of all phylogenetic speculation can endow evolutionary theory with empirical content.

According to Naef (1917, 1919, 1931), the natural system obtains through logical analysis of the hierarchy of types. The transcendental (ideal) hierarchy of types is initially grasped by the intuition of the expert. Yet in order to render comparative (i.e., systematic) morphology a true science, intuition must be replaced by rational analysis (Naef 1932, 12). The hierarchy of types is thus to be analyzed according to the laws of logic, most prominently the Law of Excluded Middle: two organisms are either similar or not; an organism is either included or excluded from a systematic group: "there is no third alternative" (Naef 1919, 25). The natural system that results from such logical analysis of the hierarchy of types is a nested hierarchy of intentionally defined classes, as also is the hierarchy reconstructed by pattern cladists (Patterson 1982, 1988). Although the empirical success in researching this natural system is highly suggestive of the theory of descent, it must proceed independently, for the strict adherence to the empirical base requires the recognition of a second, interpretive step in which the theory of descent is adduced to provide a hyopthetical-causal explanation for the hierarchy of types from which the natural system is derived through logical analysis. Through such interpretation, the empirically secured results of systematic morphology are translated into the hypothetical language of phylogenetics, yet the empirical content of phylogenetic theory will always, and necessarily, correspond only to the detail in which the hierarchy of types has previously been worked out. In presenting similar arguments, Brady (1985) invoked Darwin's (1859, 206) famous statement, "On my theory, unity of type is explained by unity of descent." It is important to note, however, that unlike pattern cladists, Naef (1917, 1919, 1931) avoided the reification of classes (or sets) through their historical interpretation (a category mistake he claimed Haeckel [1866] was guilty of: Naef 1919; 1931, 120). His natural system derived from logical analysis of the hierarchy of types, whereas his phylogenetic system obtained from the historical interpretation not of the natural system but of the hierarchy of types—a transcendental construct that links the natural system to the phylogenetic tree (Rieppel 2011a). Naef thus employed considerably more philosophical machinery than pattern cladists would be willing to endorse, but then again the philosophy Naef implicitly appealed to is hardly one with which to bolster modern systematics.

CONCLUSION: THE INSTRUMENTALISTS IN SYSTEMATICS

The phylogenetic systematist Walter Zimmermann (1892–1980), who most closely anticipated Hennig's (1950) method of phylogenetic analysis (Donoghue and Kadereit 1992) recognized in Naef's *methodology* (rather than philosophy) a mode of character analysis similar to his own (Zimmermann 1953, 483), hence with an inherent phylogenetic dimension. Zimmermann (1937, 1943), in turn, was the most important precursor of and influence on Hennig (Donoghue and Kadereit 1992). Indeed, Hennig (1950, 14; 1966, 10) praised Zimmermann as "one of the best modern theoreticians of systematic work." In his analysis of post-Haeckelian comparative morphology, Breidbach (2003, 191) recognized in Naef's "idealistic morphology . . . an a-historical calculus" that was adopted by Hennig: "Thus, Hennig was aware of the lack of an historical dimension to his phylogenetic systematics. When this calculus was adopted by modern comparative biology, an a-historical approach was imported into modern evolutionary biology" (191). In light of the above discussion of Hennig's (1950, 1966) phylogenetic system as an enkaptic hierarchy, such an assessment of Hennig's thought is certainly in need of revision. It is true, however, that the *methodology* used to arrive at the phylogenetic system ("in practice, phylogenetic systematics most of the time starts out from a system of morphological similarity relations, and thus in some sense takes recourse to idealistic morphology": Hennig 1950, 26) is an atemporal calculus, and it is this calculus that pattern cladists uprooted from its ontological grounding, separating the epistemology of systematics from the ontology implied by evolutionary theory (Patterson 1988).

The same ahistorical methods of analysis continue to prevail in modern systematics. Mathematical tools of ever-increasing complexity are brought to bear on the analysis of character distribution across the taxa under consideration under some optimality criterion (maximum parsimony, maximum likelihood, or Bayesian posterior probability) in search of a nested hierarchy that is as timeless as the language of mathematics used in its description. Ontological commitments to the results of such analyses are increasingly eschewed, since they would imply "ontological elimination" (Devitt 1997, 121, 159) at a breathtaking rate as different phylogenies are constantly provisionally accepted or rejected relative to some statistical measures of "goodness of fit" of characters to different tree topologies (Rieppel 2005). The relevance of the early cladogenesis of cladistics that resulted in the origin of pattern cladism for modern systematics may be debatable (Brower 2000; Williams and Ebach 2008), but one of its legacies certainly is the turn away from an ontologically grounded Hennigean, or indeed Darwinian, realism toward the instrumentalism that becomes increasingly manifest in the contemporary systematics literature (Rieppel 2007b, 2007c). Such instrumentalism resolves the issue of reification of systematic groups, which at the same time lose their historicity, however.

Acknowledgments

I thank Andrew Hamilton for the invitation to contribute to this volume. Gary Nelson kindly shared his index of authors cited in Hennig (1950, 1966) with me. Michael T. Ghiselin and an anonymous reviewer offered helpful advice and criticism of an earlier draft of this chapter.

NOTES

1. The first to do so was not Hennig, however, but the paleontologist Karl Beurlen (1901–85). Given his ideological commitment to National Socialism (Rieppel 2011b), Beurlen (1936) employed the classic (Kantian) conception of time as an a priori and irreversible dimension in two ways: in support of anti-Semitic Aryan physics and as an anchor for the autonomy of biology. In contrast, Hennig (1950) acknowledged the relevance of modern physics and recognized systematization as an essential part of all natural science properly conducted.

2. Patterson's talk, delivered at the second annual Willi Hennig Society meeting on October 3, 1981, in Ann Arbor, MI, was transcribed and made available by D.M. Williams, Dept. of Botany, The Natural History Museum, London.

3. The letter forms part of the "Hull papers" curated by Andrew Hamilton at Arizona State University.

4. It was again Beurlen (1937, 131) who was the first to explicitly draw an analogy between the enkaptic hierarchy of histosystems and the phylogenetic system.

5. Freie Hochschule für Geisteswissenschaft.

6. Letter from Dr. Johannes Wirz, Naturwissenschaftliche Sektion, Forschungsinstitut am Goetheanum, dated October 14, 2011.

7. Letter from Dr. Georg Maier, Dornach, Switzerland, dated October 15, 2011. Maier also met Ron Brady as a docent at the Green Meadow Waldorf School near New York in an explicitly anthropological context.

8. www.dasgoetheanum.ch/2860.html.

9. Ronald Brady was an affiliate researcher at the Nature Institute whose philosophy was inspired by, among others, Johann Wolfgang von Goethe and Rudolf Steiner. www.natureinstitute.org/about/index.htm.

REFERENCES

Ash, G.M. 1995. *Gestalt Psychology in German Culture, 1820–1967: Holism and the Quest for Objectivity*. Cambridge: Cambridge University Press.
Bäumer, Ä. 1990. *NS-Biologie*. Stuttgart: S. Hirzel.
Beatty, J. 1982. "Classes and Cladists." *Systematic Zoolology*, 31: 25–34.
Benninghoff, A. 1935–36. "Form und Funktion. I. Teil." Zeitschrift für die *gesamte Naturwissenschaft*, 1: 149–160.
Bertalanffy, L. von. 1932. *Theoretische Biologie*. Band I. Berlin: Bornträger.
———. 1941. "Die organismische Auffassung und ihre Auswirkungen." *Der Biologe*, 10: 247–258, 337–345.
Beurlen, K. 1936. "Der Zeitbegriff in der modernen Wissenschaft und das Kausalitätsprinzip." *Kant-Studien*, 41: 16–37.
———. 1937. *Die stammesgschichtlichen Grundlagen der Abstammungslehre*. Jena: G. Fischer Verlag.
Brady, R.H. 1982a. "Dogma and Doubt." *Biological Journal of the Linnean Society*, 17: 79–96.
———. 1982b. "Theoretical Issues and Pattern Cladistics." *Systematic Zoology*, 31: 286–291.
———. 1985. "On the Independence of Systematics." *Cladistics*, 1: 113–126.
———. 1986. "Form and Cause in Goethe's Morphology." In F. Amrine, F.J. Zucker, and H. Wheeler (eds.), *Goethe and the Sciences: A Re-Appraisal*. San Diego, CA: Academic Press; Berlin: Springer, 7–31.
———. 1994. "Pattern Description, Process Explanation, and the History of Morphological Sciences." In L. Grande and O. Rieppel (eds.), *Interpreting the Hierarchy of Nature*. San Diego: Academic Press, 7–31.
———. 2001. "Getting Rid of Metaphysics." *Elemente der Naturwissenschaften*, 75: 61–78.

Breidbach, O. 2003. "Post-Haeckelian Comparative Biology–Adolf Naef's Idealistic Morphology." *Theory in Biosciences,* 122: 174–193.

Brower, A.V.Z. 2000. "Evolution Is Not a Necessary Assumption of Cladistics." *Cladistics,* 16: 143–154.

Brundin, L. 1966. "Transantarctic Relationships and Their Significance, as Evidenced by Chironomid Midges, with a Monograph of the Subfamilies Podonominae and Aphroteniinae and the Austral Heptagyiae." *Kungliga Svenska Vetenskapsakademiens Handlingar,* Fjarde Serien, 11 (1): 1–472.

———. 1968. "Application of Phylogenetic Principles in Systematics and Evolutionary Theory." In T. Ørvig (ed.), *Current Problems of Lower Vertebrate Phylogeny: Proceedings of the Fourth Nobel Symposium Held in June 1967 at the Swedish Museum of Natural History in Stockholm.* Stockholm: Almqvist and Wiskell, 473–495.

———. 1972. "Phylogenetics and Biogeography." *Systematic Zoology,* 21: 69–79.

Byers, G.W. 1969. "Phylogenetic Systematics and Transantarctic Relationships and Their Significance as Evidenced by Chironomic Midges" (review of Hennig 1966 and Brundin 1966). *Systematic Zoology,* 18: 105–107.

Cain, J. 2000. "Woodger, Positivism, and the Evolutionary Synthesis." *Biology and Philosophy,* 15: 535–551.

Carus, A.W. 2007. *Carnap and Twentieth-Century Thought.* Cambridge: Cambridge University Press.

Darlington, P.J. 1970. A Practical Criticism of the Hennig-Brundin 'Phylogenetic Systematics' and Antarctic Biogeography." *Systematic Zoology,* 19: 1–18.

Darwin, C.R. 1859. *On the Origin of Species.* London: John Murray.

Devitt, M. 1997. *Realism and Truth.* 2nd ed. Princeton: Princeton University Press.

Donoghue M.J., and Kadereit, J.W. 1992. "Walter Zimmermann and the Growth of Phylogenetic Theory." *Systematic Biology,* 41: 74–85.

Dupuis, C. 1984. "Willi Hennig's Impact on Taxonomic Thought." *Annual Review of Ecology, Evolution, and Systematics,* 15: 1–24.

Forbes, S.A. 1887. "The Lake as a Microcosm." *Bulletin of the Peoria Scientific Association,* 111: 77–87.

Forman, P. 1971. "Weimar Culture, Causality, and Quantum Theory, 1918–1927: Adaptation by German Physicists and Mathematicians to a Hostile Intellectual Environment." *Historical Studies in the Physical Sciences,* 3: 1–115.

Friederichs, K. 1927. "Grundsätzliches über die Lebenseinheiten höherer Ordnung und den ökologischen Einheitsfaktor." *Naturwissenschaften,* 15: 153–178, 182–186.

Ghiselin, M.T. 1974. "A Radical Solution to the Species Problem." *Systematic Zoology,* 23: 536–544.

———. 1997. *Metaphysics and the Origin of Species.* Albany: SUNY Press.

Gregg, J.R. 1954. *The Language of Taxonomy: An Application of Symbolic Logic to the Study of Classificatory Systems.* New York: Columbia University Press.

Haeckel, E. 1866. *Generelle Morphologie der Organismen. Zweiter Band: Allgemeine Entwickelungsgeschichte der Organismen.* Berlin: Georg Reimer.

Harwood, J. 1996. "Weimar Culture and Biological Theory: A Study of Richard Woltereck (1877–1944)." *History of Science,* 34: 347–377.

Heidenhain, M. 1923. *Formen und Kräfte in der lebenden Natur.* Berlin: Springer.

———. 1937. *Synthetische Morphologie der Niere des Menschen. Bau und Entwicklung dargestellt auf neuer Grundlage.* Leiden: E.J. Brill.

Hennig, W. 1947. "Probleme der biologischen Systematik." *Forschungen und Fortschritte,* 21–23: 276–279.

———. 1949. "Zur Klärung einiger Begriffe der phylogenetischen Systematik." *Forschungen und Fortschritte,* 25: 136–138.

———. 1950. *Grundzüge einer Theorie der phylogenetischen Systematik.* Berlin: Deutscher Zentralverlag.

———. 1953. "Kritische Bemerkungen zum phylogenetischen System der Insekten." *Beitraege zur Entomologie,* 3 (*Beilageband*): 1–85.

———. 1957. "Systematik und Phylogenese." In H.-J. Hannemann (ed.), *Bericht über die Hundertjahrfeier der Deutschen entomologischen Gesellschaft Belin.* Berlin: Akademie Verlag, 50–71.

———. 1965. "Phylogenetic Systematics." *Annual Review of Ecology, Evolution, and Systematics,* 10: 97–116.

———. 1966. *Phylogenetic Systematics.* Trans. D.D. Davis and R. Zangerl. Urbana: University of Illinois Press.

Hueck, W. 1926. "Die Synthesiologie von Martin Heidenhain als Versuch einer allgemeinen Theorie der Organisation." *Naturwissenschaften,* 14: 149–158.

Hull, D.L. 1976. "Are Species Really Individuals?" *Systematic Zoology,* 25: 174–191.

———. 1988. *Science as a Process: An Evolutionary Account of the Social and Conceptual Development of Science.* Chicago: University of Chicago Press.

———. 1999. "On the Plurality of Species: Questioning the Party Line." In R.A. Wilson (ed.), *Species: New Interdisciplinary Essays.* Cambridge, MA: MIT Press, 23–48.

Jax, K. 1998. "Holocoen and Ecosystem: On the Origin and Historical Consequences of Two Concepts." *Journal of the History of Biology,* 31: 113–142.

Kälin, J. 1931. "Über den Homologiebegriff in der vergleichenden Anatomie." *Bulletin de la Société Fribourgeoise des Sciences Naturelles,* 31: 137–147.

———. 1944. "Die Homologie als Ausdruck ganzheitlicher Baupläne von Typen." *Bulletin de la Société Fribourgeoise des Sciences Naturelles,* 37: 135–161.

Lorenzen, S. 1994. "Phylogenetische Systematik gestern, heute und morgen." *Biologie in unserer Zeit,* 24: 200–206.

Maier, G., Brady, R., and Edelglass, S. 2006. *Being on Earth: Practice in Tending Appearances.* Saratoga Springs, NY: SENSRI/The Nature Institute.

Mayr, E. 1965. "Numerical Phenetics and Taxonomic Theory." *Systematic Zoology,* 14: 73–97.

———. 1968. "Theory of Biological Classification." *Nature,* 220: 545–548.

McIntosh, R. 2011. "The History of Early British and US-American Ecology to 1950." In A. Schwartz and K. Jax (eds.), *Ecology Revisited: Reflecting on Concepts, Advancing Science.* Berlin: Springer, 277–285.

Naef, A. 1913. "Studien zur generellen Morphologie der Mollusken. 2. Teil: Das Cölomsystem in seinen topographischen Beziehungen." *Ergebnisse und Fortschritt der Zoologie,* 3: 329–462.

———. 1917. *Die individuelle Entwicklung organischer Formen als Urkunde ihrer Stammesgeschichte (Kritische Bemerkungen über das sogenannte 'biogenetische Grundgesetz').* Jena: G. Fischer Verlag.

———. 1919. *Idealistische Morphologie und Phylogenetik (zur Methodik der systematischen Morphologie).* Jena: G. Fischer Verlag.

———. 1931. "Allgemeine Morphologie. I. Die Gestalt als Begriff und Idee. (Diagnostik und Typologie der organischen Formen)." In L. Bolk, E. Göppert, E. Kallius, and W. Lubosch (eds.), *Handbuch der vergleichenden Anatomie der Wirbeltiere, erster Band.* Berlin: Urban und Schwarzenberg, 77–118.

———. 1932. "Morphologie der Tiere." In R. Dittler, G. Joos, E. Korschelt, G. Linck, F. Oltmanns, and K. Schaum (eds.), *Handwörterbuch der Naturwissenschaften,* 2nd ed. Jena: G. Fischer Verlag, 7:3–17.

Nelson, G. 1985. "Class and Individual: A Reply to M. Ghiselin." *Cladistics,* 1: 386–389.

———. 2004. "Cladistics: Its Arrested Development." In D.M. Williams and P.L. Forey (eds.), *Milestones in Systematics.* Boca Raton, FL: CRC Press, 127–147.

Nelson, G.J., and Platnick, N.I. 1981. *Systematics and Biogeography: Cladistics and Vicariance.* New York: Columbia University Press.

Nickel, G. 1996. *Wilhelm Troll (1897–1978): Eine Biographie.* Halle (Saale): Deutsche Akademie der Naturforscher Leopoldina.

Patterson, C. 1978. "Verifiability in Systematics." *Systematic Zoology,* 27: 218–222.

———. 1981. "Significance of Fossils in Determining Evolutionary Relationships." *Annual Review of Ecology, Evolution, and Systematics,* 12:195–223.

———. 1982. "Morphological Characters and Homology." In K.A. Joysey and A.E. Friday (eds.), *Problems of Phylogenetic Reconstruction.* London: Academic Press, 21–74.

———. 1988. "The Impact of Evolutionary Theories on Systematics." In D.L. Hawksworth (ed.), *Prospects in Systematics.* Oxford: Clarendon Press, 59–91.

———. 1989. "Phylogenetic Relationships of Major Groups: Conclusions and Prospects." In B. Fernholm, K. Bremer, and H. Jörnvall (eds.), *The Hierarchy of Life: Molecules and Morphology in Phylogenetic Analysis. Proceedings of Nobel Symposium 70, Held at Alfred Nobel's Björkborn, Karlskoga, Sweden, August 29–September 2, 1988.* Amsterdam: Elsevier, 471–488.

———. 2002. "Evolution and Creationism." *The Linnean,* 18: 15–32.

Phillips, D. C. 1970. "Organicism in the Late Nineteenth and Early Twentieth Centuries." *Journal of the History of Ideas,* 31: 413–432.

Platnick, N.I. 1977. "Cladograms, Phylogenetic Trees, and Hypothesis Testing." *Systematic Zoology,* 26: 438–442.

Potthast, T. 2003. “Wissenschaftliche Ökologie und Naturschutz—Szenen einer Annäherung.” In J. Radkau and F. Uekötter (eds.), *Naturschutz und Nationalsozialismus*. Frankfurt am Main: Campus, 225–254.

Rensch, B. 1947. *Neuere Probleme der Abstammungslehre: Die Transspezifische Evolution*. Stuttgart: Ferdinand Enke.

Richards, R.J. 2002. *The Romantic Conception of Life: Science and Philosophy in the Age of Goethe*. Chicago: University of Chicago Press.

Rieppel, O. 2005. “A Note on Reality, Ontology, and Pattern Cladism.” *Neues Jahrbuch für Geologie und Paläontologie, Monatshefte*, 142–150.

———. 2006. “On Concept Formation in Systematics.” *Cladistics*, 22: 474–492.

———. 2007a. “The Metaphysics of Hennig’s Phylogenetic Systematics: Substance, Events and Laws of Nature.” *Systemtatics and Biodiversity*, 5: 345–360.

———. 2007b. “The Nature of Parsimony and Instrumentalism in Systematics.” *Journal of Zoological Systematics and Evolutionary Research*, 45: 177–183.

———. 2007c. “Parsimony, Likelihood, and Instrumentalism in Systematics.” *Biology and Philosophy*, 22: 141–144.

———. 2009. “Hennig’s Enkaptic System.” *Cladistics*, 25: 311–317.

———. 2010. “Species Monophyly.” *Journal of Zoological Systematics and Evolutionary Research*, 48: 1–8.

———. 2011a. “Adolf Naef (1883–1949), Systematic Morphology and Phylogenetics.” *Journal of Zoological Systematics and Evolutionary Research*, 50: 2–13.

———. 2011b. “Karl Beurlen (1901–1985), Nature Mysticism, and Aryan Paleontology.” *Journal of the History of Biology*. doi: 10.1007/s10739-011-9238-7.

———. 2011c. “Species Are Individuals—The German Tradition.” *Cladistics*, 27: 629–645.

———. 2011d. “Willi Hennig’s Dichotomization of Nature.” *Cladistics*, 27: 103–112.

———. 2012. “Wilhelm Troll (1897–1978): Idealistic Morphology, Physics, and Phylogenetics.” *History and Philosophy of the Life Sciences*, 33: 321–342.

Schmitt, M. 2001. “Willi Hennig (1913–1976).” In I. Jahn and E. Schmitt (eds.), *Darwin & Co.: Eine Geschichte der Biologie in Porträts*. Munich: C.H. Beck, 316–343.

Stadler, F. 1997. *Studien zum Wiener Kreis. Ursprung, Entwicklung und Wirkung des logischen Empirismus im Kontext*. Frankfurt am Main: Suhrkamp.

Stevens, P.F. 1983. “Report of Third Annual Willi Hennig Society Meeting.” *Systematic Zoology*. 32: 285–291.

Thienemann, A. 1925. “Der See als Lebensgemeinschaft.” *Naturwissenschaften*, 13: 598–600.

———. 1935. “Lebensgemeinschaft und Lebensraum.” *Unterrichtsblätter für Mathematik und Naturwissenschaften*, 41: 337–350.

———. 1941. *Leben und Umwelt* (Bios, 12). Leipzig: Johann Ambrosius Barth.
Troll, W. 1928. *Organisation und Gestalt im Bereich der Blüte.* Berlin: Springer.
———. 1935–36. "Die Wiedergeburt der Morphologie aus dem Geiste deutscher Wissenschaft." *Zeitschrift fur die gesammten Naturwissenschaften,* 1: 349–356.
Tschulok, S. 1908. "Zur Methodologie und Geschichte der Deszendenztheorie." *Biologisches Centralblatt,* 4–18, 33–51, 73–96, 97–117.
———. 1922. *Deszendenzlehre.* Jena: G. Fischer Verlag.
Weber, H. 1958. Konstruktionsmorphologie. *Zoologische Jahrbücher Abteilung fuer Allgemeine Zoologie und Physiologie der Tiere,* 68: 1–112.
Williams, D. M., and Ebach, M. C. 2008. *Foundations of Systematics and Biogeography.* Berlin: Springer.
Williams, P. A. 1992. "Confusion in Cladism." *Synthese,* 91:135–152.
Woodger, J. H. 1952. "From Biology to Mathematics." *British Journal for the Philosophy of Science,* 3: 1–21.
Zangerl, R. 1948. "The Methods of Comparative Anatomy and Its Contribution to the Study of Evolution." *Evolution,* 2: 351–374.
Zimmermann, W. 1930. *Phylogenie der Pflanzen. Ein Überblick über Tatsachen und Probleme.* Jena: G. Fischer Verlag.
———. 1937. "Arbeitsweise der botanischen Phylogenetik und anderer Gruppierungswissenschaften." In E. Abderhalden (ed.), *Handbuch der biologischen Arbeitsmethoden,* 3. Abt. IX, Teil 3/II. Berlin: Urban und Schwarzenberg, 941–1053.
———. 1937–38. "Strenge Objekt/Subjekt-Scheidung als Voraussetzung wissenschaftlicher Biologie." *Erkenntnis,* 7: 1–44.
———. 1943. "Die Methoden der Phylogenetik." In G. Heberer (ed.), *Die Evolution der Organismen. Ergebnisse und Probleme der Abstammungslehre.* Jena: G. Fischer Verlag, 20–56.
———. 1953. *Evolution. Die Geschichte ihrer Probleme und Erkenntnisse.* Freiburg: Karl Alber.
Zündorf, W. 1939. "Der 'Lamarckismus in der heutigen Biologie.'" *Archiv für Rassen-und Gesellschaftsbiologie,* 33: 281–303.
———. 1940. "Phylogenetische oder Idealistische Morphologie?" *Der Biologe,* 9: 10–24.
———. 1943. "Idealistische Morphologie und Stammeslehre." In G. Heberer (ed.), *Die Evolution der Organismen. Ergebnisse und Probleme der Abstammungslehre.* Jena: G. Fischer Verlag, 86–104.

6

Cladistics at an Earlier Time

GARETH NELSON

It is the historical aspect of biology, particularly the vertebrates, which has been my major interest since student days. . . . Where did they come from?

—A. S. Romer (1894–1973)

Systematic biology today is a complex of conflicting opinion, particularly about cladistics (Ebach, Williams, and Gill 2008; Williams and Ebach 2009). It was not always so. At an earlier time, things were simpler, and the genie was still in the bottle. May we begin with Romer of the 1960s and his expanding curiosity?

> In my earlier years I, too, found much of interest in comparative studies of existing vertebrates. But increasingly I found myself turning to paleontology. Why content myself with hypothetical ancestral types when actual ancestors may be discovered? (Romer 1969, 49)

Here there is no trace of cladistics, but it once began—one of its many beginnings—when the discovery of actual ancestors was generally seen by paleontologists as an illusion of wishful thinking; and the search for ancestors, a futile quest.

Looking back today, one may doubt that, except as passing fashion, the "search for ancestors" ever had much relevance for science. A Ph.D. student of Romer from those times (Thomson 1963) and a participant in the Nobel Symposium of 1967 (see below), Keith Thomson more recently wrote, "The search for ancestors has become unfashionable and the last refuge of the publicity seekers" (1997, 491). I will not belabor that history here but refer to a summary—a little mood music—offered at that earlier time by Cain and Harrison: "the actual course of evolution . . . can be known with certainty only from a really adequate fossil record" (1960, 2). They explain in greater detail: "When a good fossil record is available, of course, the whole evolutionary dendrite [phylogenetic tree] can be worked out for that group simply by putting those forms together that are more alike phenetically, without any phyletic weighting" (3).

"Known with certainty" to "a good fossil record" and beyond to "the crucial fossil evidence" are but short conceptual steps along the road of wishful expectation, steps easy for me to take when I was in graduate school (at the University of Hawaii, 1962–1966):

> The obvious way to determine the relationship of modern forms is to trace their genealogies back through the fossil record. But for modern teleostean [fish] groups the crucial fossil evidence is almost always lacking, at least as yet. (Gosline 1959, 160)

Here William Gosline (1915–2002), my major professor, means that the evidence eventually will be forthcoming. With pursuit of this possibility in mind, I obtained a postdoctoral fellowship, beginning in September 1966, at the Swedish Museum of Natural History (Naturhistoriska Riksmuseet), described as "a mecca or magnet for vertebrate palaeontologists from all over the world" (Patterson 1990, 363). Soon after there appeared in the reading room of the paleozoology department (Paleozoologiska Avdelningen) a monograph on insects by the museum's professor of entomology, Lars Brundin (1907–1993). What caught my eye were the section headings (Brundin 1966, 3):

CONTENTS

Reading the introductory pages (7–64), my reaction was much the same as Colin Patterson's a few months later:

> I don't know if anyone reads Brundin these days, but he was my first introduction to Hennig and phylogenetic systematics, what we now call cladistics. The first 50 pages of this are still a wonderfully clear and strong statement of Hennig's ideas. I was bowled over by it and became an instant convert. (Patterson 2011, 124, written in 1995 about his experience in 1967 and published posthumously)

Thus impressed, I mentioned the monograph during the department's communal morning coffee. Shortly thereafter, Tor Ørvig (1916–1994), a resident researcher, asked me if I might like to meet Brundin. We met briefly in Ørvig's office and exchanged pleasantries, and I received a copy duly inscribed by its author. Grateful for this gift, I took it home to reread the introductory and other matters often enough for adequate understanding of the many things new to me. The copy in the department was circulating; it was the property of the library of the Royal Academy of Sciences (Kungliga Vetenskapsakademien), located across the road (Roslagsvägen) from the museum, and not to be removed from the reading room.

Soon thereafter, in November, I visited London for the purpose of discussing a possible job at the American Museum of Natural History in New York City. Also visiting London was Donn Rosen (1929–1986), a curator in the American Museum's Department of Ichthyology. During my six-day visit, Brundin's monograph was discussed at length with Rosen and Humphry Greenwood (1927–1995), zoology department, and Colin Patterson (1933–1998), palaeontology department—fish researchers at what was then known as the British Museum (Natural History), informally as the BMNH or simply the BM, which was rechristened in 1992 as the Natural History Museum. We were all tobacco smokers at the time and met regularly behind the museum building for short periods in an outdoor area known as the Colonnade. Other discussion occupied our pub lunches and evening get-togethers. Suffice it to say that the "many things new to me" in Brundin's monograph, such as I could represent them, were also new to these persons who had not yet seen his publication.

I returned to Stockholm, and after a time decided to divide my postdoctoral year, with the last six months in London at the BM. It was understood that I would return to Stockholm in June 1967 for the symposium on lower vertebrate phylogeny sponsored by the Nobel Foundation. This was to be the fourth in the series begun in 1965. Before I left for London, Ørvig asked me if it might be a good idea for Brundin to address the symposium on the subject of phylogenetic principles. I replied, "Yes, of course"—this offhand comment my only contribution to planning for that event.

By auto (a new VW Beetle) I left Stockholm for London in April 1967, stopping briefly in Copenhagen to meet the fish paleontologists Eigil Nielsen (1910–1968) and Niels Bonde at the Universitetets Mineralogisk-Geologiske Institut og Museum. I discussed Brundin with Bonde, and the many things were new to him, too. I stopped also in Paris to meet Camille Arambourg (1885–1970) and several other researchers at the Institut de Paléontologie of the Muséum National d'Histoire Naturelle.

The Stockholm symposium, with forty-seven participants from fourteen countries, was held June 12–16, 1967, in the museum, except for Brundin's presentation, which was in the Academy across the road. Attending from England with Patterson and Greenwood was Roger Miles, another fish researcher from the palaeontology department of the BM. Another participant, Hans-Peter Schultze, has recently placed this meeting in perspective derived from developments over forty years and a series of nine further symposia on lower vertebrates:

> In 1967 the first symposium was dedicated to celebration of the life-long contributions to research on lower vertebrates by E. A:son Stensiö. All participants, with the exception of L. Brundin, gave detailed morphological descriptions for which the "Swedish school" was famous, and argued for homology and relationships based on general similarities. Nevertheless, most participants had their first encounter with Hennig's phylogenetic principles during this event. R. Zangerl, translator of Hennig's 1950 book (Hennig 1966), L. Brundin, a promoter of Hennig's ideas, and G.J. Nelson who was introduced to Hennig's phylogenetic method by L. Brundin and became

> thereafter the strongest promoter of Hennig's principles, were present. Before the banquet and subsequently in the proceedings, L. Brundin, an entomologist of the Naturhistoriska Riksmuseet, Stockholm, gave a presentation on the application of Hennig's phylogenetic principles (Brundin, 1968). This had no impact on the other presentations subsequently published in the proceedings (Ørvig, 1968). (Schultz 2005, vi)

Schultze comments also on a second symposium held a few years later:

> In 1972, the second symposium was organized by the Linnean Society of London to honour E. A:son Stensiö and E. Jarvik for their outstanding contributions to the description and interpretation of fossil vertebrates. Nevertheless, times had changed. G. J. Nelson has pushed strongly for the application of Hennig's approach during the years since 1967 and the second symposium was held in a completely different atmosphere. (vi–vii)

Background for the second symposium is revealed by Patterson:

> In 1972, Humphry Greenwood, Roger Miles and I organized a Linnean Society symposium called "Interrelationships of Fishes." Our excuse was to produce a Festschrift for two honorary Foreign Members of the Linnean, two Swedish heroes: Erik Stensiö (Patterson 1990) and his colleague Erik Jarvik (Janvier 1998). A Festschrift for those two was our excuse, but our hidden agenda was cladistics, to get as many groups of fishes as possible worked over in the new cladistic framework. The symposium volume came out in 1973 (Greenwood et al. 1973). We didn't manage to raise a complete cast of cladists but I think this was the first multi-author volume, anywhere in biology, in which the overall message is cladistics. It has a certain historical significance. (2011, 124)

This significance relates to Brundin, his 1966 monograph, and his participation in the Nobel Symposium of 1967, which are fittingly summarized by Wanntorp:

> Much of the later developments in cladistics can be traced back to the influence of Lars Brundin. He did not himself give birth to cladistics, but if anything he was the doctor who brought forth the baby, gave it a healthy start in the world, and announced the event to a reluctant scientific community. (1993, 357)

Another symposium, not in Schultze's list of ten because the symposium's program embraced all vertebrates, was organized in Paris by Jean-Pierre Lehman (1914–81), Institut de Paléontologie, Muséum National d'Histoire Naturelle; he had participated in Stockholm and London. This occurred on June 4–9, 1973, one of a series of Colloques Internationaux du Centre National de la Recherche Scientifique. Daniel Goujet, also in attendance in Stockholm and London, later wrote:

> The cladistic revolution among French paleontologists started with the Paris CNRS Symposium of 1973. A year earlier, the Linnean Society of London had organized the Interrelationships of Fishes meeting, which announced the impact of cladistics on fish paleontology. During the Paris Symposium cladistics was at last put on the menu of the Parisian academic circle of paleontologists; Niels Bonde and Roger Miles were the speakers. The ideas of ancestor-descendant relationships that prevailed in Paris began to waver and Colin [Patterson] was one of the prominent activists in the revolution that was taking place in Paris among the lower Vertebrate specialists. (2000, 79)

Noteworthy is Miles's (1975) cladistic treatment of lungfishes. Also Lehman's view:

> New ideas about systematics are considered as inadequate: a) Numerical taxonomy, giving the same weight to all characters, appears subjective; b) Systematical phylogeny [cladistics] is built on a dubious postulate,—the dichotomy of the taxons—and its application to Vertebrate phylogeny does not give results fitting with those of Paleontology. (1975, 257)

And Walter Georg Kühne's (1911–91) immediate response, remarkable for its "simple question":

> The cladistians of this symposium feel obliged to their host giving them the opportunity to state their case: The rejection of Sokal's and Hennig's theories in one go, as far as Palaeontology is concerned, meets our disapproval. So useful Sokal's method is, to sort—let us say—100,000 taxa, so useless is it for Palaeontology, dealing with the display of life in time. Hennig's theory however has attracted already a great number of younger palaeontologists, and this number is increasing. The reasons for this phenomenon are easily seen: Applying Hennig's theories to palaeontological material, the precision of our statement is increased, and a simple question proffered in respect to three related taxa can be answered. This question is the one of nearer relationship of two, to the remaining third taxon. (1975, 264)

Most significant perhaps is Bonde's (1975) "très belle présentation" (Tassy 1986, 98), a detailed response from within paleontology worth reading afresh today: "Discussion of some classical problems in vertebrate evolution are reviewed from a 'cladistic' standpoint" (293). Bonde observes:

> Nearly all fossil "ancestors" when studied in some detail appear to show some specializations of their own, which exclude them from a position as ancestral to other groups (Schaeffer et al. 1972, 43), and leave them as sistergroups of the monophyletic groups supposed to be their descendants.
>
> The latter position is in fact the only one left for any "ancestor" in our formal system, because of our definition of phylogenetic relationship, and this definition is self-evident.

> The principal phylogenetic information we can get about a fossil (or recent) group, A, is that it is more closely related to some group B than to any other group, meaning that A and B share a common ancestor and are sistergroups on evidence of shared apomorph features. This is why the palaeontologist's "search for the ancestor" is a futile task, and in phylogenetic studies it ought to be replaced by "search for the sistergroup." (301)

And again:

> Especially important for palaeontology is the fact that even fossil groups can only be related to other groups if synapomorphies can be demonstrated, and this leaves out all futile "search for the ancestor" and stress on importance of "primitiveness." (313)

And he concurs with Patterson's later assessment of the significance of the London Symposium of 1972:

> The latest mile-stone is the 1972 Linnean Society symposium (Greenwood et al., 1973 on interrelationships of all groups of fishes), during which probably for the first time an entire zoological meeting was to a large extent influenced by Hennig's ideas, as witnessed by the individual contributions by Miles, Patterson, Greenwood, Nelson, Forey, and Rosen. (305)

Brundin survived to help organize another Nobel Symposium in 1988, the seventieth of the series begun in 1965 (Fernholm, Bremer, and Jörnvall 1989). In his summary of this symposium Patterson writes:

> Aside from these generalities, what has really been achieved in the 21 years between 1967 and 1988? The theoretical issues are addressed in the first section of this book, on principles of phylogenetic analysis, and in several papers on particular groups. In the mid-1960s, few thought these issues worth discussion, for method was seen as self-evident. But one highlight of the 1967 meeting was Brundin's [Brundin 1968] paper on phylogenetic principles, and among the audience was Gareth Nelson. Brundin and Nelson are identified by Hull (in this volume [Hull 1989]) as the "chief conduits of Hennig's work into the English-speaking world." The heart of Brundin's paper was one message: "phylogenetics is the search for the sister group." That message eventually got through, and among morphologists the cladists have won, in Hull's terms [Hull 1988]. Much of the hundreds of pages on systematic theory and method published by morphologists during the last 20 years is embroidery on and exploration of Brundin's message; or more precisely, given the sister group as the target, it explores the concept of synapomorphy, the relation that should find the target. (1989, 472)

From the "search for ancestors" to the "search for the sister group," did anything really change in the process? According to Brundin (1988, 366), "Little by little some paleontologists have perceived that Hennig's principles of phylogenetic systematics meant a revolution to their science." Patterson agrees (1997, 4): this revolution "began in the late 1960s, accelerated in the 1970s, and was virtually complete by the eighties." There has been little comment on Patterson's view of the revolution and its duration. Independent confirmation has recently been supplied by Prothero:

> When I gave my first professional talk at the SVP [Society of Vertebrate Paleontology] meeting in Pittsburg in 1978, I was presenting one of the few cladograms on the entire program. Only a decade later, all systematics talks at the SVP meeting were cladistic, and no one was using the old methods anymore. . . . I've been to the SVP meetings every single year since 1977 (the meeting in 2009 will be my thirty-second in a row). (2009, 18, 237)

In summary, one beginning of cladistics is the revolution in, or reformation of, paleontology, particularly of vertebrates during the 1960s and 1970s (Janvier 1986; Tassy 1991, 2004; Williams and Ebach 2004). It is marked by a shift from the search for the ancestor to the search for the sister group, and the evidence relevant thereto, evidence in the form of synapomorphies, parts of organisms (Nelson 2011): "The approach of basing interrelationships on similarities, has been replaced by a search of shared derived features (synapomorphies) indicating sister group relationships" (Schultze 2005, xv).

Acknowledgments

For comments on the manuscript I am grateful to M. C. Ebach, P. Janvier, P. Y. Ladiges, J. S. Wilkins, and D. M. Williams. For his list of ten symposia, and comment about it, I am grateful to H.-P. Schultze.

REFERENCES

Bonde, N. 1975. "Origin of 'Higher Groups': Viewpoints of Phylogenetic Systematics." *Colloques Internationaux du Centre National de la Recherche Scientifique, Problèmes Actuels de Paléontologie (Évolution des Vertébrés)*, 218: 293–324.

Brundin, L. 1966. "Transantarctic Relationships and Their Significance, as Evidenced by Chironornid Midges, with a Monograph of the Subfamilies Podonominae and Aphroteniinae and the Austral Heptagyiae." *Kungliga Svenska Vetenskapsakademiens Handlingar,* Fjarde Serien, 11 (1): 1–472.

———. 1968. "Application of Phylogenetic Principles in Systematics and Evolutionary Theory." In T. Ørvig (ed.), *Current Problems of Lower Vertebrate Phylogeny*. Stockholm: Almqvist and Wiksell, 473–495.

———. 1988. "Phylogenetic Biogeography." In A.A. Myers and P.S. Giller (eds.), *Analytical Biogeography: An Integrated Approach to the Study of Animal and Plant Distributions.* London: Chapman and Hall, 343–369.

Cain, A.J., and Harrison, G.A. 1960. "Phyletic Weighting." *Proceedings of the Zoological Society of London,* 135: 1–31.

Ebach, M.C., Williams, D.M., and Gill, A.C. 2008. "O Cladistics, Where Art Thou?" *Cladistics,* 24: 851–852.

Fernholm, B., Bremer, K., and Jörnvall, H., eds. 1989. *The Hierarchy of Life: Molecules and Morphology in Phylogenetic Analysis. Proceedings of Nobel Symposium 70, Held at Alfred Nobel's Björkborn, Karlskoga, Sweden, August 29–September 2, 1988.* Amsterdam: Elsevier.

Gosline, W.A. 1959. "Mode of Life, Functional Morphology, and the Classification of Modern Teleostean Fishes." *Systematic Zoology,* 8:160–164.

Goujet, D. 2000. "The French Connection." In P.L. Forey, B.G. Gardiner, and C.H. Humphries (eds.), "Colin Patterson (1933–1998), a Celebration of His Life." *The Linnean,* Special Issue no. 2: 79–81. London: Academic Press.

Greenwood, P.H., Miles, R.S., and Patterson, C., eds. 1973. "Interrelationships of Fishes." *Zoological Journal of Linnean Society,* suppl. 1 to vol. 53. London: Academic Press.

Hennig, W. 1966. *Phylogenetic Systematics.* Trans. Trans. D.D. Davis and R. Zangerl. Urbana: University of Illinois Press.

Hull, D.L. 1988. *Science as a Process: An Evolutionary Account of the Social and Conceptual Development of Science.* Chicago: University of Chicago Press.

———. 1989. "The Evolution of Phylogenetic Systematics." In B. Fernholm, K. Bremer, and H. Jörnvall (eds.), *The Hierarchy of Life.* Amsterdam: Elsevier Science, 3–15.

Janvier, P. 1986. "L'Impact de cladisme sur la recherche dans les sciences de la vie et de la terre." In P. Tassy (ed.), *L'Ordre et la diversité de vivant: Quel statut scientifique pour les classifications biologiques?* Paris: Fondation Diderot/Fayard, 99–120.

———. 1998. "Obituary: Erik Jarvik (1907–98): Paleontologist Renowned for His Work on the 'Four Legged Fish.'" *Nature,* 392: 338.

Kühne, W.G. 1975. "Intervention de M. Kühne (après la communication de M. Lehman)." *Colloques Internationaux du Centre National de la Recherche Scientifique, Problèmes Actuels de Paléontologie (Évolution des Vertébrés),* 218: 264.

Lehman, J.-P. 1975. "Quelques réflexions sur la phylogénie des vertebrés inférieurs." *Colloques Internationaux du Centre National de la Recherche Scientifique, Problèmes Actuels de Paléontologie (Évolution des Vertébrés),* 218: 257–264.

Miles, R.S. 1975. "The Relationships of the Dipnoi." *Colloques Internationaux du Centre National de la Recherche Scientifique, Problèmes Actuels de Paléontologie (Ėvolution des Vertébrés),* 218: 133–148.

Nelson, G. 2011. "Resemblance as Evidence of Ancestry." *Zootaxa,* 2946: 137–141.

Ørvig, T., ed. 1968. *Current Problems of Lower Vertebrate Phylogeny. Proceedings of the Fourth Nobel Symposium, Held in June 1967 at the Swedish Museum of Natural History (Naturhistoriska riksmuseet) in Stockholm.* Stockholm: Almqvist and Wiksell.

Patterson, C. 1989. "Phylogenetic Relations of Major Groups: Conclusions and Prospects." In B. Fernholm, K. Bremer, and H. Jörnvall (eds.), *The Hierarchy of Life: Molecules and Morphology in Phylogenetic Analysis. Proceedings of Nobel Symposium 70, Held at Alfred Nobel's Björkborn, Karlskoga, Sweden, August 29–September 2, 1988.* Amsterdam: Amsterdam: Elsevier, 471–488.

———. 1990. "Erik Helge Osvald Stensiö." *Biographical Memoirs of Fellows of the Royal Society,* 35: 363–380.

———. 1997. "Molecules and Morphology, Ten Years On." Unpublished manuscript (delivered orally to the meeting Molecules and Morphology in Systematics, Paris, France, March 24–28). 21 pp.

———. 2011. "Adventures in the Fish Trade." Address to the Systematics Association, December 6, 1995, edited and with an introduction by David M. Williams and Anthony C. Gill. *Zootaxa,* 2946: 118–136.

Prothero, D.R. 2009. *Greenhouse of the Dinosaurs: Evolution, Extinction, and the Future of our Planet.* New York: Columbia University Press.

Romer, A.S. 1969. "Vertebrate Paleontology and Zoology." *The Biologist,* 51: 49–53.

Schaeffer, B., Hecht, M.K., and Eldredge, N. 1972. "Phylogeny and Paleontology." In T. Dobzhansky, M.K. Hecht, and W.C. Steere (eds.), *Evolutionary Biology.* New York: Appleton-Century-Crofts, 6:31–46.

Schultze, H.-P. 2005. "The First Ten Symposia on Early/Lower Vertebrates." *Revista Brasileira de Paleontologia,* 8: v–xviii.

Tassy, P. 1986. "Construction systématique et soumission au test: Une forme de connaissance objective." In P. Tassy (ed.), *L'Ordre et la diversité de vivant: Quel statut scientifique pour les classifications biologiques?* Paris: Fondation Diderot/Fayard, 85–98.

———. 1991. *L'Arbre à remonter le temps.* Paris: Christian Bourgois.

———. 2004. "La systématique contemporaine; les modalités de sa renaissance. *Bulletin d'Histoire et d'Épistémologie des Sciences de la Vie,* 11 (1): 193–217.

Thomson, K.S. 1963. "The Snout Anatomy of Rhipidistia and Dipnoi with Reference to Tetrapod Ancestry." Ph.D. dissertation, Harvard University. Consists of inserted reprint of an article by Thomson from *Breviora,* Museum of Comparative Zoology, no. 177, Dec. 1962. (Online Library Catalogue, Harvard University.)

———. 1997. "They Must Have Come from Somewhere!" (Review of H. Gee, *Before the Backbone* [London: Chapman and Hall, 1996]). *Paleobiology,* 23: 491–493.

Wanntorp, H.-E. 1993. "Lars Brundin 30 May 1907–17 November 1993" [includes L. Brundin, "From Grimsöl to Gondwanaland—Half a Century with Chironomids," probably originally written in 1985; trans. H.-E. Wanntorp]. *Cladistics,* 9: 357–367.

Williams, D.M., and Ebach, M.C. 2004. "The Reform of Palaeontology and the Rise of Biogeography—25 Years after 'Ontogeny, Phylogeny, Paleontology and the Biogenetic Law' (Nelson 1978)." *Journal of Biogeography,* 31: 685–712.

———. 2009. "What, Exactly, Is Cladistics? Re-Writing the History of Systematics and Biogeography." *Acta Biotheoretica,* 57: 249–268.

7

Patterson's Curse, Molecular Homology, and the Data Matrix

DAVID M. WILLIAMS AND MALTE C. EBACH

There is no comprehensive history of cladistics, the theory of systematics that revolutionized comparative biology in the early 1960s and inspired much discussion concerning the interrelationships of systematics, taxonomy, and evolution, as well as the relative importance of each in the discovery of phylogenetic relationships of organisms. Possibly it is too early for any useful history to be written. Three overviews do exist, each very different in perspective (Hull 1988; Craw 1992; Williams and Ebach 2008). There are a number of pertinent reminiscences (see the contributions in Forey, Gardiner, and Humphries 2000 covering reviews for France, Denmark, and the U.K.; and Bonde 2002–3 for a further examination of the impact of cladistic thinking in Denmark; see Janvier 1996 and Nelson this volume for a view from ichthyology). Not only may it be too early to write such a history, but as yet there may not even be a consensus on what cladistics really is (or was) or what it did (or did not) achieve—or even if it is still developing and progressing (Ebach and Williams 2011; Williams and Ebach 2008). Rather than attempt such an overview, it seems more useful at this time to sketch out some key episodes in the development of cladistics, particularly those that engendered a certain measure of misunderstanding and illuminate the way (some) scientists work. In this account we present one such episode that has significance on a number of levels. Oddly, the scientific aspect—the scientific content, if you like, the most important part—is rarely included in contemporary discussion.

The episode we focus our discussion on is a presentation given by the late Colin Patterson to the American Museum of Natural History's (AMNH's) Systematics Discussion Group on November 5, 1981, over thirty years ago. The presentation was titled "Evolutionism and Creationism," an incendiary title given to him but one he agreed to address (Patterson 1994b: 174). Perhaps unwisely, Patterson began with words designed to provoke:

> I should warn you that this title was laid on me by Donn Rosen: I'm speaking on it to gratify an old friend. I've never spoken on it before, but I have been kicking non-evolutionary or anti-evolutionary ideas around for 18 months or so. Usually when I get up to talk on some subject I'm confident of one thing—that I know more about it than anyone in the room. I don't have that confidence today: I'm tackling two subjects about which I feel I know nothing at all. One of the reasons I started taking a non-evolutionary view was my sudden realisation after working, as I thought, on evolution for 20 years, that I knew nothing whatever about it: It was quite a shock to learn that one could be so misled for so long. (Patterson 2002, 15)

A little later, Patterson made his aims abundantly clear:

> Well, I'm not interested in the controversy over high school teaching, and if any militant creationists have come here looking for political ammunition, I hope they will be disappointed. . . . Anyway, I'm not talking about that controversy—this is a systematics discussion group, and I shall talk about evolutionism and creationism as they apply to systematics. And since it's a discussion group, I only want to be outrageous enough to get a discussion going. (15)

But controversy there was. (Oddly enough a paper published in the *New Scientist* covered much the same ground as the AMNH presentation but lacked the deliberately provocative introduction [Patterson 1981b]). The presentation was never officially published until recently (Patterson 2002); excerpts—transcribed, erroneously or otherwise, from an unofficial tape recording—leaked out, many of them eventually making their way on to the Internet. Patterson almost always wrote out his presentations in longhand. That written version, now preserved in the archives of the Natural History Museum, London (Forey 2002), would not have differed much from the actual presentation[1]—although samizdat recordings had been circulated (often accompanied by poorly transcribed texts), a recording accompanied by a transcription of the entire talk and the subsequent discussion was made available from a creationist source (ARN 2000).

Some of the flavor of the subsequent furor might be gained from the title chosen by the ARN: "Can You Tell Me Anything about Evolution?" It is by now well known that the fallout from Patterson's talk, the extra-scientific content, touched many other areas of human understanding: sociology, philosophy, and scientific cultural history are just three that come to mind. That creationists might exploit out of context quotations from a variety of biological sources, while reprehensible, is a well-known phenomenon that should come as no surprise: it is one of their primary tools of argument, largely political in nature; that modern biologists attempt to invoke the same is shameful (Farris 2011a, 2011b, 2011c, 2012, 2013; Nixon and Carpenter 2012a, 2012b). As we shall argue later, their motives relate to the arrested development (or retardation) of cladistics as a body of thought, or at least one version of the "two cladistics" we identify, as Patterson did before us (Nelson 2004). A primary source of confusion between these two kinds of cladistics—for philosophers and historians rather than scientists—is the lengthy account written by David Hull, *Science as a Process* (Hull 1988, although some cladists have seen the marriage of the two as significant: Schmitt 2003, 375, a view at odds with other "Hennigians," 2013; Schlee 1975; Wägele 1994, 2001, 2005; Bechly 2000; and the comments of Krell 2005, 340, below).

Our account is devoted primarily to the scientific aspect of Patterson's presentation, specifically the development of cladistics (arrested or otherwise) and its relation to systematic biology and evolution in the early 1980s. We provide an account of more recent critiques of Patterson, those that dwell in areas other than science. We deal with the scientific aspect as related matters gain some perspective once Patterson's words are understood for what they were and what they were intended to convey.

We begin with a short account of Patterson's professional life.

COLIN PATTERSON (1933–1998)

Patterson was a paleoichthyologist, the majority of his career spent at the Natural History Museum, London.[2] There, at first, he studied under Errol White (1901–85), completing his Ph.D. dissertation on fossil Acanthopterygian ("spiny-finned, ray-finned") fishes, a huge group that includes bass, perch, mackerel, swordfish (Wiley and Johnson 2007). A morphologist by training—and primarily applied to fossils—Patterson was drawn by his research on fossil fishes to the Swedish

Museum of Natural History (Naturhistoriska Riksmuseet) in Stockholm, at that time the center for such investigations, led and influenced by two Swedish paleontologists, Erik Stensiö (1891–1984) and his successor, Erik Jarvik (1907–98) (see Patterson 1990 for Stensiö, Janvier 1998 for Jarvik; see Janvier 1996 and Cloutier 2004 for summaries of that period and Williams and Ebach 2008 for a nonichthyological perspective).

At that time, the Swedish entomologist Lars Brundin (1907–93) was also employed by the Swedish Museum of Natural History. His "Transantarctic Relationships and Their Significance, as Evidenced by Chironomid Midges, with a Monograph of the Subfamilies Podonominae and Aphroteniinae and the Austral Heptagyiae" (Brundin 1966), via the eagle eye of Gareth Nelson, became *the* source of dissemination of the ideas of Willi Hennig, the German dipterist who developed his "phylogenetic systematics" (Hennig 1965, 1966), which eventually became known as cladistics (Patterson 1981b, 195; see also the chapters by Hamilton, Nelson, and Rieppel in this volume).

A brief summary of that period might be thus. In the 1960s a group of paleoichthyologists, influenced by Gareth Nelson's discovery, embraced Hennig's phylogenetic systematics via Lars Brundin's chironomid midge monograph (Brundin 1966); among other things, Brundin addressed the dominance of paleontology in phylogenetic studies, seeing that subject (in particular stratigraphy) as a handicap rather than the key; systematics is revolutionized by establishing precisely what taxon (biological) relationships are and what evidence might be bought to bear in discovering those relationships. The trajectory of achievements is beautifully captured in the two diagrams published in *Nature* in 1979 illustrating the original salmon–lungfish–cow debate (fig. 7.1, after Gardiner et al. 1979 and Halstead, White, and MacIntyre 1979): it is easy to comprehend the simplicity of the upper trio of diagrams (fig. 7.1a) relative to the complexity of the single lower diagram (fig. 7.1b). We will briefly return to these illustrations shortly.

Early in his researches, Patterson noted what he thought might be a new source of data, a source somewhat different from that of bone and muscle, a source that might be gathered in abundance: DNA sequence data. Our chapter, then, sets out to sketch Patterson's exploration of various kinds of data in the light of his understanding of Hennig's term *relationship* and what that meant for future studies in systematics and evolution, especially phylogenetic studies. This sketch is viewed through

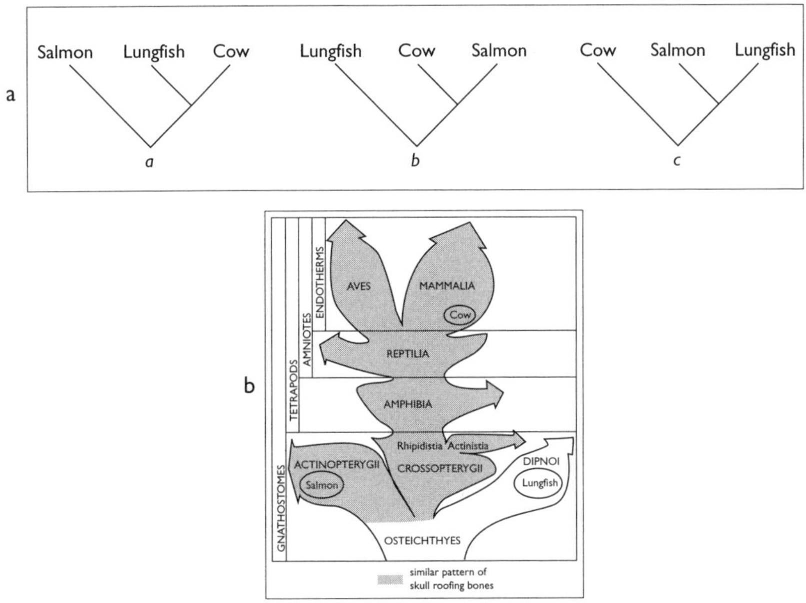

FIGURE 7.1. Diagrams illustrating the salmon-lungfish-cow debate in 1979, after Gardiner et al. 1979 and Halstead et al. 1979. See text for detailed discussion.

the lens of Patterson's American Museum of Natural History's Systematics Discussion Group presentation.

First we need to set the scene a little more and develop the idea that there are, in fact, two kinds of "cladistics," both with differing viewpoints, histories, and, probably, goals.

PHYLOGENETICS FROM THE 1960S TO THE 1990S: THE EMERGENCE OF TWO KINDS OF "CLADISTICS"

Phylogeny was not a subject for much discussion and debate in the 1950s, not quite moribund but languishing in the straightjacket of the availability of fossils, which most taxonomic groups lacked—and lacked in abundance. Nevertheless, efforts were made to recast the history of organisms in simple diagrams that were said to be a graphic representation of their history expressing specific kinds of relationships. Below we consider very briefly the development of these phylogenetic trees and then discuss the meaning of *relationship* as discussed in the late 1960s and through the 1970s. We follow this with an account of how DNA data were first understood as relevant to establishing phylogenetic relationships, homology, and, finally, how these concepts relate to the mathematization of phylogenetic trees in their construction by computer.

Phylogenetic Trees

Patterson's thesis was published by the Royal Society in 1964 and included one summary phylogenetic diagram (Patterson 1964, 472, fig. 103, reproduced here as fig. 7.2). Patterson would later describe this diagram thus:

> I ended my thesis with this awful diagram, using solid lines to show stratigraphic ranges and broken lines to show gaps in the fossil record and inferred relationships with Recent fishes. . . . I thought I'd found evidence that Recent acanthomorphs are polyphyletic, with various groups originating independently from different groups of Cretaceous beryciforms.[3] Polyphyly was the fashionable concept then, in the fifties and early sixties, and there were experts advocating polyphyly, demonstrated by fossils, for almost every major group of vertebrates. (Patterson 2011, 121; see Patterson 1967 for his early review of the problem of polyphyly in Teleosts)

Diagrams depicting polyphly in vertebrates could be found in the various summary works of Erik Jarvik (fig. 7.3, reproduced from Jarvik 1968, 510, fig. 3). The form of this diagram can be traced back to the classic spindle diagrams of Louis Agassiz (1844, reproduced here in fig. 7.4). Patterson (2011, 121) came to view Jarvik's diagram as "an extreme version of a view of evolution that was fashionable among palaeontologists—actually this is a denial of evolution." Discussion on the meaning of the various kinds of diagrams used to represent the phylogeny of organisms has created a large literature of its own (two recent useful contributions are Ragan 2009 and Tassy 2011), a dialogue that has not yet ended, even if today it extends beyond branching trees to a discussion of networks and "non-trees" as vehicles for the graphic representation of taxon relationships (see Ragan 2009 and the essays in O'Malley 2010).

The graphic representation of the phylogeny of various organisms (as opposed to their summary in a classification) began in 1866 with Ernst Haeckel's great project to represent all of life in a series of diagrams published in his *Generelle Morphologie der Organismen* (1866): the construction of the great tree of life, trunk, branches, even roots, connecting all organisms together, those still living and those extinct.

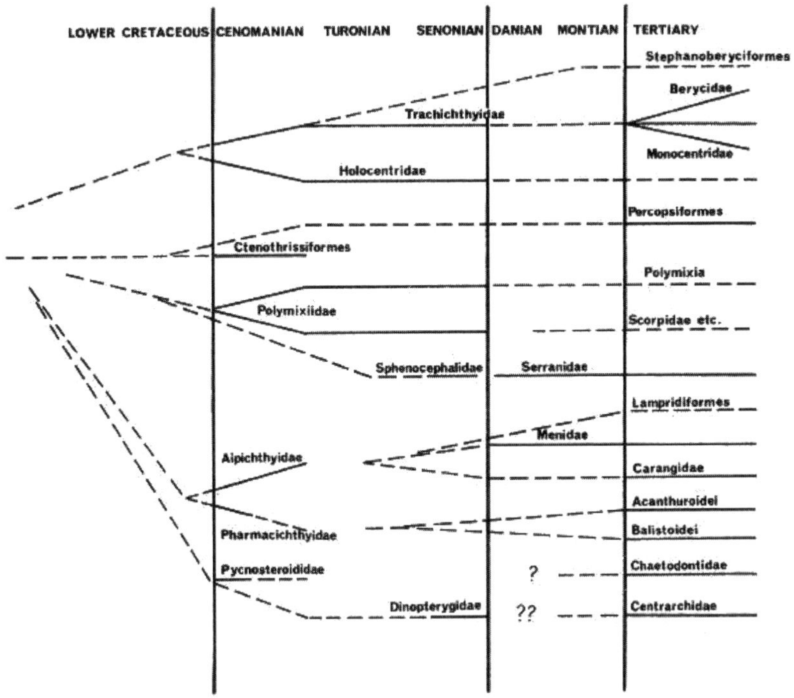

FIGURE 7.2. Summary phylogenetic diagram from Colin Patterson's thesis, published by the Royal Society in 1964. Discussion in text.

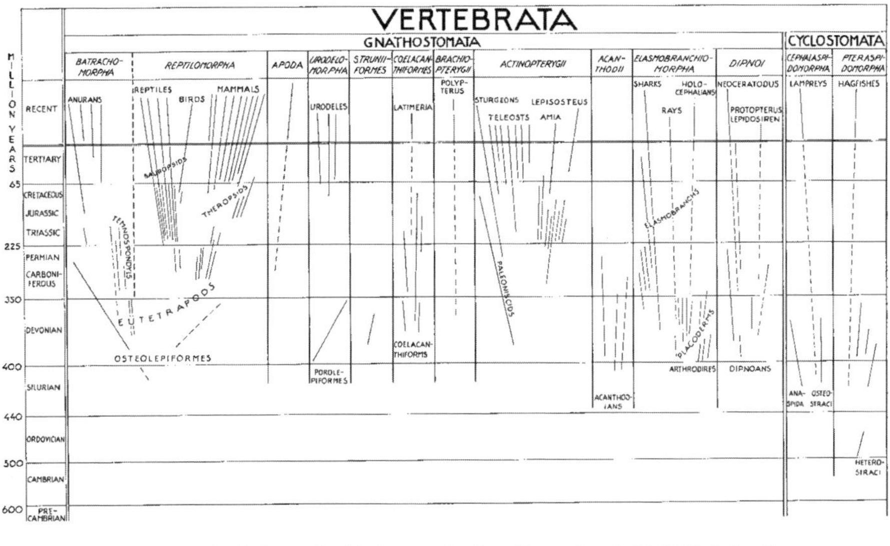

FIGURE 7.3. Diagrams depicting polyphly in vertebrates. From Jarvik 1968, 510, fig. 3.

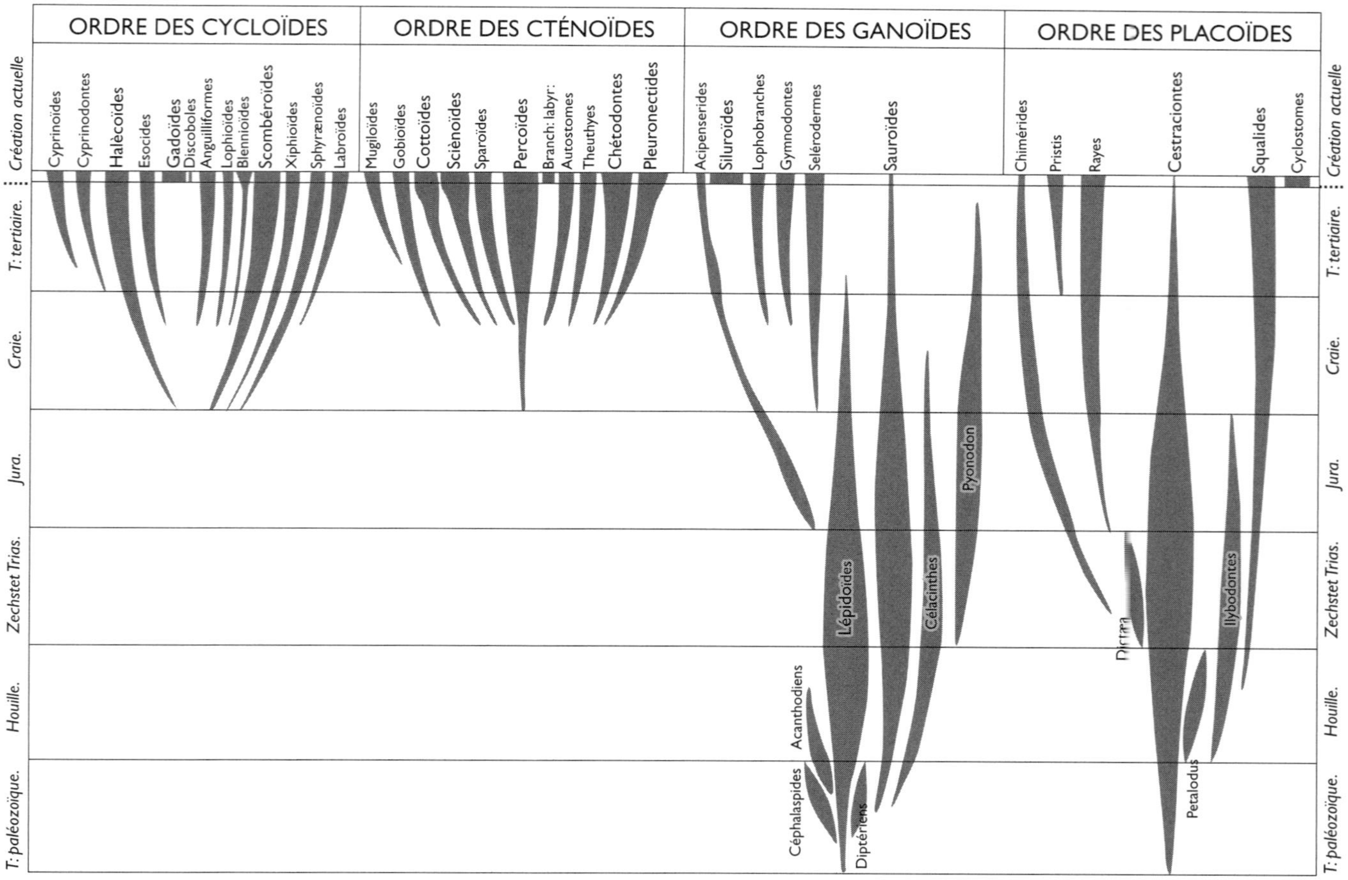

FIGURE 7.4. Classic spindle diagram of Louis Agassiz, a precursor of later diagrams by Jarvik and others. Reproduced from Agassiz 1844.

Haeckel's trees were representations of the pedigrees for life on Earth. His trees are reasonably well known and often reproduced, especially the first "Oak-like" trees (fig. 7.5a, from Haeckel 1866, Taf. 1). In the thirty years between his *Generelle Morphologie der Organismen* and the three-volume summary *Systematische Phylogenie. Entwurf eines natürlichen Systems der Organismen auf Grund ihrer Stammesgeschichte* (Haeckel 1894–96), his trees slowly changed in style, becoming simpler, more sticklike constructions (fig. 7.5b, from Haeckel 1896, Taf. 20). While the design simplified, the content was much the same: taxon names at the tips of the branches (descendants), taxon names along the branches or at the nodes (ancestors) (figs. 7.5a, b). Haeckel's trees were designed to depict the linear "relationship" of ancestor-descendant sequences, schemes that persisted well into the 1960s (fig. 7.6, after Romer 1962, 35). The example reproduced here is taken from Romer, as Brundin gave some attention to that diagram (Brundin 1966, 19, fig. 5;[4] Romer 1962, 35; Romer's tree appeared in the 1st edition of *The Vertebrate Body* [Romer 1949, 34] and remained unchanged through to the 5th edition [Romer and Parsons 1977, 37]; Pascal Tassy documents a series of these kinds of diagrams in Tassy 2011, e.g., figs. 7, 8, and 10, all the illustrations coming after Darwin; Ragan also offered a few examples: Ragan 2009, figs. 24, 25). Interestingly enough, during the development of phylogenetic diagrams it became popular to adapt Agassiz's spindle tree by simply connecting the branches together (as in Romer 1966, fig. 316; and Patterson 1977, 583, fig. 2; these diagrams have subsequently been called Romerograms: Patterson 1994a, 57; Benton 1993).

Relationships

Just a matter of days before Patterson gave his AMNH presentation, Ernst Mayr had a paper published in the October 30 issue of *Science* titled "Biological Classification: Towards a Synthesis of Opposing Methodologies" (Mayr 1981; Patterson 1994b, 174: "Fired up by Mayr's paper [1981], I gave a fairly radical talk in New York . . . "). Mayr included a diagram of relationships, a treelike diagram depicting three species, A–C, with unequal distances between each. The distances were quantified by the addition of a few lowercase letters to indicate the accumulation of hypothetical characters along certain branches (fig. 7.7a). Thus the branch leading to taxon A has two characters (*a* and *l*), the branch leading to taxon B has one character (*b*), the branch leading to taxon C has nine characters (*b–k*), and the branch leading to the node separating B + C from A has one

a

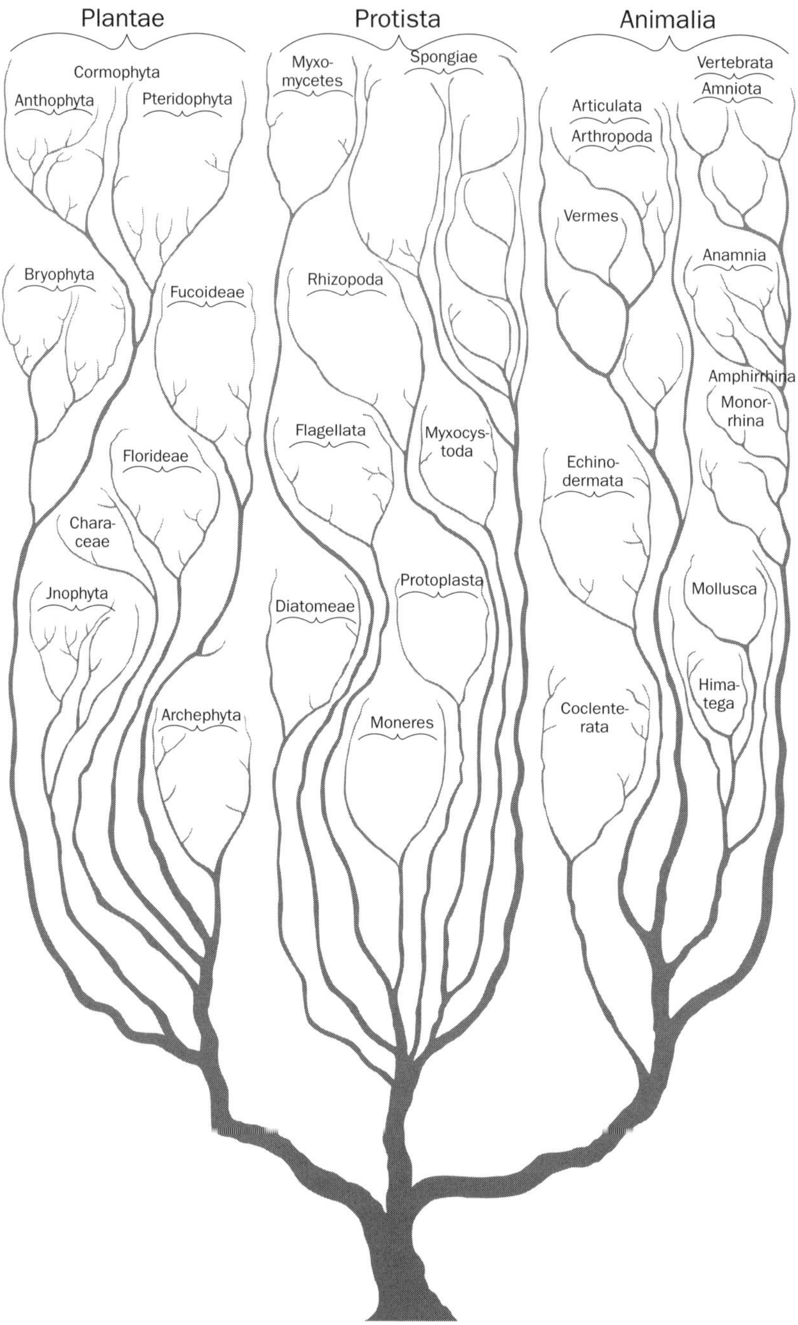

FIGURE 7.5. Haeckel's "oaklike" (a) and "sticklike" (b) representations of the pedigrees of life on Earth. This way of depicting ancestor-descendant relationships persisted into the 1960s.

b

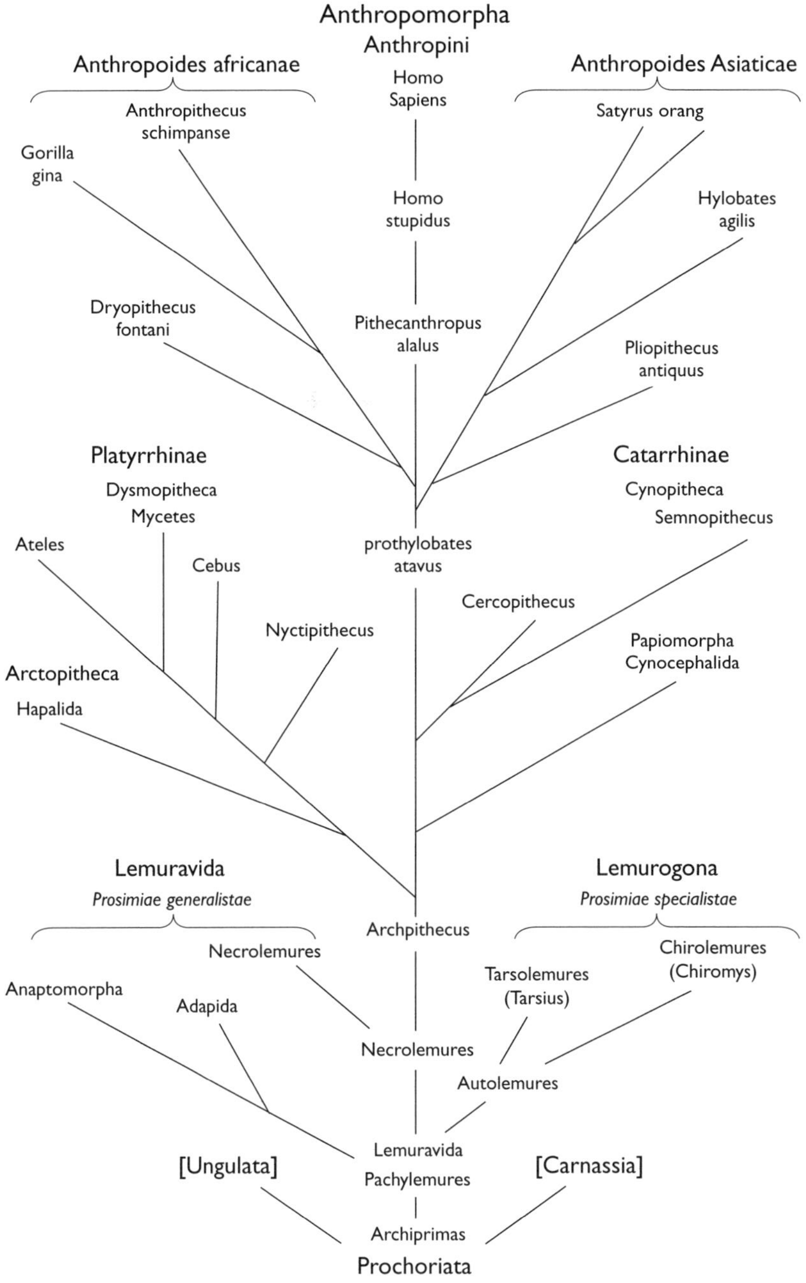
Anthropomorpha
Anthropini
Homo
Sapiens
Anthropoides africanae
Anthropoides Asiaticae
Anthropithecus
schimpanse
Satyrus orang
Gorilla
gina
Homo
stupidus
Hylobates
agilis
Dryopithecus
fontani
Pithecanthropus
alalus
Pliopithecus
antiquus
Platyrrhinae
Catarrhinae
Dysmopitheca
Mycetes
Cynopitheca
Semnopithecus
Ateles
Cebus
prothylobates
atavus
Cercopithecus
Nyctipithecus
Papiomorpha
Cynocephalida
Arctopitheca
Hapalida
Lemuravida
Prosimiae generalistae
Lemurogona
Prosimiae specialistae
Archpithecus
Necrolemures
Chirolemures
(Chiromys)
Tarsolemures
(Tarsius)
Anaptomorpha
Adapida
Necrolemures
Autolemures
Lemuravida
Pachylemures
[Ungulata]
[Carnassia]
Archiprimas
Prochoriata

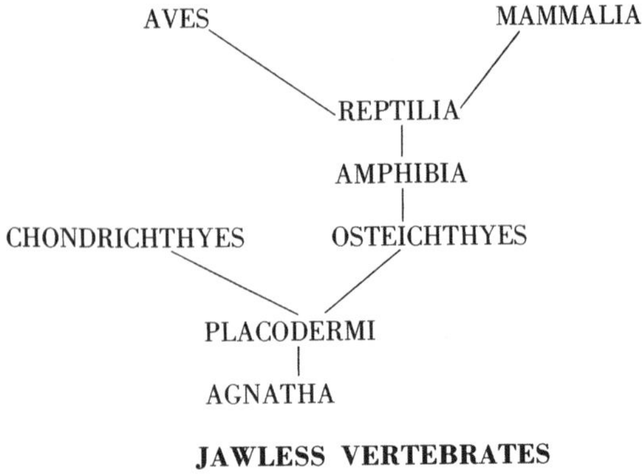

FIGURE 7.6. "Tree" of relationships between vertebrates, from Romer 1966, fig. 5. See text for discussion.

character (*a*). Note that the branches leading to taxon B and C both share character *b*. Mayr's figure legend explained the diagram:

> Cladists combine B and C into a single taxon because B and C share the synapomorph character b. Evolutionary taxonomists separate C from A and B, which they combine, because C differs by many (c through k) autapomorph characters from A and B and shares only one (b) synapomorph character with B. (Mayr 1981, 514, legend to his fig. 1)

Thus, according to Mayr, and by implication evolutionary taxonomists, the three species would be classified as (AB)C, while the cladist would classify them as A(BC). Mayr's diagram can be written in a different format so that evidence for grouping A, B, and C (character *a*) is placed at the nodes leading to all three (the most basal branch) and evidence for grouping B and C (character *b*) is placed on the node leading to that pair, while the other characters (*c–k* and *l*) are placed on their respective taxon branches (fig. 7.7b). That new kind of representation captures a key aspect of Patterson's AMNH presentation concerning the meaning of relationships. Years later, he summarized it thus:

> What we all learned from Hennig back in those early days boiled down to just one thing, what relationship means. No one had put it plainly before. Once you agreed what relationship meant, how to recognise it became obvious—synapomorphy, and then it was also obvious what was wrong with systematics as we'd been practising it in the 50's and early 60's, when everyone was preoccupied with polyphyly. Our mistake was thinking in terms of origins rather than relationships—Darwin may well be to blame for that preoccupation. Anyway, 'origins' has been a dirty word to me ever since, a symptom either of ignorance or of creationism. (Patterson 2011, 124; see also Patterson 1977, 612)

a

b

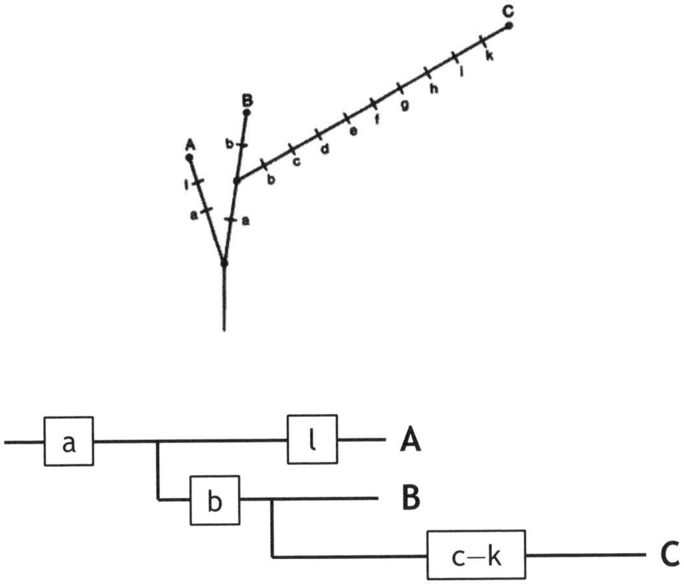

FIGURE 7.7. Treelike diagram after Mayr 1981 that was analyzed by Patterson 1994b (a); and a rewritten version of the diagram that emphasizes sister group relationships rather than origins (b). See text for detailed discussion of this figure and its context.

Evidence (in the form of synapomorphies *a* and *b*) and conclusions (in the form of taxa, B+C) could be directly connected via statements of relationship. This can be appreciated by another diagram from Hennig (fig. 7.8, after Hennig 1957, 66, Abt. 8). This diagram represents the simplest statement of relationships among three taxa A, B and C. The branch points indicate the relationship, the evidence is based on the parts of the organisms—the homologues—that suggest the relationship; the relationship is one of homology (Nelson 2011). The shift from ancestor-descendant relationships (phylogeny) to sister group relationships (classification) is universally accepted today: "the broad impact of evolutionary theory on relationship 'changed the meaning of 'affinity' (from 'resemblance' to 'blood relationship') (Bather 1927, p. lxxxii)" (Patterson 1988, 64), with the shifting of "phylogeny as historical process to systematics as hierarchy" (Patterson 1989, 471).

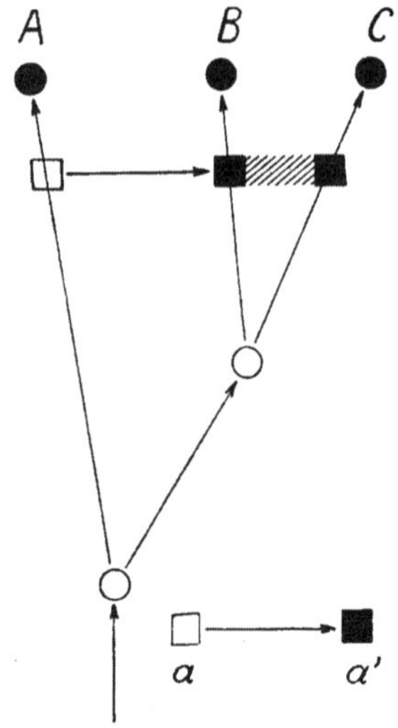

FIGURE 7.8. A statement of relationships between three taxa, as given by Henning 1957, 66, Abt. 8. See text for details.

Relationships and DNA

Mayr's 1981 diagram was a modified version of an earlier one, which contrasted a similar set of hypothetical taxa. In the earlier version the taxa were separated by differing amounts of time (Mayr 1969, 256, fig. 10–17, reproduced here as fig. 7.9a), a diagram criticized and declared wrong by George Gaylord Simpson (1975, 14). But it was another of Mayr's diagrams that attracted Patterson's attention (see Patterson 1981b, 197, 198, fig. 3; Patterson 1982a, 305, fig. 5; Patterson 1988, fig. 4.1; Patterson 2002, fig. 5). In this diagram Mayr added hypothetical percentage genetic differences along the branches (Mayr 1974, 103, fig. 1, left, reproduced here as fig. 7.10a) rather than vague notions of "divergence times" (Mayr 1969, 256, fig. 10–17) or "unit" characters (Mayr 1981, 514, legend to his fig. 1). To explain that particular diagram, Mayr wrote:

> This situation is best illustrated by a diagram [fig. 7.10a]. There will be a maximal genetic difference of 25% between the genomes of B and C [10% plus 15%], but of 60 to 70% between C and D. The cladist will say that C is more nearly related to D than to B, the evolutionist and the pheneticist that C is much closer to B than to D. (Mayr 1974, 102)

In this account, Mayr invoked a particular kind of data, suggesting that "relationship means the inferred amount of shared genotype" (Mayr 1974, 103) rather than "inferred recency of common ancestry" (103, citing Hennig's view, Hennig 1966, 74, which followed Bigelow 1956).

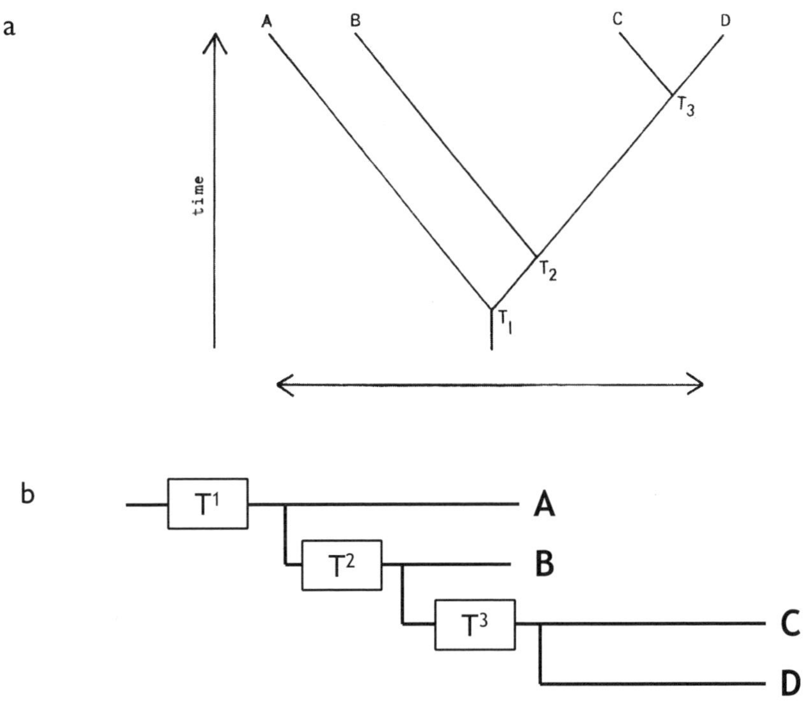

FIGURE 7.9. Earlier version of Mayr's 1981 diagram (from Mayr 1969), in which taxa are separated by differing amounts of time (a); and rewritten version that emphasizes sister group relationships (b). See text for discussion.

Many years before, Mayr had already offered the view that "when a biologist speaks of phylogenetic relationship, he means relationship in gene content" (Mayr 1965, 79; see Patterson 1977, 584, who notes that George Gaylord Simpson agreed that "genetic affinities" revealed phylogenetic relationships: Simpson 1975, 7). What appeared to be emerging was the "DNA point of view" on which genes have a special place in determining evolutionary relationships, usurping paleontology of that privileged position.

Yet Mayr presented no evidence to support this view. He simply asserted such would be the case, supporting his contention with a discussion of the relationship between a father and his child with a hypothetical scheme of inheritance of genes (Mayr 1974, 102). The thrust of his argument, in this instance, was not too dissimilar from that made by Zuckerkandl and Pauling (1965b) in their original justification of the superiority of molecules as definitive sources of phylogenetic information; elsewhere

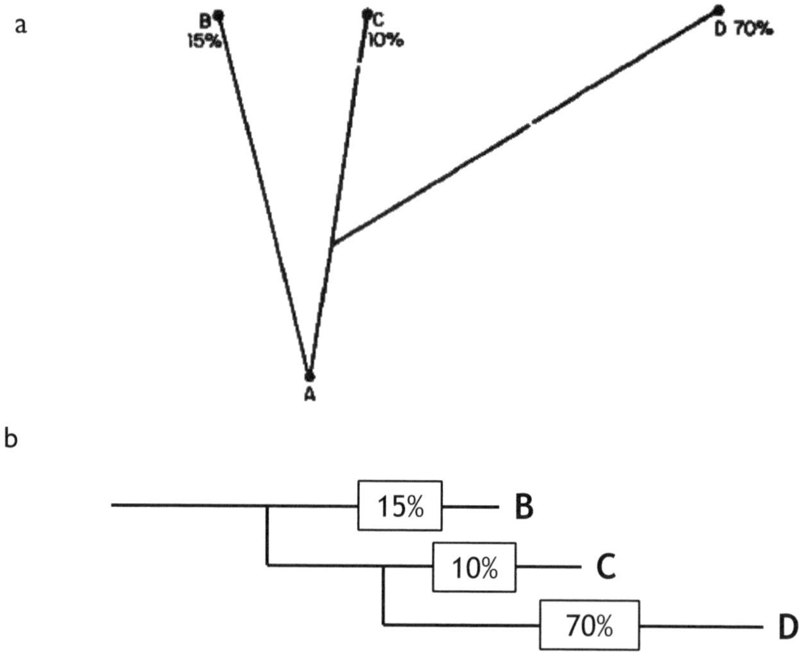

FIGURE 7.10. Diagram of relationships between four taxa to which Mayr added hypothetical percentage genetic differences along the branches, from Mayr 1974 (a); and version rewritten to emphasize sister group relationships (b). Further discussion in text.

Zuckerkandl and Pauling presented their concept of the informostat: "An organism is, by virtue of its genome, what one might call an informostat, by analogy with a chemostat or a thermostat. It keeps nearly constant the information that it contains and that it passes on" (1965a, 98).

In the discussion concerning the salmon-lungfish-cow relationships, Patterson and his colleagues took note of Mayr's view that "relationship means the inferred amount of shared genotype" (Gardiner et al. 1979).[5] Philosophical and theoretical meanderings to one side, Mayr's claim on behalf of molecules could be understood as an empirical one, something that might be explored. Initially, Patterson and his colleagues suggested that conclusions from sequence (genetic) data known at that time (and it was fairly meager) supported conclusions derived from a cladistic perspective of relationships:

> Overall similarity may be modernised or dignified by calling it 'genes in common' or 'shared genotype', yet when we have access to comparative information on genes, as in globin sequences (Goodman et al. *Nature* 253, 603;

> Romero-Herrera et al. *Nature* 261, 452; 1974), the biochemists present their information in the form of cladograms, and their use is the same genealogical concept of relationship as Hennigians. (Gardiner et al. 1979, 176)

The phrasing of their final statement is of interest: "the biochemists *present their information in the form of cladograms*, and their use is the *same genealogical concept of relationship* as Hennigians." The statements say nothing of how those relationships might be found. They deal with *representation* rather than *discovery*. Thus figures 7.7b, 7.9b, and 7.10b compare directly to Hennig's diagram (see fig. 7.8). That is, biochemists represented taxon relationships as sister groups, not ancestor-descendant sequences. This is crucial for the development of cladistics, as there were clearly two "kinds": the first deals with the concept of relationship and its *representation*; the second, which we have yet to explore here, deals with the discovery of relationships, with method. Certainly, the search for a *method* was being vigorously pursued in the 1960s by a host of different teams, but that search was linked to the development and introduction of computers into biology and the attempt to mathematize taxonomy: in short, the development of numerical taxonomy (the history of this endeavor has received more attention from historians than that dealing with relationships and the revolution in palaeontology: see Vernon 1988, 1993, 2001; Hagen 1999, 2000, 2001, 2003, 2010; Searls 2010; Strasser 2010; Sterner this volume). But that was a search for technique, not (necessarily) a philosophy. While Zuckerkandl and Pauling (1965a, 1965b) attempted to privilege molecular data by virtue of its "basic" nature, efforts to render it intelligible were explorations of methodology, of computer program dexterity. And it is here that Hull (1988) mixed the two, attempting to contrast developments in phenetics with those of cladistics, a division that might not seem best when teasing apart the fundamentals for recognizing progress in systematics.

A common thread through all of Patterson's papers was his discussion of relationships and *homology*, the latter being understood as the evidence for determining the former. Homology was the key.

Homology

In 1980 the Systematics Association, a London-based organization whose main role is the promotion and development of systematics,[6] organized a symposium entitled "Problems of Phylogenetic Reconstruction." Patterson was invited to speak on the subject of "morphological characters and homology." In an as yet unpublished talk he reflected on that event:

> Last year I was asked to write a paper on homology for a Systematics Association symposium on "Problems of Phylogeny Reconstruction." I accepted because I had always felt that the concept of homology raised vague problems that I had tried to kick back under the rug, but I wondered how on earth there could be anything to say about homology, a relation that has been under discussion for well over 150 years. (Patterson 1981a)

The proceedings of the symposium, held in Cambridge (April 8–10), were eventually published in two parts: *Problems of Phylogenetic Reconstruction* (Josey and Friday 1982, published as a book in the Systematics Association Special Volume series) and a series of papers published under the title *Methods of Phylogenetic Reconstruction* in a special issue of the *Zoological Journal of the Linnean Society* (Patterson 1982b). For the purposes of our chapter there is no need to dissect or explore in detail Patterson's views on homology and homology testing, ideas that have, in any case, already received a great deal of attention (e.g., De Pinna 1991; Rieppel 1988; Wiley 1981). Here we wish to focus on the evidence, the data, and the source of the data.

Late in his career Haeckel wrote of his thirty years of progress, from *Generelle Morphologie der Organismen* to *Systematische Phylogenie:*

> The first rough drafts of pedigrees that were published in the *Generelle Morphologie* have been improved time after time in the ten editions of my *Natürliche Schopfungsgeschichte* (1868–1902). A sounder basis for my phyletic hypotheses, derived from a discriminating combination of the three great records—morphology, ontogeny, and palaeontology—was provided in the three volumes of my *Systematische Phylogenie.* (Haeckel 1909, 148)

Haeckel's "three great records" are the evidence on which he attempted to discover (or reconstruct) taxon relationships. Today, we recognize four "great records": morphology, ontogeny, paleontology, and molecules (with a fifth, rarely exploited in this context: biogeography; Parenti and Ebach 2009). But regardless of "kinds" of evidence, it must speak to something. And under the cladistic paradigm, sister group relationships, rather than ancestor-descendant relationships, are the target (see above).

Here, then, the significant aspect is how Patterson understood the integration of Haeckel's "three great records [of evidence]—morphology, ontogeny, and palaeontology." With respect to evidence, it should be clear that the primary thrust of the cladistic revolution in the 1960s was the reformation of paleontology, meaning that fossils could—and should—be understood as a branch of morphology (Scott 1896): "The only difference between a collection of fossils and one of recent animals is that one set has been dead somewhat longer than the other" (a statement attributed to T.H. Huxley, cited in Scott 1896, 178; also in *Natural Science* 9 [1896]: 202).

But what of the relation between morphology and ontogeny, and how does that relate to homology? Patterson offered a different perspective from the usual evolutionary version. If transformation is the stuff of evolution, then ontogenetic relations (the "unfolding of form"), because they reveal actual transformations—one homologue (a part) observed changing into another homologue (another part), suggest "that characters do not transform in ontogeny, but that ontogenetic characters are themselves the characters" (Patterson 1988, 74)—they are the only direct evidence of change (Nelson 1973, 1978; Patterson 1996, 147); and if evolution could be understood as ontogeny writ large (Patterson 1983, 27, 28), then our understanding of evolution, in this sense, is an extrapolation of ontogeny (Patterson 1983). Thus for morphological characters the only source of direct evidence for transformation is ontogeny; the stratigraphic record and comparative biology (morphology) are indirect sources of evidence, based on assumptions, if you will (see also Nelson 1973, 1978). But what of the fourth and newest and most abundant "great record," molecules, the source of evidence that held some promise for Mayr and most practitioners today? According to Patterson:

> There is one class of phylogenies . . . that does differ from ontogeny: molecular phylogenies, derived from nucleotide or protein sequence data. The transformations postulated in molecular phylogenies are gene duplications and changes in nucleotides that are nowhere matched in ontogeny. (1983, 26)

What are the implications? This was the central topic explored in the AMNH presentation—explored relative to tree building, to method, the second "cladistics."

Cladistics and Tree Building

For Patterson, then, there was a crucial difference between morphological and molecular homologies and its impact was on building phylogenies:

> Molecular sequences are linear, or one-dimensional, and do not themselves display hierarchical organisation. In building phylogenies from homologous sequences, methods such as parsimony (minimum evolution), compatibility or likelihood . . . must be used to link the sequences by hypothetical ancestral sequences. (Patterson 1983, 26; Patterson 1989, 485)

He was more forthright concerning those implications in the AMNH presentation:

> So what about this tree and the numbers on the branches, and so on—where do they come from? Well, they don't pop out of the data, so I suppose they come from massaging the data with a theory—or with a computer programme based on a theory; and the theory is evolutionary theory, descent with modification. (Patterson 2002, 27)

Elsewhere, in another different but unpublished talk, the message was even more direct:

> Of course, you are all familiar with phylogenetic trees built up from comparisons of amino acid sequences. . . . So what puts the hierarchy into molecular data? . . . [T]he DNA sequences . . . are fed into the computer, and in the computer they are massaged by a programme. . . . [T]he computer does what it is told, and produces a tree. It will give you a tree from completely random data, because that's what it is for, the programme is designed to makes trees. (Patterson 1981c)

Patterson's comment is significant in many ways: tree-building programs, being "designed to make trees," are so assumption laden that they point toward a black-box approach to systematics. That is, there seemed to be no escape from using assumption-laden computer programs to determine treelike relationships. But where do we search for those assumptions? And "what does the tree tell us about—is it telling us something about nature, or something about evolutionary theory—I'll leave you to decide" (Patterson 2002, 27). In the same vein he wrote, "Parsimony is a necessary property of methods of analysis, not of nature" (Patterson 1988, 79). We pass over the view that Patterson (and other pattern cladists) were suggesting an assumption-free approach to systematics. This is clearly nonsense ("But I know of no cladist so naive as to believe that there are any scientific statements whatsoever that are theory-free": Platnick 1985, 88; see also Platnick 1986), even if some persist in using it as a characterization (Delaporte 2012).

Many years later, as a result of examining in detail a dataset for a group of fishes (Patterson and Johnson 1997a, 1997b), Patterson noted, with some justified frustration, "I really feel that in adopting this modern version of cladistics [numerical cladistics] we may be replacing one pernicious black box, evolutionary systematics, with another, the matrix" (Patterson 2011, 131), a view eloquently expanded on by Mooi and Gill (2010).

"CREATIONISTS": SELECTIVE CITATION AND AUTO-EDITING

In our view, Patterson identified—but did not specifically state it as such[7]—what we have been referring to as the "two cladistics": the first effected the revolution in palaeontology (Williams and Ebach 2004, 2008); and the second, the numerical version, the exploration of, or extension to, numerical taxonomy (Sneath and Sokal 1973; Felsenstein 2004), in this context specifically the development of Wagner parsimony as a "cladistic" proxy (Farris 1970). Patterson did outline the basics of a method of character analysis based on his understanding of homology, a method he likened to character compatibility (Patterson 1982a, 1988). His method naturally shared a number of similarities with existing numerical methods, specifically Wagner parsimony and the related compatibility approach (for a contemporary review of methods, see Felsenstein 1982; see also Felsenstein 2004 for a book-length treatment of progress in numerical taxonomy and Williams, Ebach, and Wheeler 2010 for commentary on the development of that intellectual lineage). From a "science culture" point of view, contemporary critique of Patterson's ideas is limited; in cladistic circles it can be reduced to three: creationism (by association), selective citation (avoiding the literature), and auto-editing (doctoring the literature).

Creationism

Arguments critical of Wagner parsimony often solicit responses tinged with a certain amount of political posturing, of limited scientific relevance but enormous sociological significance. For example, Kluge and Farris, when commenting on a methodology that did not absorb all of their teachings on the "correct" interpretation of parsimony, wrote, "No one but a creationist could think it realistic to exclude transformational considerations from the process of grouping" (Kluge and Farris 1999, 208).[8] Rather than offer any comments on the veracity of that statement, we invite reflection on an earlier piece by Farris:

> Consider as an analogy Halstead's (1980) charge that cladistics is a communist conspiracy.[9] On its face it seems simply a mindless lie, but in fact it has quite a different purpose than to represent an actual state of affairs. Halstead, being an evolutionary taxonomist, despises cladists. He may very well also dislike Marxists, but he need not; it only matters that others are apt to dislike them. *By associating the two, Halstead hoped to suppress cladistics.* (Farris 1985, 198; emphasis ours)

Associating creationism with "anything other than Wagner parsimony" might indeed suppress consideration of those ideas too.[10] But Farris supplies our conclusion also, in sentences immediately following those above:

> That might be hoped to work on the general public, *but for arguing to biologists it is much more effective to make a charge that biologists will consider as utterly damning, and anti-evolutionism fits the requirement perfectly.* Once the idea is established that there are people called pattern cladists who deny the place of evolution in systematics, every worker who dislikes any cladistic view or result has the ideal weapon. (Farris 1985, 198; emphasis ours)[11]

Selective Citation

Later, in this political battle of wills, another device, selective citation, was used, again by Farris, in a review of Kitching et al. (1998), a primer on the practical aspects of cladistics, its second edition outlining the rudiments of three-item analysis, an approach to systematic data that moved beyond binary representation of conventional data matrices (Nelson and Platnick 1991) and Wagner parsimony. Here Farris targets Patterson once again, noting that he (Patterson) wished to avoid using outgroups when comparing characters to determine what is primitive and what is derived: "he [Patterson] devised a technique based on 'letting the characters speak for themselves' . . . " (Farris 2000, 427).

Farris cited another passage from Patterson: "it is not necessary to polarize characters, or to sort them into primitive and derived, to find a cladogram that is unambiguously rooted" (Patterson 1988, 76). He then notes that Patterson's resulting cladogram (1988, fig. 4.2) was indeed rooted, "but as Patterson's method did not distinguish symplesiomorphy from synapomorphy, it behaved like phenetic clustering" (Farris 2000, 427).

Where, then, might one look for "proper" information on how to distinguish symplesiomorphy from synapomorphy, how to distinguish primitive from derived characters? Farris suggested Nixon and Carpenter (1993) because of their "excellent discussion of the advantages of outgroup rooting" (Farris 2000, 427). One might examine Nixon and Carpenter's paper:

> Perhaps the most striking aspect of this subject [outgroup comparison] is that the logically inconsistent treatment of ingroup and outgroup data proposed by Maddison et al. (1984) has largely escaped attention for 10 years. *Meacham (1984) argued that character polarity need not be determined before cladistic analysis,* citing Farris (1970) regarding parsimony of undirected networks. (Nixon and Carpenter 1993, 417; emphasis ours)

Many years later Nixon and Carpenter repeated this claim: "In Farris' approach, *character polarity is unimportant prior to a cladistic analysis*" (Nixon and Carpenter 2012a, 162; emphasis in original).

Colless (1985) concurred with Meacham (1984) and cited Farris (1972) to the effect that simultaneous analysis of the ingroup and outgroup together would properly implement outgroup analysis and rooting (Nixon and Carpenter 1993, 418).

Nixon and Carpenter offered a reason for the conspicuous lack of reference to Colless and Meacham in the general literature: "Perhaps because Meacham advocated clique analysis and Colless advocated phenetics, their papers are seldom cited" (1993, 418).

But Patterson did cite both Meacham and Colless (and Nixon and Carpenter did not cite Patterson in 1993 or, many years later, in Nixon and Carpenter 2012a):

> Outgroup analysis is still subject to the charge of circularity (Colless, 1984 [1985]). General congruence might be called letting the characters speak for themselves, a phrase used by Meacham (1984) in arguing against a priori directing of characters, and recommending undirected cladistic analysis. (Patterson 1988, 74)

Evidently, Colless, Patterson, and Nixon and Carpenter (among others) did see something of value in Meacham's work, the source of the phrase "let the characters speak for themselves" (Meacham 1984, 35: "Direction of character state transitions we are more sure of can suggest the direction of those we are less sure of. It lets the characters speak for themselves"), a phrase Farris found so obnoxious; and it was Meacham who supported unconstrained analysis of both ingroup and outgroup. Nixon and Carpenter offered the following:

> The most pervasive misunderstanding of outgroups and their use in cladistics is thus tied to the concept of "character polarization" and centers on the

> belief that we must know "primitive" and "derived" states before we attempt analysis. (Nixon and Carpenter 1993, 419)
>
> Polarity in fact need not be decided before analysis, contrary to common misconception. (Nixon and Carpenter 1993, 420)

Which all sounds remarkably like Patterson: "demanding that the information within the character set dictates the polarity of characters" (Patterson 1988a, 74)

In matters of principle, there seems no difference in Patterson's line of argument and that of Nixon and Carpenter; however, seen politically, Nixon and Carpenter side with Farris and "selectively eliminate" appropriate literature. But a difference does remain, as neither Colless, nor Meacham, nor Patterson considered Wagner parsimony the "perfect" method for implementing cladistics (a suspect proxy). Indeed, Patterson's method, which he called "General Congruence," was designed to "let the characters speak for themselves." But rather than phenetics, he compared it to compatibility, a different approach to parsimony.

Identifying "true" motives is a tricky game, but, as we have noted, the use of creationist jibes linked to pattern cladistics betrays argument beyond mere evidence, even logic, and locates itself in the arena of politics.

Auto-Editing

By "auto-editing," we mean editing out the apocryphal literature from a politically motivated (and usually false) message, namely, that three-taxon analysis[12] is kin to pattern cladistics, which is in turn creationist.

Learning something about pattern cladistics must be a difficult endeavor for anyone wishing to do so. What literature should be examined? Reference to Farris's papers is often made, rather than the primary literature written by pattern cladists themselves. In most cases, if not all, this leads to erroneous assumptions based on initial false claims. Years after the discussions above (under "Selective Citation"), Nixon and Carpenter wrote:

> In reality, there are almost no cladists who can be categorised as pattern cladists, and as Farris (2011) has observed, the true pattern cladists may be those few remaining (e.g., Williams and Ebach, 2008) who adhere to the method termed "three taxon analysis" of Nelson and Platnick (1991), mixed with Patterson's (1988) "pattern analysis." (Nixon and Carpenter 2012a, 165)

Never mind whether there is any substance to these claims (as before, there is none), our purpose here is to draw attention to the *style* of argument.[13] The citation of Farris (2011) as an accurate account of what pattern cladistics might be is merely politically motivated. The association of a viewpoint with a particular methodology, erroneously linking "three taxon analysis" with "pattern analysis," is a further dose of political chicanery. The final charge mustered is Farris's creationist "argument":[14]

> This confusing explanation of the cladistic method, divorcing hypotheses of homology from evidence, viewing characters/states as vacuous "hypotheses of grouping" without reference to homology, then using such grouping information to infer homology, is beyond comprehension for most evolutionary biologists, and has led to the view in some quarters that the few remaining pattern cladists are in league with creationists (Farris 2011). (Nixon and Carpenter 2012a, 165)

"In league with creationists": absurd though that is, Nixon and Carpenter might indeed cite whom they believe to be pattern cladists, but they do not refer to them directly; they refer, as ever, to Farris (2011): Farris appears to be the canon that cladists refer to while totally ignoring the apocryphal literature, namely, the primary literature. Once a society or self-regulated group of people collectively decide in print on what must be right versus what is unacceptable, a canon of literature is created. This might reflect the democratization of science, but a majority view is not necessarily the correct view and in any case should certainly not be canonized. In order to understand any debate in any field, one needs to address and understand the many sides of the argument—for there are always many. Unfortunately this no longer occurs, and possibly has not since the earlier canonization of the evolutionary taxonomy trio Simpson, Mayr, and Stebbins, in contrast to the new "evolutionary taxonomy" trio, Nixon, Carpenter, and Farris.

How does auto-editing affect the way we do science? It would be safe to assume that most students do not read the philosophical literature that justifies any particular method. That might seem an enormous task, so why should they? Usually, a program manual would suffice. For example, to use the new computer program TNT one would not need to read Hennig (1966) to understand how it operates: reading the manual, following others, and receiving the program gratis from the Willi Hennig Society is sufficient (Goloboff, Farris, and Nixon 2003). Once TNT produces a result, then the practitioner is satisfied that the program has done what the manual promised—delivered the phylogenetic relationships between the taxa of interest, based on the transformation of character states at various nodes of the tree. Evidence for this, namely, homology, does not need to be identified or investigated further because the literature states that the result is supported. Furthermore, popular belief that the program actually works justifies this action. In other words, a program that produces a result similar to the one the practitioner is expecting is greater than actual evidence. And if not, then collect more and more data and redo the analysis, until everything stabilizes.

If the practitioner reads a large body of literature that details, as well as criticizes, the history and theory behind a particular method, then he or she will most certainly make a different choice. Not only is this impractical, but it requires the practitioner to have far more knowledge of theory than is actually necessary to use any program. So the question remains: Does one need to know the literature and theory that support a methodology? If a methodology is found lacking in terms of generating satisfactory results, then it is generally rejected in favor of one that does. But when a popular method does not produce reliable results, a credible explanation needs to be made. These are the canons.

Papers in which theory is "created" to support a popular method becomes a canon, and, by definition, anything that is critical becomes apocryphal (Nixon and Carpenter 2012a, 2012b). Now the method has a theory that is used in support of an implementation that may be flawed. Such theories corrode evidence even further. Former errors that showed a methodology to be faulty become forms of evidence that now support it. If this practice is followed every time an error occurs, a totally "artificial" theory is created, one divorced entirely from the history of systematics (examples are Nixon and Carpenter 2012a, 2012b; and Farris 2011, 2012a, 2012b, 2012c).

CLADISTS BEHAVING BADLY

Why have both sides decided to fight their battles at the level of methodology? Three-item analysis, for all it is worth, is a theory of classification, applicable to any scientific field (e.g., historical biogeography) (Williams and Ebach 2008), while Wagner parsimony is a highly specialized technique applicable to a very specific area, namely, phylogenetic systematics (Farris 1970). Why a simple technique used in phylogenetics is pitched against a theory of classification may appear strange to someone outside the community. Within, however, the conflict is obvious: Numerical cladistics created a social clique based on and united by a particular technique, namely, Wagner parsimony, on which everything else—history, theory, methodology—is judged. Those histories (e.g., Hull 1988), theories (e.g., three-item), and methods (e.g., compatibility) that do not fit this worldview are rejected as being "noncladistic" ("Counterfeit Cladistics," Farris 2012a). Even today, the term *cladistics* is defined by its application (Wagner parsimony) rather than the history, theory, and methodology preceding it, meaning that much of the debate since the 1980s has had nothing to do with logic, reason, or method but with schoolyard bullying and name-calling, such as "pattern cladists are in league with creationists" (Farris 2011).

PATTERSON'S CURSE

We have largely concentrated on cladists' reaction to Patterson's AMNH presentation, and to his development of those ideas in subsequent publications. Patterson took his ideas much further, suggesting that molecules and morphology could be integrated to form a unified approach to the systematics of organisms, forming a comprehensive way to discovering phylogeny and thereby to understanding evolution: "The impact of neutral theory and the molecular clock on systematics is to make phenetic and cladistic methods equivalent at the level of DNA" (Patterson 1988, 83), summed up as the twin pillars "clocks and clades" (also Patterson 1994b, 188). A starting point, and an acceptable challenge removed from the model-based (assumption-based) analysis of data, was the question of the data's relevance, directly addressing a twentieth-century concern applied to Haeckel's "three great records—morphology, ontogeny, and palaeontology," with molecules as the fourth, a source of data unknown to Haeckel. The question—the data's relevance—was profound, and remains so: If the stratigraphic record is no longer the definitive arbiter of phylogenetic relationships, what is? What could function that way? Or could anything function that way, any source of data? (Nelson 1989).

And what of phenetics, the apparent bête noir of systematics, the theory that simply will not go away (Lewens 2011)? Patterson did come to understand numerical cladistics as a kind of phenetics: " we have learned that one factor (perhaps the only one) distinguishing cladistics from phenetics and eclectics [evolutionary systematics] is the cladist's refusal to treat absence as a character" (Patterson 1994b,

176). That comparison may have been born from a knowledge and familiarity with developments in numerical cladistics, the role of ontogeny in character determination, and his explorations with DNA, this entirely new kind of data. But beyond "kinds" of data, the equivalence between numerical cladistics and phenetics is not entirely unjust and their equivalence with Hennig's work somewhat misplaced, points raised elsewhere: "Since Hennig's times, cladistics has developed rapidly, but mainly as a numerical methodology, which Hennig's Phylogenetic Systematics never was (Schlee 1969)" (Krell 2005, 339). Actually, the approach and procedure of modern cladistics differ substantially from Hennig's methodology, a fact that is not widely recognized though occasionally explained. It has even been shown (Wägele 2001, 185f.) that cladistic methods did not derive primarily from Hennig's methodology but from the numerical taxonomy of Sokal and Sneath (1963), a procedure that Hennig and his collaborators rigorously rejected (Hennig 1971; Schlee 1975; see also Krell 2005, 340; Wägele 1994; Bechly 2000).

Historically and theoretically, there is more to Hennig's phylogenetic systematics than a single (proxy) methodology (Wagner parsimony) and the political posturing designed to suppress any advances in that body of thought made under his name. At the very least, Patterson offered a constellation of ideas worthy of exploration.

In our view, and with the benefit of hindsight, Patterson's journey was within the phenetic paradigm, as no one at that time had thought to extend the cladistic revolution to the characters themselves, to the concept of homology (Nelson 1989, 1994). We believe Patterson was mistaken: mistaken about "clocks and clades." But the problems rest not with Patterson's reasoning but with fundamentals concerning the evidence and the data and their representation (Nelson 1996; Williams 2002; Williams and Ebach 2008). These problems have been misinterpreted as methodological (Farris 2011, 2012a, 2012b; Nixon and Carpenter 2012a, 2012b) when in fact they are fundamental (Williams and Ebach 2008).[15]

To seriously consider that pattern cladists are in league with creationists—or perhaps even the devil—is simple slander and not worthy of further consideration. Rather, the association has been a curse for the late Colin Patterson and a bane for all of systematics, retarding its progress. Patterson himself retreated from the limelight, having suffered too many personal attacks from professional colleagues. David Hull, perhaps the first to identify the bully-boy tactics in systematics, was himself a victim. Ousted for being gay at a Willi Hennig Society meeting, Hull too could bear a grudge but certainly not in print.

We hope the history of cladistics will be told in full once its protagonists fade away and the dirty deeds are finally uncovered. Until then, enjoy the show—what's left of it: "Cladistics should have been purer than this" (Farris 1985, 200). Indeed.

Acknowledgments

We are grateful to the following for comments and suggestions: Andrew Hamilton, Gareth Nelson, Francisco Vergara-Silva, and John Wilkins.

NOTES

1. Early on in the presentation Patterson says, "If it appears that I'm reading this stuff and sometimes seem surprised by what I find there, well, that's true, I am reading it." This statement appears in the transcript of the presentation issued by the Access Research Network (ARN 2000).

2. There are numerous obituaries, but the most useful are Bonde 1999; Fortey 1999; Forey 2004; Nelson 2007; and the various papers in Forey, Gardiner, and Humphries 2000.

3. Beryciformes is an order of actinopterygian fishes with many fossil representatives.

4. "Still in 1962 we are told by Romer that the Osteichthyes have given rise to Amphibia, that Reptilia are the descendants of Amphibia, and that Aves and Mammalia arose from Reptilia. . . . It [Romer's figure] visualises a successive development from one evolutionary level to another and interprets the actual groups as evolutionary grades. It visualises phenomena connected with phylogeny, not phylogeny itself" (Brundin 1966, 19).

5. Patterson later explored Mayr's claim in more detail and, after an examination of the available sequence data, concluded that "Mayr's interpretation of evolutionary relationships is refuted . . . [and relationships determined by] common ancestry and inferred common genotype lead to the same interpretation" (Patterson 1981b, 199).

6. The original aims of the Systematics Association were given in these two articles: *Nature*, July 24, 1937, 163; *Nature*, August 7, 1937, 211–212.

7. Patterson did later contrast two significant events of 1967: Brundin's presentation at the 1967 Nobel Symposium (Brundin 1968) and the Cavalli-Sforza and Edwards and Fitch and Margoliash papers on the construction of phylogenetic trees (Patterson 1989, 472).

8. It should be noted here that Farris invented Wagner parsimony, a method of interpreting systematic character data in terms of "special similarity" and so was not exactly an unbiased commentator.

9. L. Beverley Halstead attempted to compare cladistics to certain aspects of Marxism, a foolish comparison that solicited a vast correspondence in the journal *Nature* from 1979 until 1981 ("if the cladistic approach becomes established as the received wisdom, then a fundamentally Marxist view of the history of life will have been incorporated into a key element of the educational system of this country": Halstead 1980, 208). Halstead's original target was the Natural History Museum's new exhibition policy, but it extended to cladistics in general, the debate gaining a notable event with the discussion of the relationships of the salmon-lungfish-cow debate (see Schafersman 1985; Bonde 1999, 257; and Williams and Ebach 2008 for a review of the controversy).

10. This line of argument is a variation on the rather better known *reductio ad Hitlerum*, whereby an opponent's view is associated with those Hitler supposedly held. This version might be called *reductio ad ex nihilo*, whereby an opponent's view is associated with those of creationists, a line of argumentation, remarkably, shared by Halstead and Farris.

11. Historians have a valuable document in Farris 1985. He concludes the section we cite thus: "It is clear, then, why the myth of pattern cladistics has become so popular. Anti-cladists find it useful" (Farris, 2013). And, later, so did he.

12. Three-taxon analysis is a relatively new method that explores the structure and representation of systematic data rather than various effects of different computer algorithms. First specifically outlined by Nelson and Platnick (1991), a certain amount of discussion took place in the pages of *Cladistics*, the house journal of the Willi Hennig Society. A larger review giving some historical context can be found in Williams and Ebach 2008.

13. The style is similar when Farris addresses the issue of phenetics (Farris 2012c).

14. We place the word *argument* in quotes as no argument is ever made. Farris simply follows the tactic he assigned to Halstead years before. Here Farris's prose speaks for itself: "In a posthumous publication of a talk from 1981, Patterson (2002, 26 [the transcription of the AMNH talk addressed in our chapter]) characterized his method as 'what a creationist makes of it [the data].' Perhaps reluctant to be perceived as a mindless fanatic, he employed less forthright language in publication while still alive, using Newspeak to pass off creationism as 'cladistics' (Patterson, 1980, 1982, 1988)" (Farris 2011, 211 n. 11; other similar comments litter this tawdry contribution); "Well, ppa [an acronym attributed to Patterson as "Patterson's Pattern Analysis"] actually is phenetic, but Williams and Ebach have never been known to call Patterson a pheneticist, probably (if perhaps paradoxically) because they admire his creationism" (Farris 2012a, 228). Nixon and Carpenter cannot resist either: "Strangely, proponents of 3-ta [three-item analysis] believe they are avoiding unnecessary assumptions about evolution, and that method is now closely tied to 'pattern cladistics' and often favorably cited by creationists and others who deny evolution" (Nixon and Carpenter 2012a, 226). Favorably cited by creationists? No references are supplied to support that view.

15. A tactic repeated when dealing with phenetics: Farris 2012c.

REFERENCES

Agassiz, L. 1844. *Recherches sur les Poissons Fossiles.* Vol. 1. Text. Neuchâtel: Petitpierre.

ARN (Access Research Network). 2000. "Can You Tell Me Anything about Evolution? November 1981, Presentation at the American Museum of Natural History by Colin Patterson." Audio CD and annotated transcript.

Bather, F.A. 1927. "Biological Classifications: Past and Future." *Quarterly Journal of the Geological Society,* 83: lxii–civ.

Bechly, G. 2000. "Mainstream Cladistics versus Hennigian Phylogenetic Systematics." *Stuttgarter Beiträge zur Naturkunde Series A,* 613: 1–11.

Benton, M.J., ed. 1993. *The Fossil Record 2.* London: Chapman and Hall.

Bigelow, R.S. 1956. "Monophyletic Classification and Evolution." *Systematic Zoology,* 5: 145–146.

Bonde, N. 1999. "Colin Patterson (1933–1998): A Major Vertebrate Palaeontologist of This Century." *Geologie en Mijnbouw,* 78: 255–260.

Bonde, N., with J. Høeg and O. Seberg. 2002–3. "Introduktion af fylogenetisk systematik. Kladismens fremme I Danmark og andre lande." *Dansk Naturhistorisk Forening, Årskrift,* 13: 8–34.

Brundin, L. 1966. "Transantarctic Relationships and Their Significance, as Evidenced by Chironomid Midges, with a Monograph of the Subfamilies Podonominae and Aphroteniinae and the Austral Heptagyiae." *Kungliga Svenska Vetenskapsakademiens Handlingar,* Fjarde Serien, 11 (1): 1–472.

———. 1968. "Application of Phylogenetic Principles in Systematics and Evolutionary Theory." In T. Ørvig (ed.), *Current Problems of Lower Vertebrate Phylogeny: Proceedings of the Fourth Nobel Symposium Held in June 1967 at the Swedish Museum of Natural History (Naturhistoriska riksmuseet) in Stockholm.* Stockholm: Almqvist and Wiksell, 473–495.

Cloutier, R. 2004. "Hans-Peter Schultze's Contribution to Our Understanding of Lower Vertebrate Evolution." In G. Arratia, H.V.H. Wilson, and R. Cloutier (eds.), *Advances in the Origin and Early Radiation of Vertebrates.* Munich: Verlag Dr Friedrich Pfeil, 11–28.

Colless, D.H. 1985. "On the Status of Outgroups in Phylogenetics." *Systematic Zoology,* 34: 364–366.

Craw, R.C. 1992. "Margins of Cladistics: Identity, Difference and Place in the Emergence of Phylogenetic Systematics, 1864–1975." In P. Griffiths (ed.), *Trees of Life: Essays in Philosophy of Biology.* Australasian Studies in History and Philosophy of Science, 11. Amsterdam: Kluwer Academic, 65–107.

De Pinna, M.C.C. 1991. "Concepts and Tests of Homology in the Cladistic Paradigm." *Cladistics,* 7: 367–394.

Delaporte, P. 2012. "The Systemist Emergentist View of Mahner and Bunge on 'Species as Individuals': What Use for Science and Education?" *Science and Education,* 21: 1535–1544.

Ebach, M.C., and Williams, D.M. 2011. "A Devil's Glossary for Biological Systematics." *History and Philosophy of the Life Sciences,* 33: 251–258.

Farris, J.S. 1970. "Methods for Computing Wagner Trees." *Systematic Zoology,* 19: 83–92.
———. 1972. "Estimating Phylogenetic Trees from Distance Matrices." *American Naturalist,* 106: 645–668.
———. 1985. "The Pattern of Cladistics." *Cladistics,* 1: 190–201.
———. 2000. "Paraphyly, Outgroups and Transformations." *Cladistics,* 16: 425–429.
———. 2011. "Systemic Foundering." *Cladistics,* 27: 207–221.
———. 2012a. "Counterfeit Cladistics." *Cladistics,* 28: 227–228.
———. 2012b. "3ta Sleeps with the Fishes." *Cladistics,* 28: 422–436.
———. 2012c. "Fudged 'Phenetics.'" *Cladistics,* 28: 231–233.
———. 2013. "'Taxic Homology' is Neither." *Cladistics, 29: 1–3.*
Felsenstein, J. 1982. "Numerical Methods for Inferring Evolutionary Trees." *Quarterly Review of Biology,* 57: 379–404.
———. 2004. *Inferring Phylogenies.* Sunderland, MA: Sinauer Associates.
Forey, P.L. 2002. "Systematics and Creationism" (Introduction to Patterson's "Evolution and Creationism.") *The Linnean,* 18: 13–14.
———. 2004. "Patterson, Colin (1933–1998)." In H.C.G. Matthew and B. Harrison (eds.), *Oxford Dictionary of National Biography, from the Earliest Times to the Year 2000.* Oxford: Oxford University Press in association with the British Academy.
Forey, P.L., Gardiner, B.G., and Humphries, C.J., eds. 2000. *Colin Patterson (1933–1998): A Celebration of His Life. The Linnean,* Special Issue No. 2, Linnean Society of London.
Fortey, R.A. 1999. "Colin Patterson. 13 October 1933–9 March 1998." *Biographical Memoirs of Fellows of the Royal Society,* 45: 367–377.
Gardiner, B.G., Janvier, P., Patterson, C., Forey, P.L., Greenwood, P.H., Miles, R.S., and Jefferies, R.P.S. 1979. "The Salmon, the Lungfish and the Cow: A Reply." *Nature,* 277: 175–176.
Goloboff, P., J.S. Farris, and Nixon, K. 2003. *TNT (Tree Analysis Using New Technology).* Tucumán, Argentina: Published by the authors.
Haeckel, E. 1866. *Generelle Morphologie der Organismen: Allgemeine Grundzüge der organischen Formen-Wissenschaft, mechanisch begründet durch die von C. Darwin reformirte Decendenz-Theorie.* Berlin: G. Reimer.
———. 1894–96. *Systematische Phylogenie: Entwurf eines natürlichen Systems der Organismen auf Grund ihrer Stammesgeschichte.* 3 vols. Vol. 1 [1894], *Systematische Phylogenie der Protisten und Pflanzen;* Vol. 2 [1895], *Systematische Phylogenie der Wirbellosen Thiere (Invertebrata);* Vol. 3 [1896], *Systematische Phylogenie der Wirbelthiere (Vertebrata).* Berlin: G. Reimer.
———. 1909. "Charles Darwin as an Anthropologist." In A.C. Seward (ed.), *Darwin and Modern Science: Essays in Commemoration of the Centenary of the Birth of Charles Darwin and of the Fiftieth Anniversary of the Publication of "The Origin of Species."* Cambridge: Cambridge University Press, 137–157.
Hagen, J.B. 1999. "Naturalists, Molecular Biologists, and the Challenges of Molecular Evolution." *Journal of the History of Biology,* 32: 321–341.
———. 2000. "The Origins of Bioinformatics." *Nature Reviews Genetics,* 1: 231–236.

———. 2001. "The Introduction of Computers into Systematic Research in the United States during the 1960s." *Studies in the History and Philosophy of the Biological and Biomedical Sciences,* 32: 291–314.

———. 2003. "The Statistical Frame of Mind in Systematic Biology from Quantitative Zoology to Biometry." *Journal of the History of Biology,* 36: 353–384.

———. 2010. "Waiting for Sequences: Morris Goodman, Immunodiffusion Experiments, and the Origins of Molecular Anthropology." *Journal of the History of Biology,* 43: 697–725.

Halstead, L.B. 1980. "Museum of Errors." *Nature,* 288: 208.

Halstead, L.B., White, E.I., and MacIntyre, G.T. 1979. "L.B. Halstead and Colleagues Reply." *Nature,* 277: 176.

Hennig, W. 1957. Systematik und Phylogenese. In H. von Hannemann (ed.), *Bericht über die Hundertjahrfeier der Deutschen Entomologischen Gesellschaft, Berlin. 30 September bis 5 Oktober 1956.* Berlin: Akademie-Verlag, 50–71.

———. 1965. "Phylogenetic Systematics." *Annual Review of Entomology,* 10: 97–116.

———. 1966. *Phylogenetic Systematics.* Trans. D.D. Davis and R. Zangerl. Urbana: University of Illinois Press [Reprinted 1979, 1999].

Hull, D.L. 1988. *Science as Process: An Evolutionary Account of the Social and Conceptual Development of Science.* Chicago: University of Chicago Press.

Janvier, P. 1996. *Early Vertebrates.* Oxford: Clarendon Press.

———. 1998. "The Palaeontologist Renowned for His Work on the 'Four-Legged Fish.' Obituary: Erik Jarvik (1907–98)." *Nature,* 392: 338.

Jarvik, E. 1968. "Aspectsl of Vertebrate Phylogeny." In T. Ørvig (ed.), *Current Problems of Lower Vertebrate Phylogeny: Proceedings of the Fourth Nobel Symposium Held in June 1967 at the Swedish Museum of Natural History (Naturhistoriska riksmuseet) in Stockholm.* Stockholm: Almqvist and Wiksell, 496–527.

Josey, K.A., and Friday, A.E. 1982. *Problems of Phylogenetic Reconstruction.* Oxford: Academic Press.

Kitching, I., Forey, P.L., Humphries, C.J., and Williams, D.M. 1998. *Cladistics: The Theory and Practice of Parsimony Analysis.* Systematics Association Publications, 11. Oxford: Oxford University Press.

Kluge A.G., and Farris, J.S. 1999. "Taxic Homology = Overall Similarity." *Cladistics,* 15: 205–212.

Krell, F.T. 2005. "A Hennigian Monument on Vertebrate Phylogeny" (Book review of G. Mickoleit, *Phylogenetische Systematik der Wirbeltiere* [Phylogenetic Systematics of Vertebrates]. Munich: Verlag Dr. Friedrich Pfeil, 2004). *Systematics and Biodiversity,* 3: 339–341.

Lewens, T. 2011. "Pheneticism Reconsidered." *Biology and Philosophy,* 27: 159–177.

Mayr, E. 1965. "Numerical Phenetics and Taxonomic Theory." *Systematic Zoology,* 14: 73–97.

———. 1969. *Principles of Systematic Zoology.* New York: McGraw-Hill.

———. 1974. "Cladistic Analysis or Cladistic Classification?" *Zeitschrift für Systematik und Evolutionsforschung*, 12: 94–128.

———. 1981. "Biological Classification: Towards a Synthesis of Opposing Methodologies." *Science*, 214: 510–516.

Meacham, C.A. 1984. "The Role of Hypothesized Direction of Characters in the Estimation of Evolutionary History." *Taxon*, 33: 26–38.

Mooi, R.D., and Gill, A.C. 2010. "Phylogenies without Synapomorphies—A Crisis in Fish Systematics: Time to Show Some Character." *Zootaxa*, 2450: 26–40.

Nelson, G.J. 1973. "The Higher Level Phylogeny of Vertebrates." *Systematic Zoology*, 22: 86–90.

———. 1978. "Ontogeny, Phylogeny, Paleontology and the Biogenetic Law." *Systematic Zoology*, 27: 324–345.

———. 1989. "Species and Taxa: Systematics and Evolution." In D. Otte and J. Endler (eds.), *Speciation and Its Consequences*. New York: Sinauer, 60–81.

———. 1994. "Homology and Systematics." In B.K. Hall (ed.), *Homology: The Hierarchical Basis of Comparative Biology*. San Diego, CA: Academic Press, 101–149.

———. 1996. *Nullius in Verba*. New York and Melbourne: Published by the author. Reprinted in *Journal of Comparative Biology*, 1 (1966): 141–152; *Botanical Review*, 71 (2005): 355-387.

———. 2004. "Cladistics: Its Arrested Development." In D.M. Williams and P.L. Forey (eds.), *Milestones in Systematics*. Boca Raton, FL: CRC Press, 127–147.

———. 2007. "Patterson, Colin." *New Dictionary of Scientific Biography*, 7: 30–34.

———. 2011. "Resemblance as Evidence of Ancestry." *Zootaxa*, 2946: 137–141.

Nelson, G., and Platnick, N.I. 1991. "Three Taxon Statements: A More Precise Use of Parsimony?" *Cladistics*, 7: 351–366.

Nixon, K.C., and Carpenter, J.M. 1993. "On Outgroups." *Cladistics*, 9: 413–426.

———. 2012a. "More on Homology." *Cladistics* 28: 225–226.

———. 2012b. "On Homology." *Cladistics*, 28: 160–169.

O'Malley, M.A., ed. 2010. Special Issue: *The Tree of Life. Biology and Philosophy* 25 (4).

Parenti, L.R., and Ebach, M.C. 2009. *Comparative Biogeography: Discovering and Classifying Biogeographical Patterns of a Dynamic Earth*. Berkeley: University of California Press.

Patterson, C. 1964. "A Review of Mesozoic Acanthopterygian Fishes, with Special Reference to those of the English Chalk." *Philosophical Transactions of the Royal Society, London, Ser. B*, 247: 213–482.

———. 1967. "Are the Teleosts a Polyphyletic Group?" *Colloques Internationale du Centre National de la Recherche Scientifique*, 163: 93–109.

———. 1977. "The Contribution of Paleontology to Teleostean Phylogeny." In M.K. Hecht, P.C. Goody, and B.M. Hecht (eds.), *Major Patterns in Vertebrate Evolution*. New York: Plenum, 579–643.

———. 1981a. "Homology and Phylogeny." Unpublished presentation.
———. 1981b. "Significance of Fossils in Determining Evolutionary Relationships." *Annual Review of Ecology and Systematics,* 12: 195–223.
———. 1981c. [No title.] Unpublished presentation.
———. 1982a. "Cladistics and Classification." *New Scientist,* 94: 303–306. Reprinted in J. Cherfas (ed.), *Darwin Up to Date.* London: New Science Publications, 1982, 35–39.
———. 1982b. "Morphological Characters and Homology." In K.A. Joysey and A.E. Friday (eds.), *Problems of Phylogenetic Reconstruction.* London: Academic Press, 21–74.
———. 1983. "How Does Ontogeny Differ from Phylogeny?" In B.C. Goodwin, N. Holder, and C.C. Wylie (eds.), *Development and Evolution.* Cambridge: Cambridge University Press, 1–31.
———. 1988. "The Impact of Evolutionary Theories on Systematics." In D.L. Hawksworth (ed.), *Prospects in Systematics.* Oxford: Clarendon Press, 59–91.
———. 1989. "Phylogenetic Relations of Major Groups: Conclusions and Prospects." In B. Fernholm, K. Bremer, and H. Jörnvall (eds.), *The Hierarchy of Life: Molecules and Morphology in Phylogenetic Analysis. Proceedings of Nobel Symposium 70 Held at Alfred Nobel's Björkborn, Karlskoga, Sweden, August 29–September 2, 1988.* Amsterdam: Elsevier, 471–488.
———. 1990. "Erik Helge Osvald Stensiö." *Biographical Memoirs of Fellows of the Royal Society,* 35: 363–380.
———. 1994a. "Bony Fishes." In D.R. Prothero and R.M. Schoch (eds.), *Major Features of Vertebrate Evolution.* Short Courses in Paleontology, 7, 57, Paleontological Society. Knoxville: University of Tennessee.
———. 1994b. "Null or Minimal Models." In R. Scotland, D.J. Siebert, and D.M. Willams (eds.), *Models in Phylogeny Reconstruction.* Systematics Association Special Volume 52. Oxford: Oxford University Press, 173–192.
———. 1996. "Comments on Mabee's 'Empirical Rejection of the Ontogenetic Polarity Criterion.'" *Cladistics,* 12: 147–167.
———. 2002. "Evolutionism and Creationism." *The Linnean,* 18: 15–32.
———. 2011. "Adventures in the Fish Trade" (address to the Systematics Association, December 6, 1995), edited and with an introduction by David M. Williams and Anthony C. Gill. *Zootaxa,* 2946: 118–136.
Patterson, C., and Johnson, G.D. 1997a. "Comments on Begle's 'Monophyly and Relationships of Argentinoid Fishes.'" *Copeia,* 401–409.
———. 1997b. "The Data, the Matrix, and the Message: Comments on Begle's 'Relationships of the Osmeroid Fishes.'" *Systematic Biology,* 46: 358–365.
Platnick, N.I. 1985. "Philosophy and the Transformation of Cladistics Revisited." *Cladistics,* 1: 87–94.
———. 1986. "On Justifying Cladistics." *Cladistics,* 2: 83–85.
Ragan, M.A. 2009. "Trees and Networks before and after Darwin." *Biology Direct,* 4: 43. doi:10.1186/1745-6150-4-43.
Rieppel, O. 1988. *Fundamentals of Comparative Biology.* Basel: Birkhäuser.

Romer, A.S. 1949. *The Vertebrate Body*. London: W.B. Saunders.
———. 1962. *The Vertebrate Body*. 3rd ed. London: W.B. Saunders.
———. 1966. *Vertebrate Paleontology*. 3rd ed. Chicago: University of Chicago Press.
Romer, A.S., and Parsons, T.S. 1977. *The Vertebrate Body*. 5th ed. Philadelphia: Saunders.
Schafersman, S.D. 1985. "Anatomy of a Controversy: Halstead vs. the British Museum (Natural History)." In L.R. Godfrey (ed.), *What Darwin Began: Modern Darwinian and Non-Darwinian Perspectives on Evolution*. Boston: Allyn and Bacon.
Schlee, D. 1969. "Hennig's Principle of Phylogenetic Systematics, an 'Intuitive, Statistico-Phenetic Taxonomy'?" *Systematic Zoology*, 18: 127–134.
———. 1975. "Numerical Phyletics: An Analysis from the Viewpoint of Phylogenetic Systematics." *Entomologica Scandinavica*, 6: 139–208.
Schmitt, M. 2003. "Willi Hennig and the Rise of Cladistics." In A. Legakis, S. Stenthourakis, R. Polymeni, and M. Thessalou-Legaki (eds.), *The New Panorama of Animal Evolution*. Sofia: Pensoft, 369–379.
———. 2013. *From Taxonomy to Phylogenetics: Life and Work of Willi Hennig*. Leiden: Brill.
Scott, W.B. 1896. "Palaeontology as a Morphological Discipline." *Biological Letters of the Marine Biology Laboratory*, 1895: 43–61.
Searls, D.B. 2010. "The Roots of Bioinformatics." *PLoS Computational Biology*, 6(6): e1000809. doi:10.1371/journal.pcbi.1000809.
Simpson, G.G. 1975. "Recent Advances in Methods of Phylogenetic Inference." In W.P. Luckett and F. Szalay (eds.), *Phylogeny of the Primates*. New York: Plenum Press, 3–19.
Sneath, P.H.A., and Sokal, R.R. 1973. *Numerical Taxonomy*. San Francisco: Freeman.
Strasser, B.J. 2010. "Collecting, Comparing, and Computing Sequences: The Making of Margaret O. Dayhoff's *Atlas of Protein Sequence and Structure*, 1954–1965." *Journal of the History of Biology*, 43: 623–660.
Tassy, P. 2011. "Trees before and after Darwin." *Journal of Zoological Systematics and Evolutionary Research*, 49: 89–101.
Vernon, K. 1988. "The Founding of Numerical Taxonomy." *British Journal for the History of Science*, 21: 143–158.
———. 1993. "Desperately Seeking Status: Evolutionary Systematics and the Taxonomists' Search for Respectability, 1940–1960." *British Journal for the History of Science*, 26: 207–227.
———. 2001. "A Truly Taxonomic Revolution? Numerical Taxonomy 1957–1970." *Studies in History and Philosophy of Biological and Biomedical Sciences*, 32: 315–341.
Wägele, J.W. 1994. "Review of Methodological Problems of 'Computer Cladistics' Exemplified with a Case Study on Isopod Phylogeny (Crustacea: Isopoda)." *Zeitschrift für zoologische Systematik und Evolutionsforschung*, 32: 81–107.
———. 2001. *Grundlagen der phylogenetischen Systematik*. [2 Aufl.] Munich: Verlag Dr Friedrich Pfeil.

———. 2005. *Foundations of Phylogenetic Systematics.* (Translated from the German second ed. by C. Stefen and J.-W. Wägele and revised by B. Sinclair.) Munich: Verlag Dr Freidrich Pfeil.

Wiley, E.O. 1981. *Phylogenetics: The Theory and Practice of Phylogenetic Systematics.* New York: Wiley Interscience.

Wiley, E.O., and Johnson, G.D. 2007. "*Acanthopterygii* Johnson and Patterson 1993." http://tolweb.org/Acanthopterygii/15094/2007.01.09 in the Tree of Life Web Project, http://tolweb.org/.

Williams, D.M. 2002. "Precision and Parsimony." *Taxon,* 51:143–149.

Williams, D.M., and Ebach, M.C. 2004. "The Reform of Palaeontology and the Rise of Biogeography—25 Years after 'Ontogeny, Phylogeny, Paleontology and the Biogenetic Law' (Nelson 1978)." *Journal of Biogeography,* 31: 685–712.

———. 2008. *The Foundations of Systematics and Biogeography.* New York: Springer.

Williams, D.M., Ebach, M.C., and Wheeler, Q.D. 2010. "Beyond Belief: The Steady Resurrection of Phenetics." In D.M. Williams and S. Knapp (eds.), *Beyond Cladistics.* Berkeley: University of California Press, 169–197.

Zuckerkandl, E., and Pauling, L. 1965a. "Evolutionary Divergence and Convergence in Proteins." In V. Bryson and H.J. Vogel (eds.), *Evolving Genes and Proteins.* New York: Academic Press, 97–166.

———. 1965b. "Molecules as Documents of Evolutionary History." *Journal of Theoretical Biology,* 8: 357–366.

8

History and Theory in the Development of Phylogenetics in Botany

BRENT D. MISHLER

This chapter is meant to be a brief intellectual history of cladistic approaches to botanical systematics. As such it focuses on the development of ideas, and the forces constraining them, rather than on biography or bibliography. Botany started from a different place, and went in different directions, than did zoology. The empirical concerns were different, the goals of classification were different, beliefs in the underlying pattern of nature were different, and so on. Even the sociological interactions among botanists were different from those among zoologists. There are still differences in all these things today, although at the end of this chapter I'll suggest some ways they should be resolved and complementary strengths adopted across systematics.

EARLY DIFFERENCES IN THE HISTORY OF BOTANICAL AND ZOOLOGICAL SYSTEMATICS

The deep history of botanical systematics developed differently from zoological systematics in various ways (see the masterful treatment in Stevens 1994). In the earliest days of scientific approaches to systematics, botany had a strongly practical emphasis, because of the uses of plants in medicine that stretches back to the Middle Ages and across all of the world's cultures. Botany was an important subject for a physician; most of the early plant systematists were trained M.D.s. To facilitate uses of herbs in medicine, inventories of plants of a region (floras) became a strong emphasis in the field. And to make these floras useful for medicine, a premium was placed on providing tools for sure identification. Descriptions emphasized characteristics useful for either recognition or use rather than some deeper theoretical meaning.

This emphasis on floras remains to this day: it represents the majority of research efforts of current plant systematists worldwide. There are major ongoing floristic studies in many regions and states throughout the United States and in most countries of the world. There is little counterpart in the zoological world; relatively few investigators make it their goal to find, describe, and provide identification aids for all the animals of some specific geographic or political area. The effects of this obsession with flora writing and the wealth of specimens that has resulted are both beneficial and limiting to the field of botanical systematics. On the one hand, botanists have a very detailed knowledge of plant distributions, which has resulted in important online databases that have for example become widely used in recent predictive studies of the effects of climate change. On the other hand, a regional approach to systematics, rather than a focus on natural groups throughout their total range, produces many flaws in the resulting taxonomy. The terminal taxa might be correctly distinguished in the local area, but it is often difficult to name them correctlywithout study of additional specimens and types outside the region. Thus it is very difficult to make global comparisons or classifications until a truly worldwide monograph of a group is done, and the number of such studies has been few and declining through the twentieth century given an academic climate that favors publication of more, shorter studies.

Partly because of the emphasis on practical use and identification, in their classifications botanists emphasized the finer-scale units (species and genera). Plants were put together in classifications by criteria allowing easy identification and/or their practical uses, and botanists often didn't worry about theoretical frameworks for higher-level classification. When they did, the conceptual framework or metaphor used was most often conceived of as a map, with higher taxa having direct relationships in many directions (Stevens 1984, 1994, 2000). Botanists spoke of a given family X as approaching family Y at this end of its included variation and approaching family Z at another end. Virtually never was the chosen metaphor a tree.

Free from many of these practical constraints, zoology by contrast had a stronger theoretical bent from the beginning; systematists worried more about the nature of the underlying framework that connected living things together. Among many theoretical frameworks and metaphors considered (O'Hara 1991), including a divine plan, circles of five, and a great chain of being among others, zoologists most often thought of relationships in terms of a treelike hierarchy rather than a map. Before Darwin this was not an evolutionary tree, of course (with very few exceptions). The hierarchy initially was thought to be due to a divine plan, with higher taxa like phyla being major ideas in the mind of God and progressively lower taxa being smaller elements of the Plan (Agassiz 1859).

The development of the "Natural System" in the late 1700s and early 1800s (as discussed by Stevens 2000; Glimour 1937, 1940) was primarily in botany. In this rejection of the single-character Linnaean approach (which was deemed "unnatural" or "artificial"), taxa were recognized by overall resemblance in many characters. Interestingly the switch to the natural system had already happened before the Darwinian revolution, and since several of Darwin's close friends and correspondents, such as Hooker and Gray, were architects of the natural system, it was part of the intellectual background in which Darwin worked. The demonstration that there is a hierarchy of obvious natural groups in living things was prime evidence for Darwin to present in favor of evolution in the *Origin.*

Thus, as has been noted by many (e.g., de Queiroz 1988), the Darwinian revolution caused no fundamental revolution at the time in systematics in general, and especially in botany. After Darwin there was a shift in the language systematists used: lip service was now given to evolution. Instead of similarities being part of a creator's plan, they were now said to be inherited from common ancestors, but systematists' fundamental approaches remained the same: classification was done by grouping organisms together by overall resemblance ("phenetics"). Zoologists tended to apply the phenetic approach to building their trees, botanists to building their maps.

This same phenetic approach was enhanced by the use of computers in biology in the late 1950s and 1960s (see Sterner this volume). Not surprisingly perhaps, many of the developers of the new objective algorithmic approaches were botanists or bacteriologists (Sneath and Sokal 1962, 1973). Clustering algorithms made it possible to apply the Natural System to many characteristics in a repeatable manner. Initially there was no intent to apply a historical concept of homology, or to interpret the phenetic clusters as lineages (e.g., Sneath and Sokal 1973). However, many workers, especially zoologists, couldn't resist the temptation to consider cluster diagrams as evolutionary trees.

There was a considerable influence of operationalist philosophies of science in the development of numerical phenetics. The supposedly theory-free nature of the endeavor was viewed as an advantage over a competing approach, the so-called evolutionary systematics that developed from the Modern Synthesis of the 1940s. The Modern Synthesis was primarily engineered by zoologists, and in contrast to classical phenetics, classification was not based on overall similarity in *all* characters but rather overall similarity in "evolutionarily important" characters, that is, characters considered to be of adaptive importance. Note that while numerical phenetics and evolutionary systematics did use a different set of characters and considered themselves violently opposing camps philosophically (e.g., Mayr vs. Sokal), they shared the same basic methodology of grouping organisms by overall similarity. Both were fundamentally phenetic approaches; the advent of a truly evolutionary approach awaited Hennig and the cladists (see below).

One impact of the modern synthesis in zoological systematics was a change in focus from higher classification to populations and species. This restriction of focus to the species level, and the relegation of higher taxa to the back burner, brought botanical and zoological thinking together in this particular regard, although many differences remained in how classification was regarded theoretically. Despite differing views of the underlying ontology and appropriate metaphors for representing relationships, botanists and zoologists continued to practice their systematic work as either intuitive or numerical pheneticists.

The distinction between plant and animal studies is probably most glaring on the topic of species; the history of ideas here has never run parallel in botany and zoology. Some botanists, notably the "biosystematists," primarily located in California and England (Clausen, Keck, and Hiesey 1939; Stebbins 1950; Grant 1971), did participate in the development of the Modern Synthesis and absorbed from zoologists an emphasis on the biological species concept. However, even the botanists who ostensibly embraced the BSC tempered their views considerably with considerations of rampant hybridization between species. One gets the impression that some botanists, including Stebbins for instance, only gave lip service to the BSC so they would be accepted by zoologists like Mayr and allowed to join the Modern Synthesis club. The BSC, in its Mayrian state, was not truly favored by any botanists. By far the prevailing view of the nature of species in botany, as clearly laid out by Levin (1979, 2000), was and is as the smallest phenetic clusters separated by gaps in character variation on all sides. Workers today still commonly refer to "species boundaries"—a holdover of the map metaphor—and decide if an unknown plant is a new species by examining whether it fits within the range of variation of known species. Of course, most zoological advocates of the BSC, including Mayr, applied the same phenetic cluster methodology when describing species rather than doing breeding tests. (Mayr apparently never did a breeding test in his life!) For classification in general, clearly the main distinction between botany and zoology with respect to species is conceptual rather than empirical. It is an interesting feature of the history of science that competing schools of thought can viciously attack each other over ontological issues while differing little over the epistemological methods they apply in day-to-day work (Hull 1988).

THE PICTURE IN BOTANY BY THE 1970S

The situation by the 1970s in systematic botany was dire. Other types of botanists were questioning whether the loose, theory-free, intuitive approach being taken was scientific. Even systematists were not sure: when I started graduate school in the late 1970s, one of my advisers told me that systematics was not a science with objective principles but rather an art acquired over many years of careful apprenticeship. Herbaria were in trouble across the United States: many lost curatorial faculty, and a number were orphaned. The field was teetering on the brink of intellectual extinction. It was perceived as too lacking in rigorous methods and too loosely connected to evolutionary biology.

Phylogenetics came along in the nick of time to save systematic botany, but before turning to that important development, it will be important to understand several attitudes that plagued the field by conspiring against a rigorous phylogenetic approach. To set the stage for the discussion to follow, here is a list of prevailing attitudes of the 1970s that needed to be cleared away for phylogenetic progress to be made.

Overconcern about reticulation. Many botanists of the mid-twentieth century had become so convinced of the prevalence of hybridization that they felt there was no phylogenetic tree of plants to discover. Instead, there was a network (or map!) of genetic relationships going in all directions. Botanists such as Anderson (1949), Stebbins (1950), and Grant (1971) suggested that many angiosperm species were entirely or partly of hybrid origin, which made it more difficult for a phylogenetic tree to be accepted as the metaphor for classification. Instead, these concerns supported continued use of the traditional map metaphor.

The wrong metaphors. The long tradition of representing relationships in a maplike fashion, where a taxon has "boundaries" in all directions, culminated in this time period. Spectacular two-dimensional diagrams were produced by prominent botanists to show their ideas about relationships of plants (figs. 8.1, 8.2). These were argued to represent cross sections of evolutionary trees, but then of course without being able to see branches, no true phylogenetic relationships could be shown. This lack of specificity was considered a benefit by the authors of these diagrams (it was hard to prove them wrong!), but the lack of explicit hypotheses of relationships and methods to test them was clearly holding the field back.

When phylogeny was attempted, the tendency was to extend the map metaphor back in time and to link up extant taxa in ladderlike series (fig. 8.3). The botanists who took such approaches viewed some taxa as primitive and others as derived progressively along adaptive lines (Bessey 1915; Cronquist 1968). Trends in adaptive value were evaluated a priori and considered to be general—resulting in lists of "dicta" (e.g., Bessey 1915). For example, woodiness is primitive, herbaceous derived; many flower parts is primitive, few flower parts derived, and so on. Again, there was a lack of explicitness in the reasoning process for deciding a priori which characters and taxa are primitive.

Operationalism. Despite the participation of some botanists in the modern synthesis, most plant systematists in England and the United States were enamored with the operationalist philosophy of science fashionable in the 1960s.They were skeptical that one can know evolution (without enough fossils; see below) and thus preferred to produce classifications that are neutral in language and methodology. The most important virtue of a classification under this view is that it be practical—easy to use to identify plants found in the field.

Rejection of a historical concept of homology. Due partly to the philosophical biases mentioned in the previous paragraph in England and the United States and partly to the strong idealistic tradition in plant morphology and anatomy prevalent in Europe (Zimmermann 1931, 1937; Kaplan 2001; see also the chapters by Hamilton and Rieppel in this volume), the treatment of "homology" in botany was quite different from that prevalent in zoology. Either homology was neglected entirely (major works

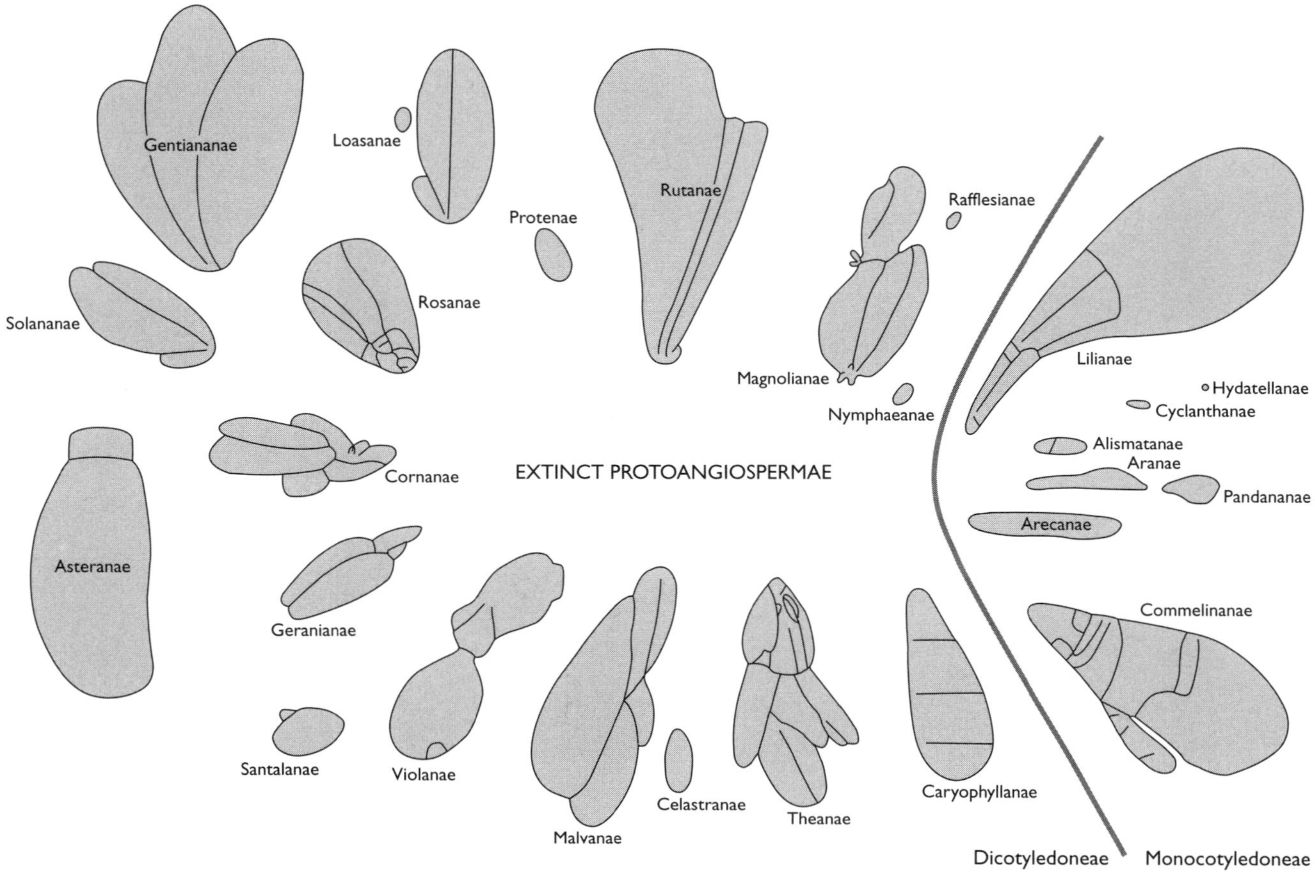

FIGURE 8.1. Phylogenetic shrub of the Angiosperms. Fig. 1, p. 249, in R. F. Thorne, "Classification and Geography of the Flowering Plants," *Botanical Review*, 58 (1992): 225–348. Springer. Used with permission.

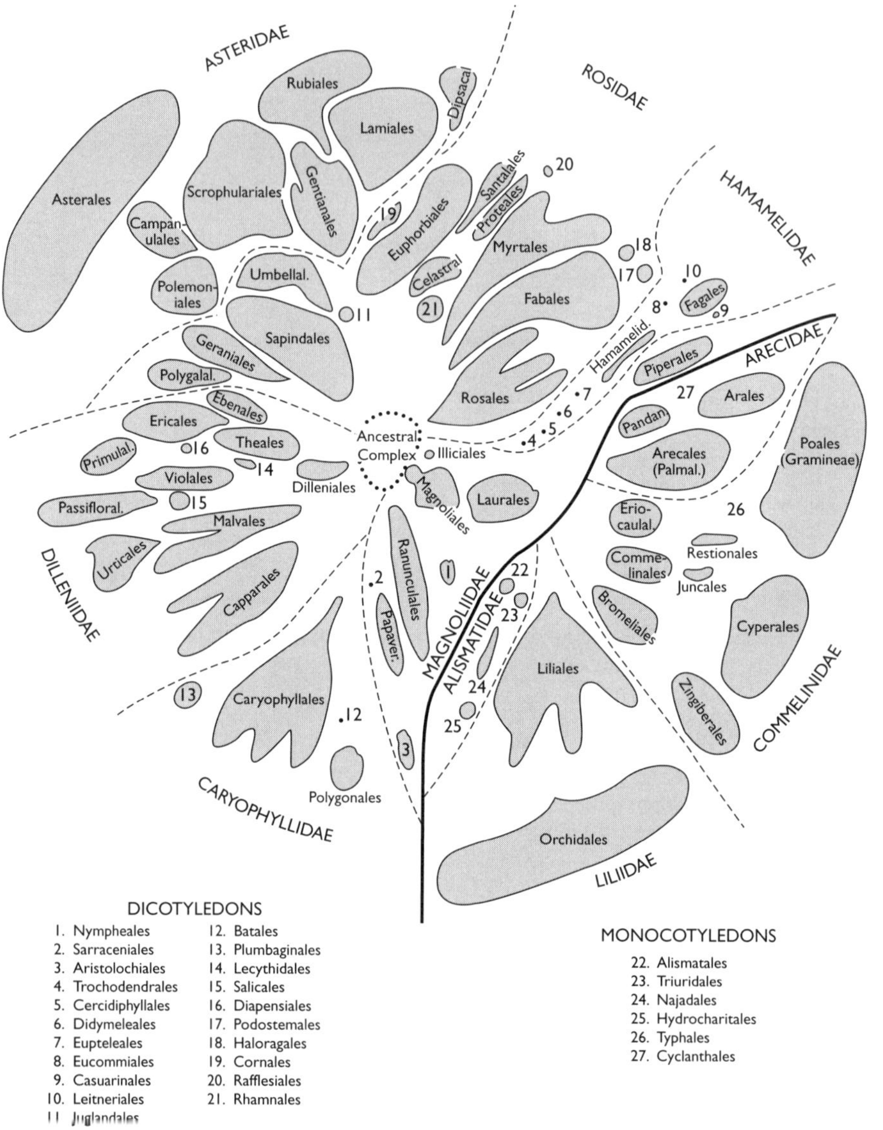

FIGURE 8.2. Diagram showing the relative degree of specialization of the orders of angiosperms. Figure 11–1, p. 247, in G. L. Stebbins, 1974. *Flowering Plants: Evolution above the Species Level* (Cambridge, MA: Belknap Press of Harvard University Press). Used with permission.

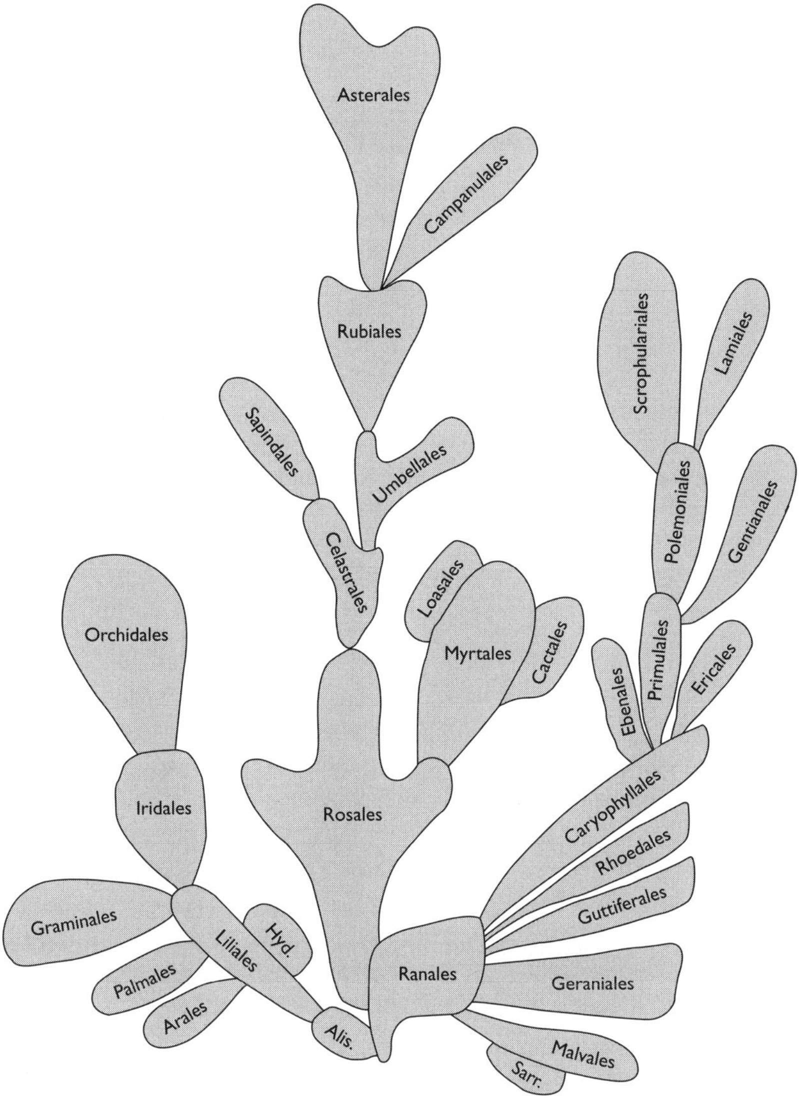

FIGURE 8.3. Bessey's cactus. *Annals of the Missouri Botanical Garden*, 2 (1915): 118.

following the operationalist tradition did not even mention the word), or it was defined based on developmental criteria following the idealistic tradition. In either case, the lack of an appropriate *historical* concept of homology greatly limited phylogenetic reasoning in botany.

Mistaken views on the role of fossils in phylogeny. Because of the lack of a historical concept of homology, botanists were missing out on the powerful phylogenetic evidence to be gained from comparative analysis of modern organisms. Many botanists felt that the fossil record carried all relevant evidence about phylogeny, that only a robust fossil record could give reliable relationships, and that evolution could be directly read from it. However, the methodology for linking fossils was poorly explored or explained: fossils were to be linked into ladders using overall similarity in some intuitive manner. But because there is a sparse fossil record for many plant groups, it was felt that there is no need to develop a rigorous methodology to build phylogenies! This completed a dubious circle of reasoning that kept the study of phylogenetic relationships deeply buried in the background of systematic botany during the 1970s.

EARLY CLADISTICS IN BOTANY

The Hennigian revolution took place in the 1970s to 1980s (Hull 1970, 1988), initially due mainly to the efforts of zoological systematists (Hennig 1965, 1966; Nelson 1973; Wiley 1981; Patterson 1982; Farris 1970, 1983; Kluge 1989). The fundamental conceptual advance was rejecting the use of overall similarity in favor of what Hennig called *special similarity.* Hennig made a distinction between shared derived similarities (*synapomorphies*) that are indicators of shared ancestry at some particular level and shared primitive similarities (*sympleisomorphies*) that are not indicators of shared ancestry (Sober 1988). Hennigian phylogeneticists (or *cladists,* to use a term originally coined as a denigration by Mayr) recognized that the fundamental organizing principle for biological classification should be phylogeny: descent with modification punctuated occasionally by branching of lineages. A strongly historical definition of homology was required (Patterson 1982). Cladists restricted the use of the formal Linnaean system to hypothesized *monophyletic groups* (i.e., groups composed of all and only descendants of a common ancestor), as evidenced by synapomorphies. The fundamental metaphor for classification should be a phylogenetic tree. These advances finally completed the Darwinian revolution in systematics (de Queiroz 1988).

The 1970s and 1980s saw the beginning of Hennigan cladistics in botany. The early adopters were young botanists, in many cases graduate students at the time, located in centers where they were exposed to teaching and research by zoological cladists. The first cladistic study in plants was by Timo Koponen (1968), working on mosses in Helsinki, Finland, under the influence of the entomologist Tuomakowsi, himself a colleague of Hennig's. Another early study was on land plant relationships by Lynne Parenti (1980), under the influence of zoological cladists at the American Museum of Natural History. Mishler and Donoghue published early plant cladistic studies with colleagues (Mishler and Churchill 1984, 1985; Doyle and Donoghue 1986) partly because of the influence of the fish systematist William Fink at Harvard. Other examples of early applications of cladistics in botany were Bremer and Wanntorp (1978) in Sweden, Humphries (1979) and Hill and Crane (1982) in England, Funk (1981) in Ohio, and Sluiman (1985) in the Netherlands. A bibliography of early plant cladistic studies can be found in Funk and Wagner (1981).

As was true of cladistics as a whole in that pre-PCR (polymerase chain reaction) era, early studies were solely morphology based and were non-numerical at first. The methodology employed early was manual "Hennigian argumentation": the tree was built by adding one character at a time, basically with pencil and paper. It is important to note that while major advances were soon to come in both sources of data (as molecular characters became available) and numerical analysis methods (as algorithms developed rapidly), the initial relationships discovered using only morphological data, and simple manual data analyses have stood the test of time. Later studies have greatly resolved relationships and added support to the early results, but the major branches in the plant portions of the tree of life were apparent right from the start. Thus the real revolution in plant systematics was caused by theoretical progress. The coming flood of molecular data and the rapid development of computer algorithms were important developments, but they were icing on the cake of the theoretical advances.

I don't want to imply that the young botanists adopting cladistic methods, initially developed by zoologists, were doing so passively. Botanists began to make important contributions to theory and method in cladistics. As befitting the clear biological differences between plants and animals, botanists entered enthusiastically into the debate over how species should be conceptualized in cladistics (e.g., Mishler and Donoghue 1982).

Likewise, botanists insisted that hybridization was real and needed to be studied using cladistic methods (Wagner 1983; Funk 1985; McDade 1990, 1992). Botanists contributed to the development of computer algorithms, particularly those for studying character evolution (Maddison, Donoghue, and Maddison 1984), to assessment of character polarity (Stevens 1980), and to the debates over the role of ontogenetic information in polarization of characters (Mishler 1986, 1988; Blackmore and Crane 1988). Botanists were also heavily involved in working out how fossils should be used in cladistic analyses (Doyle and Donoghue 1987; Donoghue et al. 1989) and in the development of comparative methods to apply cladistics to studies of evolution and ecology (Donoghue 1989; Wanntorp et al. 1990). And of course, botanists were fully involved in considerations of how to apply the newly available DNA sequence data in cladistic analyses (e.g., Mishler et al. 1988).

THE ADVENT OF MOLECULAR PHYLOGENETICS (STARTING IN THE LATE 1980s)

The invention of the polymerase chain reaction and its transfer to systematics labs in easy-to-use protocols in the late 1980s came at a perfect time to reinforce the earlier conceptual advances in phylogenetic reconstruction and further rescue the reputation of systematics that had sunk so low in the 1960s. Combined with rapid advancement in the sophistication of computational techniques, which allowed the use of many more characters in an analysis than was possible by manual techniques, systematics moved from the periphery to the very center of biological research. Phylogenetic research in systematics gave it the full luster of the highest scientific standards of the time; systematists were rigorously testing hypotheses and using cutting-edge laboratory technology. Furthermore, the development of phylogenetic comparative methods (e.g., Donoghue 1989; Funk and Brooks 1990; Wanntorp et al. 1990; Brooks and McLennan 1991; Harvey and Pagel 1991; Miles and Dunham 1993; Martins 1996) and the rapid proliferation of these methods into virtually all areas of biology made phylogenetic expertise a valuable academic skill, enhancing employment opportunities for systematic botanists. The institutions of plant systematics—herbaria and botanical gardens—which provide much-needed resources and expertise for both morphological and molecular studies, underwent a renaissance in status and funding.

Not everything was rosy as systematics developed into the 1990s, however. DNA-based studies were quicker to do than morphology-based ones, and took much less training. A student could be taught how to PCR and sequence DNA in a few weeks; it took much longer to train a student in techniques of microscopy and anatomy and how to conduct fieldwork and use collections properly. Because DNA-based research was more expensive than traditional "muddy boots" systematics, and regarded as the cutting edge, young faculty naturally concentrated their grant-raising efforts in this area. It was a good route to tenure; large grants could be acquired and publications produced more quickly. The net result of these facts was a tendency to go overboard in the enthusiasm for molecular data and leave out morphology entirely. Too many students were trained in molecular lab skills but not well rounded enough, lacking understanding of the morphology, anatomy, ecology, and systematics of their study group. Too many faculty members felt forced to concentrate on molecular phylogenetics and evolution, to the exclusion of producing systematic results and taxonomic treatments. These trends led to limits on the ability to integrate different kinds of data and synthesize results.

Another difficulty that developed in the early 1990s because of the glut of new data was a lack of coordination across the field. Many genes were being sequenced but from different examplar organisms in different terminal taxa. The morphological studies that were being done often concentrated on yet other taxa. Thus the possibility of compiling large, comparable datasets for multiple genes and morphology was severely limited—the data matrix was full of holes. This situation was brought about because individual labs were working in isolation and even in secrecy until eventual publication, due to worries about being scooped and maintaining competitive advantages in grant seeking. No single lab with a standard National Science Foundation grant could address the big phylogenetic questions that needed to be tackled, and yet there were barriers keeping labs from working collaboratively. The field of systematics needed to make the transition into "big science" much as had previously occurred in physics, astronomy, and molecular biology (particularly genomics)—all areas where the scope of the questions, and the expense and the range of expertise needed to address them, had outstripped the ability of single labs.

The solution to this difficulty clearly lay in broad-scale phylogenetic collaboration, and botanists led the way, taking advantage of a greater degree of existing collaboration and trust among labs than was present in systematic zoology. For example, a large number of labs collaborated in 1993 to produce what was then the largest cladistic analysis ever undertaken, with five hundred taxa of seed plants (Chase et al. 1993).

Most of the botanists working on phylogeny sought ways to collaborate by organizing sampling of standard exemplar taxa for multiple genes with the goal of assembling truly comparable datasets across green plants. The result was the Green Plant Phylogeny Research Coordination Group (known as "Deep Green"), jointly funded by the U.S. National Science Foundation, Department of Energy, and Deptartment of Agriculture. This collaboration began working in 1994 to facilitate interactions among many research labs through a series of workshops, meetings, and sharing of data and information on public websites. This effort culminated at the 1999 International Botanical Congress in St. Louis where the group sponsored eight symposia with seven speakers each, presenting phylogenetic results across the green plants at several nested levels and attracting national media attention. The success of this unusual coordination effort in turn stimulated the formation of two general funding programs by the National Science Foundation: the Research Coordination Network (RCN) program and the Assembling the Tree of Life (AToL) program. Both programs funded a number of further plant collaborations such as the RCN's "Deep Gene" and "Deep Time" and a number of plant AToL grants, as well as many collaborations in animals and microbes.

Thus botanists led the way toward a big science approach to phylogenetics, a trend that has only continued to accelerate. For example, massive collaborative analyses have been done for all green plants (Mishler et al. 1994), liverworts (Forrest et al. 2006), ferns (Pryer et al. 2001), and angiosperms (Qui et al. 2005). There are active research collaborations across most branches of the plant parts of the tree of life today, with botanists increasingly involved in comparative genomics and evolutionary studies that take advantage of the ever more detailed and well supported phylogenetic trees available.

THOUGHTS ON THE FUTURE

Where does systematic botany go from here? In my opinion, it would be good if the traditional differences with systematic zoology became a thing of the past, and we moved forward by adopting complementary concepts and approaches from both sources. Basically, we need to realize plants and animals aren't really that different after all. A common path is needed.

As has been traditional in zoology, the tree of life is the appropriate metaphor for all organisms: animals, plants, and microbes. There is, however, not a purely diverging tree. Reticulation caused by various processes that result in a horizontal transfer across distinct branches has been clearly shown to be a real phenomenon. Recent work suggests that reticulation, once a concern mostly of botanists, is a potential concern in any group of organisms, and general methods need to be developed to take the possibility of reticulation into account. As has been studied for a long time in plants, reticulation occurs through distinctly different processes (e.g., hybrid speciation, introgression, and horizontal gene transfer). Each of these processes will need distinctly different analytical approaches to sort out and distinguish them from other, nonreticulation processes (such as lineage sorting) that produce similar patterns. And of course, despite misguided arguments to the contrary (Dolittle 1999), phylogenetic methods initially assuming a purely diverging tree are necessary for the discovery of reticulation.

Another area where both plant and animal phylogenetics need to move ahead in parallel is in devoting more effort to selecting data to use for particular questions. There has been an overemphasis on analysis methods; most theoretical work in phylogenetics has been directed toward the issue of how to turn a data matrix into a phylogenetic tree rather than the issue of what kind of data should go into a particular matrix. Phylogenetic analyses at different scales require different datasets, and we need more objective means for selecting the appropriate data. For example, fine-scale "shallow" analyses need rapidly evolving markers but need to pay particular attention to the likelihood of reticulation or other sources of incongruence by using several completely independent data sources. Deep analyses looking for relatively short periods of shared history in the distant past can be confused by artifacts such as long-branch attraction, and handicapped by loss of signal, but are less affected by reticulation. The latter type of analyses needs slowly and episodically evolving markers such as morphological and anatomical characters. Genomic structural data, including, for example, changes in gene order and arrangement, show considerable promise for studies of this type (Mishler 2000, 2005)

Thus the trend toward emphasizing molecular data to the exclusion of morphological data needs to be reversed. We need to reintegrate morphology and molecules into combined matrices for most phylogenetic studies. As detailed in table 8.1, a rigorous morphological character matrix is needed to achieve most of the goals of phylogenetics, including incorporating information from fossils, getting the tree right, and interpreting character evolution rigorously.

TABLE 8.1 TEN REASONS TO INCLUDE OBJECTIVELY ANALYZED MORPHOLOGICAL DATA IN PHYLOGENETIC DATA MATRICES

1. *The greater complexity of morphological data may often allow better homology assessments.* Unlike DNA sequences, which are one-dimensional strings of data (unless you have secondary or tertiary structure models to use in alignment), morphology is complex and three-dimensional, plus it has a time component (ontogeny).
2. *Morphological data have many potential character states.* An important parameter determining whether data might be subject to "long-branch attraction" problems is the number of potential character states. False reconstructions are only a problem when parallel changes to the same character state occur, a phenomenon that is most frequent with binary data and rare with many available states.
3. *Data can be gathered from many specimens, cheaply and quickly.* A systematist can base his or her conclusions on samples from thousands of semaphoronts.
4. *We need to be able to identify clades easily in the field.* Morphological apomorphies are easier to apply in field keys and in photo ID guides.
5. *We need to objectively determine morphological apomorphies for clades.* A real analysis is needed to show what the apomorphies at a particular level are. It is not rigorous to inspect a purely molecular tree and hang morphological characters onto branches intuitively, although this is unfortunately done frequently.
6. *Morphology gives another independent dataset*, distinct from organellar and nuclear genes. Comparing the topology of morphological datasets to those derived from specific genes can help one discover reticulation, lineage sorting, etc.
7. *Morphological characters might actually help get the best-supported answer!* Even in cases where the topology of the total evidence tree is the same as with the molecules alone, support values such as bootstrap values often go up. And sometimes the total-evidence topology has novel, highly supported branches, synergistically supported by the combined data.
8. *Episodic patterns of change.* Despite common misconceptions to the contrary, clocklike markers are actually undesirable for reconstructing deep, short branches. Such markers continue to click along, changing at a regular rate until all the signal marking the deep branch is gone. The best marker for such deep branches is like the clock on the *Titanic:* it ticks once and stops forever. Slow change with long periods of stasis works best for these cases, i.e., the pattern shown by some morphological and anatomical features, as well as by structural features of the genome, such as inversions.
9. *Better sampling of the tree of life.* Good sampling is extremely important for reconstructing the correct phylogenetic tree. We need to break down those long branches. Over 99% of the lineages that have existed on the tree of life are extinct, and the only feasible way to get information about them is by adding fossils, which in turn requires morphology.
10. *Calibrating molecular clocks and dating of lineages.* In order to include fossils, we must have morphological characters in the matrix, and therefore optimized to the cladogram. Fossils do not come with a taxon ID in the fossil record; they simply come with some morphological characters from which we might be able to infer phylogenetic placement and classification. The fossil must therefore be attached to the cladogram based on its characters, then (and only then) can we infer that its sister group is at least as old as the age of the fossil.

A final area where general progress is needed is classification. Classifications are rightly or wrongly used by many people for many purposes, both within and outside biology, including biodiversity inventories, identification guides, conservation planning, ecological models of food webs and community structure, macroevolutionary studies of origination and extinction, predictions of the effect of climate change, coevolutionary studies, and many more. The number of uses is increasing. The availability of comprehensive online databases such as the Paleontology Database, GBIF, and Genbank has greatly increased the ability (and temptation) to conduct broad meta-analyses using taxonomic names as a bridge. The current state of taxonomy is woefully inadequate to support most of these uses: the application of most current taxonomic names found in lists or databases is quite ambiguous. We don't know which names refer to monophyletic clades rather than some paraphyletic or polyphyletic assemblage. Furthermore, even in the best-case scenario where a name does refer to a clade, we don't precisely know which clade (because of a limitation with the current codes of nomenclature) and we can't assume that two clades given the same rank have anything in common, such as age, included diversity, or distinctness in characteristics (de Queiroz and Gauthier 1992, 1994).

In a nutshell, taxonomic practice has not kept up with the recent developments in phylogenetic theory discussed above. We are still using codes of nomenclature that were developed in preevolutionary times to attempt to name evolving entities. The Darwinian revolution may have been completed for systematics by the development of Hennigian cladistics, but it has not yet been completed for taxonomy per se. The developing Phylocode (www.ohio.edu/phylocode/) will attempt to rectify the problems caused by the existing codes in three main ways that will benefit users by making it clear what taxonomic names represent: (1) all taxa named under the Phylocode will be postulated clades; (2) multiple specifiers are used to precisely define the clade being named, unlike the current system, which with only one type specimen does not allow clear specification of an ancestral node; (3) ranks are eliminated, removing the temptation to make inappropriate comparisons. Phylocode names will be hierarchically indexed in a database (REGNUM), along with their associated data, and will allow users to approach their goals appropriately given the modern evolutionary worldview.

REFERENCES

Agassiz, L. 1859. *An Essay on Classification.* Boston, MA: Harvard University Press.

Anderson, E. 1949. *Introgressive Hybridization.* New York: Wiley and Sons.

Bessey, C.E. 1915. "Phylogenetic Taxonomy of Flowering Plants." *Annals of the Missouri Botanical Garden,* 2: 109–164.

Blackmore, S., and Crane, P.R. 1988. "The Systematic Implications of Pollen and Spore Ontogeny." In C.J. Humphries (ed.), *Ontogeny and Systematics.* New York: Columbia University Press.

Bremer, K., Humphries, C.J., Mishler, B.D., and Churchill, S.P. 1987. "On Cladistic Relationships in Green Plants." *Taxon,* 36: 339–349.

Bremer, K., and Wanntorp, H.-R. 1978. "Phylogenetic Systematics in Botany." *Taxon,* 27: 317–329.

Brooks, D.R., and McLennan, D.A. 1991. *Phylogeny, Ecology, and Behavior: A Research Program in Comparative Biology.* Chicago: University of Chicago Press.

Cantino, P.D., and de Queiroz, K. 2007. *International Code of Phylogenetic Nomenclature.* Version 4b. Available at www.ohiou.edu/phylocode/.

Chase, M.W., Soltis, D.E., Olmstead, R.G., Morgan, D.R., Les, D.H., Mishler, B.D., Duvall, M.R., Price, R.A., Hills, H.G., Qiu, Y.-L., Kron, K.A., Rettig, J.H., Conti, E., Palmer, J.D., Manhart, J.R., Sytsma, K.J., Michaels, H.J., Kress, W.J., Karol, K.G., Clark, W.D., Hedren, M., Gaut, B.S., Jansen, R.K., Kim, K.-J., Wimpee, C.F., Smith, J.F., Furnier, G.R., Strauss, S.H., Xiang, Q.-Y., Plunkett, G.M., Soltis, P.S., Swensen, S.M., Williams, S.E., Gadek, P.A., Quinn, C.J., Eguiarte, L.E., Golenberg, E., Learn, G.H., Jr., Graham, S.W., Barrett, S.C.H., Dayanandan, S., and Albert, V.A. 1993. "Phylogenetics of Seed Plants: An Analysis of Nucleotide Sequences from the Plastid Gene rbcL." *Annals of the Missouri Botanical Garden,* 80: 528–580.

Clausen, J., Keck, D.D., and Hiesey, W.M. 1939. "The Concept of Species Based on Experiment." *American Journal of Botany,* 26: 103–106.

Cronquist, A. 1968. *The Evolution and Classification of Flowering Plants.* Boston, MA: Houghton Mifflin.

de Queiroz, K. 1988. "Systematics and the Darwinian Revolution." *Philosophy of Science,* 55: 238–259.

de Queiroz, K., and Gauthier, J. 1992. "Phylogenetic Taxonomy." *Annual Review of Ecology, Evolution, and Systematics,* 23: 449–480.

———. 1994. "Toward a Phylogenetic System of Biological Nomenclature." *Trends in Ecology and Evolution,* 9: 27–31.

Dolittle, W.F. 1999. "Phylogenetic Classification and the Universal Tree." *Science,* 284: 2124–2128.

Donoghue, M.J. 1989. "Phylogenies and the Analysis of Evolutionary Sequences, with Examples from Seed Plants." *Evolution,* 43: 1137–1156.

Donoghue M.J., Doyle, J.A., Gauthier J., Kluge A., and Rowe, T. 1989. "The Importance of Fossils in Phylogeny Reconstruction." *Annual Review of Ecology, Evolution, and Systematics,* 20: 431–460.

Donoghue, M.J., and Kadereit, J.W. 1992. "Walter Zimmermann and the Growth of Phylogenetic Theory." *Systematic Biology,* 41: 74–84.
Doyle, J.A., and Donoghue, M.J. 1986. "Seed Plant Phylogeny and the Origin of the Angiosperms: An Experimental Cladistic Approach." *Botanical Review,* 52: 321–431.
———. 1987. "The Importance of Fossils in Elucidating Seed Plant Phylogeny and Macroevolution." *Review of Palaeobotany and Palynology,* 50: 63–95.
Farris, J.S. 1970. "Methods for Computing Wagner Trees." *Systematic Zoology,* 19: 83–92.
———. 1983. "The Logical Basis of Phylogenetic Analysis." In V. Funk and D.R. Brooks (eds.), *Advances in Cladistics.* Vol. 1. New York: New York Botanical Garden, 7–36.
Forrest, L.L., Davis, E.C. , Long, D.G., Crandall-Stotler, B.J., Clark, A., and Hollingsworth, M.L. 2006. "Unraveling the Evolutionary History of the Liverworts (Marchantiophyta): Multiple Taxa, Genomes and Analyses." *Bryologist,* 109: 303–334.
Funk, V.A. 1981. "Special Concerns in Estimating Plant Phylogenies. In N. Platnick and F. Funk (eds.), *Advances in Cladistics.* Vol. 2. New York: New York Botanical Garden, 73–86.
———. 1985. "Phylogenetic Patterns and Hybridization." *Annals of the Missouri Botanical. Garden,* 72: 681–715.
Funk, V.A., and Brooks, D.R. 1990. *Phylogenetic Systematics as the Basis of Comparative Biology.* Washington, DC: Smithsonian Institution Press.
Funk, V.A. and Wagner, W.H., Jr. 1982. Bibliography of Botanical Cladistics: I. 1981. *Brittonia,* 34: 118-124.
Gilmour, J.S.L. 1937. "A Taxonomic Problem." *Nature* 139, 1040–1042.
———. 1940. "Taxonomy and Philosophy." In J. Huxley (ed.), *The New Systematics.* Oxford: Oxford University Press, 461–474. Reprinted in 1971 by the Systematics Association.
Grant, V. 1971. *Plant Speciation.* New York: Columbia University Press.
Harvey, P.H., and Pagel, M.D. 1991. *The Comparative Method in Evolutionary Biology.* Oxford: Oxford University Press.
Hennig, W. 1965. "Phylogenetic Systematics." *Annual Review of Entomology,* 10: 97–116.
———. 1966. *Phylogenetic Systematics.* Trans. D.D. Davis and R. Zangerl. Urbana: University of Illinois Press.
Hill, C.R., and Crane, P.R. 1982. "Evolutionary Cladistics and the Origin of Angiosperms." In K.A. Joysey and A.E. Friday (eds.), *Problems of Phylogenetic Reconstruction.* London: Academic Press, 269–361.
Hull, D.L. 1970. "Contemporary Systematic Philosophies." *Annual Review of Ecology and Systematics,* 1: 19–54.
———. 1988. *Science as a Process: An Evolutionary Account of the Social and Conceptual Development of Science.* Chicago: Chicago University Press.
Humphries, C. 1979. "A Revision of the Genus *Anacyclus* L. (Compositae: Anthemideae)." *Bulletin of the British Museum (Natural History), Botany Series,* 7: 83–142.

Kaplan, D.R. 2001. "The Science of Plant Morphology: Definition, History, and Role in Modern Biology." *American Journal of Botany,* 88: 1711–1741.

Kenrick, P., and Crane, P.R. 1997. *The Origin and Early Diversification of Land Plants.* Washington, DC: Smithsonian Institution Press.

Kluge, A.J. 1989. "A Concern for Evidence and a Phylogenetic Hypothesis of Relationships among *Epicrates* (Boidae, Serpentes)." *Systematic Zoology,* 38: 7–25.

Koponen, T. 1968. "Generic Revision of the Mniaceae Mitt. (Bryophyta)." *Annales Botancini Fennici,* 14: 429–439.

Levin, D.A. 1979. "The Nature of Plant Species." *Science,* 204: 381–384.

———. 2000. *The Origin, Expansion and Demise of Plant Species.* Oxford: Oxford University Press.

Maddison, W.P., Donoghue, M.J., and Maddison, D.R. 1984. "Outgroup Analysis and Parsimony." *Systematic Zoology,* 33: 83–103.

Martins, E.P. 1996. "Phylogenies, Spatial Autoregression, and the Comparative Method: A Computer Simulation Test." *Evolution,* 50: 1750–1765.

Mayr, E. 1965. "Classification and Phylogeny." *American Zoologist,* 5: 165–174.

McDade, L. 1990. "Hybrids and Phylogenetic Systematics. I. Patterns of Character Expression in Hybrids and Their Implications for Cladistic Analysis." *Evolution,* 44: 1685–1700.

———. 1992. "Hybrids and Phylogenetic Systematics II. The Impact of Hybrids on Cladistic Analysis." *Evolution,* 46: 1329–1346.

Miles, D. B. and Dunham, A.E. 1993. "Historical Perspectives in Ecology and Evolutionary Biology: The Use of Phylogenetic Comparative Analyses." *Annual Review of Ecology and Systematics,* 24: 587–619.

Mishler, B.D. 1986. "Ontogeny and Phylogeny in Tortula (Musci: Pottiaceae)." *Systematic Botany,* 11: 189–208.

———. 1988. "Relationships between Ontogeny and Phylogeny, with Reference to Bryophytes." In C.J. Humphries (ed.), *Ontogeny and Systematics.* New York: Columbia University Press, 117–136.

———. 2000. "Deep Phylogenetic Relationships among 'Plants' and Their Implications for Classification." *Taxon,* 49: 661–683.

———. 2005. "The Logic of the Data Matrix in Phylogenetic Analysis." In V.A. Albert (ed.), *Parsimony, Phylogeny, and Genomics.* Oxford: Oxford University Press, 57–70.

Mishler, B.D., Bremer, K., Humphries, C.J., and Churchill, S.P. 1988. "The Use of Nucleic Acid Sequence Data in Phylogenetic Reconstruction." *Taxon,* 37: 391–395.

Mishler, B.D., and Churchill, S.P. 1984. "A Cladistic Approach to the Phylogeny of the 'Bryophytes.'" *Brittonia,* 36: 406–424.

———. 1985. "Transition to a Land Flora: Phylogenetic Relationships of the Green Algae and Bryophytes." *Cladistics,* 1: 305–328.

Mishler, B.D., and Donoghue, M.J. 1982. "Species Concepts: A Case for Pluralism." *Systematic Zoology,* 31: 491–503.

Mishler, B.D., Lewis, L.A., Buchheim, M.A., Renzaglia, K.S., Garbary, D.J., Delwiche, C.F., Zechman, F.W., Kantz, T.S., and Chapman, R.L. 1994.

"Phylogenetic Relationships of the 'Green Algae' and 'Bryophytes.'" *Annals of the Missouri Botanical Garden,* 81: 451–483.
Nelson, G. 1973. "Classification as an Expression of Phylogenetic Relationships." *Systematic Zoology,* 22: 344–359.
O'Hara, R.J. 1991. "Representations of the Natural System in the Nineteenth Century." *Biology and Philosophy,* 5: 255–274.
Parenti, L.R. 1980. "A Phylogenetic Analysis of Land Plants." *Biological Journal of the Linnean Society,* 13: 225–242.
Patterson, C. 1982. "Morphological Characters and Homology." In K.A. Joysey and A.E. Friday (eds.), *Problems of Phylogenetic Reconstruction.* London: Academic Press, 21–74.
Pryer, K.M., Schneider, H., Smith, A.R., Cranfill, R., Wolf, P.G., Hunt, J.S., and Sipes, S.D. 2001. "Horsetails and Ferns Are a Monophyletic Group and the Closest Living Relatives to Seed Plants." *Nature,* 409: 618–622.
Qiu, Y.-L., Dombrovska, O., Lee, J., Li, L., Whitlock, B.A., Bernasconi-Quadroni, F., Rest, J.S., Davis, C.C., Borsch, T., Hilu, K.W., Renner, S.S., Soltis, D.E., Soltis, P.S., Zanis, M.J.J., Cannone, J., Gutell, R.R., Powell, M., Savolainen, V., Chatrou, L.W., and Chase, M.W. 2005. "Phylogenetic Analyses of Basal Angiosperms Based on Nine Plastid, Mitochondrial, and Nuclear Genes." *International Journal of Plant Sciences,* 166: 815–842.
Sluiman, H.J. 1985. "A Cladistic Evaluation of the Lower and Higher Green Plants (Viridiplantae)." *Plant Systematics and Evolution.* 149: 217–232.
Sneath, P.H.A., and Sokal, R.R. 1962. "Numerical Taxonomy." *Nature,* 193: 855–860.
———. 1973. *Numerical Taxonomy: The Principles and Practice of Numerical Classification.* San Francisco: Freeman.
Sober, E. 1988. *Reconstructing the Past.* Cambridge, MA: MIT Press.
Soltis, P. S., Soltis, D.E., and Chase, M.W. 1999. "Angiosperm phylogeny inferred from multiple genes: A research tool for comparative biology." *Nature,* 402: 402–404.
Stebbins, G.L. 1950. *Variation and Evolution in Plants.* New York: Columbia University Press.
———. 1974. *Flowering Plants: Evolution above the Species Level.* Cambridge, MA: Belknap Press.
Stevens, P.F. 1980. "Evolutionary Polarity of Character States." *Annual Review of Ecology, Evolution, and Systematics,* 11: 333–358.
———. 1984. "Metaphors and Typology in the Development of Botanical Systematics 1690–1960, or the Art of Pouring New Wine in Old Bottles." *Taxon,* 33: 169–211.
———. 1994. *The Development of Biological Systematics.* New York: Columbia University Press.
———. 2000. "Botanical Systematics, 1950–2000: Change or Progress, or Both?" *Taxon,* 49: 635–659.
Thorne, R.F. 1968. "Synopsis of a Putative Phylogenetic Classification of Flowering Plants." *Aliso,* 6: 57–66.
———. 1992. "Classification and Geography of the Flowering Plants." *Botanical Review,* 58: 225–348.

Wagner, W.H. 1983. "Reticulistics: The Recognition of Hybrids and Their Role in Cladistics and Classification." In N.I. Platnick and V.A. Funk (eds.), *Advances in Cladistics.* Vol. 2. New York: Columbia University Press, 63–79.

Wanntorp, H.-E., Brooks, D.R., Nilsson, T., Nylin, S., Ronquist, F., Stearns, S.C., and Wedell, N. 1990. "Phylogenetic Approaches in Ecology." *Oikos,* 57: 119–132.

Wiley, E.O. 1981. *Phylogenetics: The Theory and Practice of Phylogenetic Systematics.* New York: Wiley-Blackwell.

Zimmermann, W. 1931. "Arbeitsweise der botanischen Phylogenetik." In E. Abderhalden (ed.), *Handbuch der biologischen Arbeitsmethoden,* Abt. IX, Teil 3/II. Berlin: Urban and Schwarzenberg, 941–1053.

———. 1937. "Strenge Objekt/Subjekt-Scheidung als Voraussetzung Wissenschaftlicher Biologie." *Erkenntnis,* 7: 1–44.

PART THREE

Technology, Concepts, and Practice

9

Well-Structured Biology

Numerical Taxonomy's Epistemic Vision for Systematics

BECKETT STERNER

The history of twentieth-century systematics is full of periodic calls for revolution and the battles for dominance that followed. At the heart of these disputes has been a disagreement over the place of evolutionary theory in the field: some systematists insist that it is central to their methodology, while others argue that evolution can only be studied from an independent foundation (Hull 1988; Vernon 2001; Felsenstein 2001, 2004). Despite the importance of these battles, another trend across the twentieth century is increasingly relevant as it assumes center stage in current methodology: the value and costs of integrating mathematics and computers into the daily research practices of biology (Hagen 2001, 2003; Vernon 1993; Agar 2006; Hine 2008; Hamilton and Wheeler 2008; Strasser 2010). These two features of the history of systematics are related, and we can easily observe today how mathematics forms the common language within which systematists pose their fundamental disagreements. Indeed, the fates of revolution and of mathematics in twentieth-century systematics were inseparable. Moreover, their interdependence exemplifies a general relation between scientific change and the symbolic formalization of a body of practices. I will argue for these two theses from the historical perspective of the founding and articulation of numerical taxonomy (NT) roughly between 1950 and 1970, a period that marks the first major attempt to revolutionize the whole of systematics based on the mathematization of the key concepts and techniques in the field.

The central issue will prove to be understanding the character of a particularly *mathematical* revolution in a biological science; that is, what does it look like when a group of scientists set out to reenvision an entire field of biology in quantitative terms? It is important to note at the outset, however, that the integration of mathematics into biology has often proceeded in piecemeal fashion without such a holistic vision. Indeed, systematists began to incorporate the ideas of population sampling and other statistical methods, such t-tests for statistical significance, into their research long before NT became prominent in the 1960s (Hagen 2003). In addition, Ronald Fisher's work at Rothamsted Station beginning in the 1920s opened the door for thousands of biologists to use statistical models as a valuable way to increase the precision and complexity of experiments without undue worry about needing theoretical explanations for the experimental phenomena. Also, the advent of quantitative traits such as serotypes in mid-twentieth-century taxonomy encouraged a certain mathematical facility without requiring a radical reconception of what a character is in general. Usually these piecemeal adoptions of mathematical techniques take an instrumentalist perspective, such as in much of contemporary bioinformatics: math is a tool for getting things done rather than a reflection of any deeper truth about biology.

Nonetheless, we can clearly see historical moments when the adoption of mathematical techniques had consequences rippling out to encompass the entire field's practices. The importance of statistical thinking, for example, is unavoidable in the development of the Modern Synthesis. In fact, NT was in a sense an unintended offspring of the groundbreaking statistical work of Sewall Wright and Fisher, both of whom proved an important influence on Robert Sokal, cofounder of NT with Peter Sneath. As we will see, whereas the Modern Synthesis's mathematics failed to penetrate into systematics, Sokal and Sneath reenvisioned the field according to a nonevolutionary yet still statistical foundation. By demonstrating that mathematics had no easy favorite in the systematists' disputes over the proper place of evolution in their field, Sokal and Sneath also opened the door to future arguments over how to reconstruct evolutionary history in terms of mathematical demonstrations and formal philosophical arguments (Farris 1969).

Revolutions are not much if they are just philosophy, however, and the character of a mathematical revolution must center on the daily practices of systematists. How can we uncover the inseparable relation between the mathematical view of biology held by Sokal and Sneath and the revolutionary character of their work for systematics? Although many perspectives are possible on the historical and philosophical significance of NT, I will borrow from Jon Agar the approach of examining the history as an overarching reorganization of scientific work. While my immediate purpose is somewhat orthogonal to his, we are both working within the same field of problems.

Agar touches on the history of systematics insofar as he is pursuing a more general topic, "What difference did computers make?" (Agar 2006). Aiming to deflate or at least analyze the hyperbolic claims often made for the transformative effect of computers on science, Agar investigates their impact on a number of concrete historical cases. He argues that there are three modes in which computers can affect an ongoing scientific research project: "the computer program could closely follow the existing organization of computation (and associated work)"; computerization "could be used as an opportunity for reorganizing methods of computation (and associated work), even though it was still feasible to use manual or mechanical means"; or last, it "could be used to reorganize methods of computation with new methods that, in principle or in practice, would be impossible to accomplish by manual or mechanical means" (Agar 2006, 873).

He finds that none of his cases fit this last mode, and he speculates that it has indeed never happened. Numerical taxonomy, somewhat implicitly, he places in the second category of using computerization as an opportunity to reorganize the work process without a discontinuous leap from old practices. Keith Vernon offers an account of NT that corroborates this classification insofar as both Sokal and Sneath arrived at their mathematical innovations independent of the use of computers, although they quickly recognized them as a valuable resource (Vernon 1988). Agar limits the scope of his case studies, however, to computerization within a single laboratory or institution, and it is important that what was continuous development in Sokal and Sneath's own research appeared as a discontinuous leap from the perspective of almost any other systematist at the time. What was relatively continuous growth for a small group of innovators appeared as a universal, radical new take on the field when Sokal and Sneath went public with their research program.

Numerical taxonomy as a reorganization of work therefore has significance as a historical process in which a small group of scientists universalized their originally quasi-private methods into a research program for every systematist. I will not be able to pursue Agar's interest in the impact of computers on this process further, though. Instead, I will focus on how the symbolic formalisms of mathematics were central to Sokal and Sneath's universalization of their local practices into the larger community. That is, I will be concerned with how their effort to arrive at a mathematical method of classification that was universal to every type of organism intrinsically implicated them in a qualitative reorganization of the work of all systematists. This chapter therefore expands Agar's topic to consider how local reorganizations of work can expand beyond individual labs or institutions to communities as wholes. This sort of process—in which a provincial prototype of a research strategy is generalized into a (putatively) universal program—is a common one that happens on many different scales. For convenience, I want to label the thinking that guides this process an "epistemic vision," briefly defined as a feasible strategy for solving ill-structured problems by the principled reorganization of work. I will have a bit more to say about the nature of epistemic visions below, but the major part of this chapter will focus on elucidating the particular vision that Sokal and Sneath had rather than epistemic visions in general.

An excellent place to examine how Sokal and Sneath universalized their work into a general research program is their manifesto and bible of NT, *Principles of Numerical Taxonomy,* published in 1963 (Sokal 1963). The publication of *Principles* in 1963 is a critical intersection for the integration of mathematics into systematics and its contentious ideological debates. Although Sokal and Sneath, along with a number of interested colleagues, had already published similar arguments in a couple journal articles, the scope, provocativeness, and coherence of the book made it a flashpoint of debate as well as a critical reference source. In the book, Sokal and Sneath set out a new theory and method of classifying organisms based solely on a concept of similarity that, at least initially, excluded the use of evolutionary theory. In order to carry this out, they reenvisioned the character traits of organisms, which form the empirical observations upon which classifications are based, as quantitative variables sampled from a larger population of traits ultimately contained in the genome. In NT, characters gained essential statistical properties that guaranteed the convergence of a classification as the number of measurements increases. In this way, a required reorganization of the order of research—classify using similarity alone before proceeding to evolutionary considerations—is inseparably linked to new statistical methods.

My particular conclusions about NT's contributions to twentieth-century systematics focus on how certain aspects of the new organization of work Sokal and Sneath proposed became entrenched even as opposing schools of thought criticized NT's larger ambitions. The way that morphological characters are treated even today owes something to how NT first reorganized the process of classification to treat characters as inherently statistical in nature. This heritage depends specifically on how Sokal and Sneath distinguished two self-sustaining and autonomous lines of work within systematics: methodology, that is, the study of numerical taxonomy in general, and taxon-specific research, the study of particular groups of organisms, within which NT was supposedly the universal method of classification. The character of how they distinguished methodology from taxon-specific research remains influential today, even if the content of the method and of taxon research have changed dramatically. In other words, I believe one of NT's most important legacies for the field has been the institutional entrenchment of a distinction between types of research that hinges on particular assumptions and techniques in the process of classification.

ARTICULATING A GENERAL PROCESS OF CLASSIFICATION

The problem of a universal method for systematics, if there even was just *one,* had engaged researchers for some time before Sokal and Sneath entered the scene.[1] Systematics, such as it was before 1960, was a fractious field where entomologists might feel little communion with other zoologists, let alone botanists or microbiologists (Johnson 2009). Few ready resources were available to support systematic inquiry into the common techniques used across communities studying different taxa, and advancing a substantive perspective on the field as a whole was a considerable achievement. The efforts of Ernst Mayr, George Gaylord Simpson, and Willi Hennig were influential for Sokal and Sneath's 1963 book, *Principles,* both as scaffolding for their own ideas and as critical targets. The work of Mayr and Simpson on describing a general step-wise process for classification, in particular, is important background to understanding how Sokal and Sneath sought to reorganize classificatory work.

A leading light of the New Systematics, Ernst Mayr published in 1942 his influential contribution to the Modern Synthesis, *Systematics and the Origin of Species.* The book aimed to fix a new purpose for classifications beyond the practical needs of description and identification: to aid in the study of speciation and evolution. Mayr's stated goal was for classificatory work to be a step along the way to analyzing patterns of variation and speciation, but this aim required resources and huge collection sizes far beyond the means of most taxonomists. Mayr also included a section titled "The Procedure of the Systematist" (Mayr 1942, 11), where he notes the paucity of introductions to "the practical methods of taxonomic work" and recommends that a beginner should find a "lucid" treatise that "set an example of method" and would serve to begin the education-by-apprenticeship model characteristic of taxonomy at the time. Although Mayr does go on to describe several stages in the procedures of systematics, including collection, identification, and description, his exposition quickly leaves practical method behind for a conceptual discussion of types, nomenclature, and rank. Nonetheless, his effort to organize the work of systematics into a series of consecutive stages was one of several important attempts to which Sokal and Sneath would respond in *Principles*.

Publishing *Principles of Animal Taxonomy* almost two decades later, in 1961, G.G. Simpson went beyond Mayr in articulating the actual practice of classification, although he also embraced its open-ended nature. In the section "What Do Taxonomists Do?" Simpson writes that "in general, he does the following things, more or less in this sequence but usually with a good deal of overlapping, jumping, and backtracking." I summarize the steps here: the taxonomist (1) selects the organisms, (2) records observations on them, (3) sorts them into different units (local populations, species, etc.), (4) compares the units' characteristics, (5) interprets their relationships, (6) makes inferences about an evolutionary pattern, (7) translates these conclusions into hierarchic rankings, and finally (8) selects appropriate names for the units (Simpson 1961, 108–109). Simpson was a proponent of the artistic side of taxonomy, and while he adds new detail to the description of classificatory work, he also leaves considerable creative freedom to the taxonomist regarding how to achieve the final product (Hagen 2003).

A final example of methodological description will help flesh out the background to NT's criticisms. In Willi Hennig's 1950 book, translated into English much later, in 1966, he discusses "reciprocal illumination" as the method of systematics. Since both a whole and its parts have specific properties and coherences, one can investigate the parts and then learn something about the whole by recalibrating one's results to its nature, and vice versa. As Hennig describes it, "The phylogenetic affiliations are, so to speak, the whole. With its parts—the relations of morphological, ecological, physiological, geographic, etc. similarity, which reflect the phylogenetic affiliations of their bearers—[the whole] is subjected to the method of checking, correcting, and rechecking" (Hennig 1966, 23). Thus one can use geographic distribution as a way of checking the reliability of phylogenetic relationships (Hennig 1966, 148). Hennig quotes the ethnographer Mühlmann to help refute a common objection against reciprocal illumination: "The objection that this is circular reasoning has been discussed by Dilthey; it applies logically, but not to practical investigation" (Hennig 1966, 22). To paraphrase, while Hennig believed that a logical translation of his scientific research might appear to be deriving its premises from its conclusions, this circularity does not apply in the face of the empirical uncertainty of practice. That is, logical circularity in the world of precise deductive methods does not necessarily imply a poor research strategy in the world of messy, ambiguous ideas.

Learning from exemplars, applying artistic expertise, and reinterpreting data on the go are characteristic modes of work based on practical expert knowledge about particular organisms. Each of these modes is well fitted to coping with the open-ended diversity of concrete specifics about which little is known beforehand. Consequently, from this perspective there is no single procedure for classifying that can be simultaneously perfectly fitted for arthropods and fungi because the specifics of each are at least as important as their commonalities. Moreover, any two entomologists, for example, could make classifications using different procedures and principles. As a result, Mayr's goal of studying variation and speciation helped harmonize the work of some systematists but still left a lot of leeway (Johnson 2005, 2009). In addition, although Simpson's list of steps is helpful, almost all the practical knowledge of the expert is contained in the "overlapping, jumping, and backtracking" of the actual process. As long as the structure of systematic method was so open-ended, it was difficult even for leading researchers such as Mayr or Simpson to provide much helpful guidance.

Indeed, a variety of research groups across the world struggled to make do under the umbrella of the New Systematics. When there was no evolutionary evidence about the history of traits, the evidence was ambiguous, or a simple classification was desired independent of evolutionary hypotheses, no exemplars of practice led the way for researchers. As Vernon has documented, the shortcomings of the New Systematics appear to have produced a remarkable, independent agreement among critics about what was best for classificatory work to proceed outside the Modern Synthesis. As a result, Sokal and Sneath's research program was not the only one to advance a controversial emphasis on segregating evolutionary theory from the classification process: similar ideas sprang up in numerous locations around the world in the time between 1950 and 1963, when *Principles of Numerical Taxonomy* was first published. I will use the general label "phenetic" methods to describe their resulting proposed solutions.[2]

DETERMINING THE METHODOLOGICAL ORDER OF TAXONOMY

While the Modern Synthesis in general would be hamstrung without statistical concepts, it was not the research program's statistics with which the pheneticists took issue; rather, the particular problem of phenetics was how to classify without making use of evolutionary theories. Statistics would in fact prove critical to the phenetic approach in general, insofar as it provided new, more formalized exemplars for how to infer classifications. The point of phenetics, putting inferences about similarity before historical genealogy, went hand-in-hand with a new perspective on what material went into the start of classificatory work and how it could be used. Phenetics, therefore, represents a first step for NT as a reorganization of work; it was both a general response to intellectual conditions at the time and the ground upon which Sokal and Sneath developed their particular approach.[3]

For our purposes, what is most important about the phenetic approach is how it articulates an alternative problem situation that encompasses the same general process as Simpson's eight steps but differs on the start, end, means, and goal of the process. In order to describe this new structuring of classification, it will help to bring in Herbert Simon's discussion of well- versus ill-structured problems (Simon 1973). Simon and others developed this distinction (independently of systematics) in order to study the tractability of various challenges for artificial intelligence, and the general scope of their description of problem structures offers a useful and appropriate perspective for our situation here. I'll proceed in this section by first summarizing the relevant history of phenetics' independent origins and then examining them in more depth in relation to the new structure placed on classification.

In the 1950s, working outside an evolutionary perspective was a thoroughly "ill-structured problem." There were few answers to important questions such as, How should a researcher select and characterize the data that went into the classification? What techniques should he use to go from initial observations to groupings of taxa? How could he know when the classification was finished? Without a general, uniform structure to the problem of nonevolutionary classification, it remained baffling in complexity, offering too many choices with evaluative criteria that were too weak to determine a clear answer. The innovations of phenetics would contribute several key insights, including that all characters that were not logically redundant were acceptable, and they should each be weighted equally in judging similarity.

Phenetic methods developed independently across a set of three initial groups comprising five researchers in all: A. J. Cain and G. A. Harrison, Sokal and C. D. Michener, and Sneath.[4] These three groups published the seminal papers of what would become NT in a thirteen-month period, between June 1957 and July 1958. The reasons for this striking simultaneity of independent innovations deserve further research, but Vernon finds that the innovations were produced by reactions of outsiders to the general state of taxonomy at the time. The interests of each individual help us to understand the scope of and motivations behind what became a radical change in systematic practice, so it is worth summarizing their motivations and the problems they faced.

Of Cain and Harrison, Harrison was the outsider to taxonomy and Cain an expert who dissented from aspects of the Modern Synthesis. Cain had already spent considerable time studying the geographic variation of the land snail *Cepaea nemoralis* with the geneticist P. Sheppard, and this research served as a basis for his arguments against the Modern Synthesis's emphasis on drift as opposed to selection. Also, the research highlighted the problems with using phylogenetic methods in data-poor case studies as there were few fossils available for study. Cain leveraged his difficulties into general arguments against evolutionary classifications that used poor data and unanalyzed notions of taxonomic importance.

When Harrison, a budding biological anthropologist, took a course from Cain at Oxford, his interest in craniometry sparked a collaboration that led to their innovative theories about taxonomy. Harrison was interested in the Mount Carmel skulls but struggled with how anthropologists lacked a justifiable way to identify clear evolutionarily significant and measurable characters. Over the course of several years, Cain and Harrison arrived at the idea that any characters not correlated with each other by necessity were acceptable, and that weighting each character equally would allow the calculation of affinities without evolutionary input. Thus they separated out the judgment of affinity, "phenetics," from the judgment of evolutionary history, phylogenetics.

Sneath started off even further from evolutionary biology than Harrison: he trained in medicine and was working in bacteriology when he first struggled with the problems of taxonomy. He was preparing an M.D. thesis in London on clinical cases of infection by *Chromobacterium* when he began to wonder if his classification of thirty-eight strains into two broad groups meant the groups were different species. He quickly discovered the chaotic state of bacterial taxonomy at the time, especially the seemingly arbitrary emphasis on some characters above others and unjustified attempts at phylogenetic classifications. In consultation with the botanist J. S. L. Gilmour, Sneath worked to articulate a nonphylogenetic method for classifying bacteria that had a more logical way to assess quality and method. As with Harrison and Cain, the failings of taxonomy's tool kit, including the phylogenetic approach, inspired Sneath to attempt a reorganization of classification to make it more coherent, justifiable, and self-complete with respect to evolution.

In contrast, Michener and Sokal were motivated by a different dynamic opposing the objectivity of mathematical methods to the subjectivity of evolutionary methods. Sokal arrived in the Department of Entomology at the University of Kansas after a fellow graduate student at the University of Chicago had converted him to applying mathematical methods in biology. Sokal had taken a biometry course from Sewall Wright at Chicago and over time learned or taught himself cutting-edge methods in statistics. At Kansas, however, he was a colleague of five other entomologists, all of whom were traditional taxonomists with little appreciation for mathematics. Sokal in fact had no initial interest in taxonomy—his dissertation was on ecological variation—but he was spurred by the tenor of the department to make a bet one day at lunch that he could do a better job of classifying organisms with mathematics than the traditional approach.

Michener, the department heavyweight, provided Sokal with data on 122 characters of 98 bee species and then helped him evaluate the results of the new mathematical procedure Sokal created. Michener was an exemplar of the New Systematics, and he produced authoritative work on bees that relied on the traditional "by eye" approach to classification. While Michener knew little mathematics himself, it made a certain intuitive sense to him, and he was open to Sokal's innovations where many systematists were not. In fact, when Sokal's results turned out to make a slight improvement on Michener's prior results, he embraced them and coauthored a lengthy paper with

Sokal. What neither of them expected, however, was that Sokal's approach did not rely on evolution; as Vernon relates, it took a couple years before they realized the study was purely one of affinity, not of history. In parallel to Cain and Harrison, Michener and Sokal therefore concluded that taxonomy was in fact based on a judgment of similarity first and evolution second. They ultimately diverged, however, on whether evolutionary considerations should be incorporated into the method, and Sokal preferred keeping the static analysis of similarity pure.

The three groups therefore originated in a common concern of the outsider: to make use of the products of systematics, that is, its classifications or phylogenies, without having to choose among the potentially idiosyncratic commitments and methods of systematics' practitioners. In order to better interpret the consequences of phenetics for the problem situation of nonevolutionary classification, it will help to borrow Simon's work on describing the relation between well- and ill-structured problems. Simon listed six criteria that go into making a particular problem well versus ill structured, although ultimately the distinction is one of degree rather than kind. I quote them at length because they indicate very richly what sorts of knowledge, techniques, and criteria are necessary to define a process of calculation that does not involve Simpson's "overlapping, jumping, and backtracking."[5] Our immediate interest in this section is point 4, although 1, 2, and 3 are also relevant for considering NT's distinctive merits in later sections.

1. There is a definite criterion for testing any proposed solution, and a mechanizable process for applying the criterion.
2. There is at least one problem space in which can be represented the initial problem state, the goal state, and all other states that may be reached, *or considered*, in the course of attempting a solution of the problem.
3. Attainable state changes (legal moves) can be represented in a problem space, as transitions from given states to the states directly attainable from them. But considerable moves, whether legal or not, can also be represented—that is, all transitions from one considerable state to another.
4. Any knowledge that the problem solver can acquire about the problem can be represented in one or more problem spaces.
5. If the actual problem involves acting upon the external world, then the definition of state changes and of the effects upon the state of applying any operator reflect with complete accuracy in one or more problem spaces the laws (laws of nature) that govern the external world.

6. All of these conditions hold in the strong sense that the basic processes postulated require only practicable amounts of computation, and the information postulated is effectively available to the processes–i.e., available with the help of only practicable amounts of search. (Simon 1973, 183)

The common situation facing all three groups was that of desiring a classification whose *method of production* was intelligible even to outsiders. Harrison, Sneath, and Sokal each in his own way found the guildlike mastery of systematists obstructive or questionable. One of the greatest merits of setting universal rules about the admissibility and weighting of characters—that is, any logically nonredundant character is acceptable and all are weighted equally—was to explicitly delimit what prior knowledge could be used to make a classification. Under these strictures, the set of characters available for a classification therefore fulfilled the dictum, per item 4, that "any knowledge that the problem solver can acquire about the problem can be represented in one or more problem spaces." Methodological rules therefore ensured the sufficiency of the pheneticists' abstracted representation of what initial knowledge went into their classifications.

Where Simpson sought a general description of systematists' methodology but explicitly recognized its ultimately ill-structured nature, pheneticists pursued a similar program but in the critical vein, emphasizing a need for new rigor. My interest in this chapter is how that rigor manifested itself in the reorganization of classificatory work. Each of the different groups, beginning from independent experiences of frustration with contemporary systematics, arrived at analogous resolutions that allowed them to proceed in solving what they had initially set as their problem.[6] Broadly speaking, these resolutions to otherwise narrow problems took on greater significance for all of systematics because they were posed as a critique and positive alternative to the "whole" of the field already envisioned by Mayr, Simpson, and many others in the New Systematics. Only through the test of time had it become clear that the practical gaps in the methodological prescriptions of the New Systematics were systematic in nature, in that the gaps included a general class of problems rather than merely a hodgepodge of disparate cases.[7] Moreover, these problems proved sufficiently fundamental in nature that their local solutions supported further research by pheneticists to generate a new view of the field.

In the introduction I offered a name for the sort of thinking that guides this fairly common process of moving from local to general problems. I termed them "epistemic visions," defined as a feasible strategy for solving ill-structured problems by the principled reorganization of work. Now that we have seen a concrete instance of how putting judgments of similarity before evolutionary history helped make classification a better-structured problem for the early pheneticists, we can say more about epistemic visions in general, although a full discussion would take too much space. The local problems the early pheneticists faced in developing their thoughts were relevant to the whole field in part because their challenge required finding a new *way* to make what is a nearly universal product of every systematics research project: some sort of branching diagram that organized the species (or other units) into mutual relationships. The key concepts of the field are intrinsic to carrying out this work, in that it is only through ideas such as character, species, similarity, and homology that the scientists can *know* what they are *doing*.[8] Hence articulating a different way of classifying intrinsically implicates a shift in concepts because it is through these concepts that the method becomes generally intelligible. In our present context, we can uncover these linked changes in concepts and practices through careful attention to the particular epistemic obstacles systematists faced, the universal assumptions they made about their objects of study to justify certain procedures, and also how they represented their work process as a whole (more on this last item later). Simon's characterization of well- and ill-structured problems therefore offers a heuristic frame for us to track what about a work process is being reorganized.

In general, an epistemic vision refers to a plan of action, a strategy, that a person has thought through as a way of making some previously insoluble class of problems tractable. Often, as was the case for the early pheneticists, one will stumble onto a solution for a local version of the general problem first, often not even appreciating its true scope, and only then, upon reflection, begin to ask how the local situation can be understood as a more general one.[9] Through this process of reflection, the scientists revise their conceptual understanding of their work in such a way as to end up claiming that something similar to the local solution would succeed more broadly. Their reflection thus essentially involves them in *envisioning how* to demonstrate in action the truth of this new conceptual understanding. For Sokal and Sneath, their vision of a new sort of taxonomy engaged them in a multipronged research program whereby they aimed to demonstrate the correctness of their approach using theoretical arguments, concrete comparisons against older methods, simulations, and mathematical proofs.

There are a host of important questions that are raised by the place of epistemic visions in scientific change, but I will focus here only on its relation to the significance of Sokal and Sneath's reorganization of classificatory work. Beyond the general phenetic move to classifications based on similarity using any nonredundant character with equal weight, Sokal and Sneath articulated a particular approach with its own unique consequences: they added enough further structure to the problem to separate out two different types of work that were tied together in a functional relationship. Numerical taxonomy supported *methodology* as a full-time research project, in addition to the traditional production of concrete classifications of various organisms. In the rest of this chapter I articulate the further structure Sokal and Sneath added to the process of classification and describe what parts of it have become entrenched, forming NT's major historical contribution to methodology in systematics.

BRACKETING METHOD OFF FROM TAXON-SPECIFIC RESEARCH

As contrasted to the descriptions of method offered within the New Systematics that we saw above, the methodological innovations of NT were twofold. The first was rigorous enforcement of classification as a strictly linear process, born out of the general phenetic move to put classifying *before* inferring phylogeny and instantiated in practice almost exclusively by computer programs. Second, Sokal and Sneath devoted additional attention to specific steps in the overall process: the three stages in the middle of the process were where they made their statistical and methodological contribution, as is reflected by the technical sophistication and length of treatment received by those stages in *Principles of Numerical Taxonomy*. It was in the second innovation that NT made its distinctive mark on the history of systematics: the particular way it articulated and justified a methodology universal to all taxa, drawing on conceptual resources from statistics, logical empiricism, and genetic determinism. While many of the historical specifics of Sokal and Sneath's vision have dropped away, the boundaries and relations of methodology to taxon-specific work have remained largely unchanged, but this organization is neither necessary nor unproblematic for the field. Indeed, some systematists appear to be struggling with this legacy of NT in just the past few years.

Sokal and Sneath's first innovation in reorganizing the work of taxonomy was to insist strictly that even in practice one must proceed

TABLE 9.1 COMPARISON OF THE PROCEDURES OF SIMPSON AND SOKAL AND SNEATH

Step	Simpson		Step	Sokal and Sneath
1.	Select specimens	=	1.	Select specimens
2.	Record observations	≈	2.	Discover and measure characters
3.	Sort into units	X		
4.	Compare units	→	3.	Code characters
5.	Interpret relations	→	4.	Calculate affinity
6.	Infer evol. patterns	→	5.	Cluster specimens
7.	Make rankings	≈	6.	Extraction of data
8.	Select names	=	7.	Identify specimens

NOTE: Sokal and Sneath have transformed the middle of the process, steps 4–6 (denoted by arrows in the middle column), while leaving the extremes relatively untouched. Step 3 was eliminated because Sokal and Sneath treated all clustering problems as independent of which units in the Linnaean hierarchical ranking system are used.

linearly instead of reciprocally. In particular, NT split classification into stages not unlike those of Simpson but required that each stage depend on only the output of the previous stage.[10] Later stages could not directly affect the execution of earlier stages unless the taxonomist decided to start the process anew with different specimens, characters, or algorithms. In addition, their insistence on the equal weighting of characters, at the least the first time through the classification process, precluded much of the backtracking and reciprocal illumination Simpson and Hennig found so valuable.

Figure 9.1 shows the emphatically linear process Sokal and Sneath articulated, which was enforced in part by normative strictures in *Principles* and by the linear, input-output nature of computer programs. If we compare the steps of NT's process to Simpson's, we see that the steps given in figure 9.1 closely mirror the eight given by Simpson but show important differences. See table 9.1 for a summary. Step 3 in Simpson, the forming of specimens into groups, has disappeared for Sokal and Sneath because their general procedure was intended to be the same no matter to what hierarchical level the clustering applied, for example, genera or varieties. They introduced the term *operational taxonomic unit* (OTU; see steps 3–5 in fig. 9.1) in order to side step the loaded issues of species and ranks. Thus the expert knowledge about geographic variation, species concepts, or any determinate ecological relation between the specimens is hidden or diffused into the remaining stages in NT's procedure.

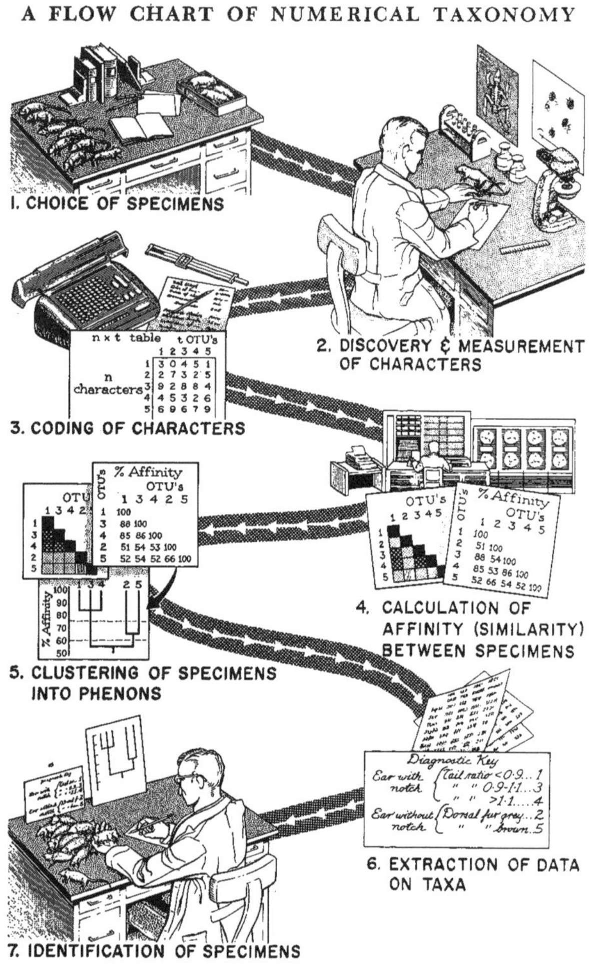

FIGURE 9.1. The seven stages of carrying out a classification process in numerical taxonomy. Sokal and Sneath's innovations compared to previous descriptions of taxonomic procedure lie in stages 3 through 5.

In addition, step 6 in Simpson, the inference of evolutionary pattern, has been transformed into a nonevolutionary analog. The replacement is justified from Sokal and Sneath's point of view because such evolutionary conclusions putatively could be based only on an already complete classification. Thus the inference of evolutionary patterns, especially in a reciprocal method, invited a vicious circle. Hence instead of stage 6 in Simpson we have stage 5 of Sokal and Sneath, where the taxonomist infers clusters by affinities, that is, measures of similarity between the characters of specimens.

The heart of NT's technical innovation as of 1963 was stages 3–5, which reflect in order its statistical approach to taxonomic characters, the articulation of affinity as a mathematical function, and the use of clustering algorithms to produce a branching classification diagram. The corresponding three chapters in *Principles* (chaps. 5–7) transform steps in Simpson's account from a "descriptive generalization" to a technically involved, explicit procedure (Sokal 1963, 21). In contrast, Sokal and Sneath have changed almost nothing about the extreme stages (1 and 8 in Simpson), and their stance toward stages 2 and 7 was concerned only to ensure that the bracketing off of methodology was secured, leaving a wide range of decisions about character choice and ranking to the particular systematist. For example, while Sokal and Sneath did have points to make about the selection of specimens and character measurement, these issues receive a much briefer discussion in *Principles* that also lacks extensive mathematical symbolization (e.g., their take on nomenclature, the logical independence of characters, or species as a useful choice for operational taxonomic units). In addition, the technical issues of how to build diagnostic keys for identification embodied by step 6 in figure 9.1 receive little discussion in *Principles* and were treated by other authors outside NT.[11]

Focusing on this second innovation of NT for now, we will see that the transformation of the *middle* of the process of classification carried a far-reaching consequence for the organization of work. On the one hand, Sokal and Sneath's work opened well-structured research problems about the statistical nature of characters, measuring affinity, and clustering OTUs. These problems could and did demand the full attention of many researchers who made careers out of studying methods of classification, irrespective of any knowledge about particular organisms.[12] On the other hand, methodology was now sharply distinguished from the work of producing actual classifications, which were the end products of the great majority of taxonomists.[13] These systematists' work specialized on particular organisms, and their results rarely aimed for a universal impact on the work of other taxonomists. I will call the former group methodologists and the latter group taxon specialists.[14] This differentiation of research work demonstrates how clearly Mayr's 1942 suggestion about learning taxonomy through exemplars had been challenged by the claim of numerical taxonomists to a method universal for all taxa.

The intended universality of method in NT is evident in a number of ways. One simple example is how Sokal and Sneath redefine classification:

> We have adapted Simpson's definition (1961, p. 9), which restricted itself to zoological classification, to more general usage. . . . We have restricted [this definition] to organisms, since this book is primarily intended for the biological taxonomist. However, there are many methods of classification, including numerical taxonomy, which are equally applicable to concepts and entities other than organisms. (Sokal and Sneath 1963, 3)

In other words, NT is a general method of classification that extends beyond living organisms and is most certainly not restricted to any group of them. This universality is also reflected in their use of *operational taxonomic unit*, with which Sokal and Sneath referred to data taken from taxonomic units at any level of the Linnaean hierarchy (Sokal and Sneath 1963, 120–121). The issue is again evident in the fact that Sokal and Sneath looked for a definition of characters that would "allow the broad treatment needed for a *general* theory of systematics" (Sokal and Sneath 1963, 63; my emphasis). Hence NT is a universal method in the sense that its concepts, principles, and procedures are putatively applicable to all living organisms (and more), and although particular procedures may prove more apt in a given study than others, they are still general in form.

Numerical taxonomy's putative universality for all kinds of organisms depended in practice on two important objects that bracket off the taxon specialists' work from the methodologists'. That is, the objects are the transitions where the process moves up into the abstractions of the computational methods and then back down into the concreteness of the organisms, and they are the character matrix and dendrogram, prominently visible in steps 3 and 5 in figure 9.1. These two objects function as boundary objects between the lines of work of methodologists and taxon specialists (Star and Griesemer 1989). For NT, the character matrix and dendrogram are boundary objects that join together but also hold apart the expertises of the methodologist and the taxon specialist.

In a more concrete sense, the boundary objects help articulate ill-structured problems into well-structured ones by making the problems' structure and solutions mostly independent of each other. (This is analogous to Simon's [1962] near decomposability concept.) By establishing the character matrix and dendrogram as pivotal objects between lines of work, Sokal and Sneath can claim to legitimately have abstracted

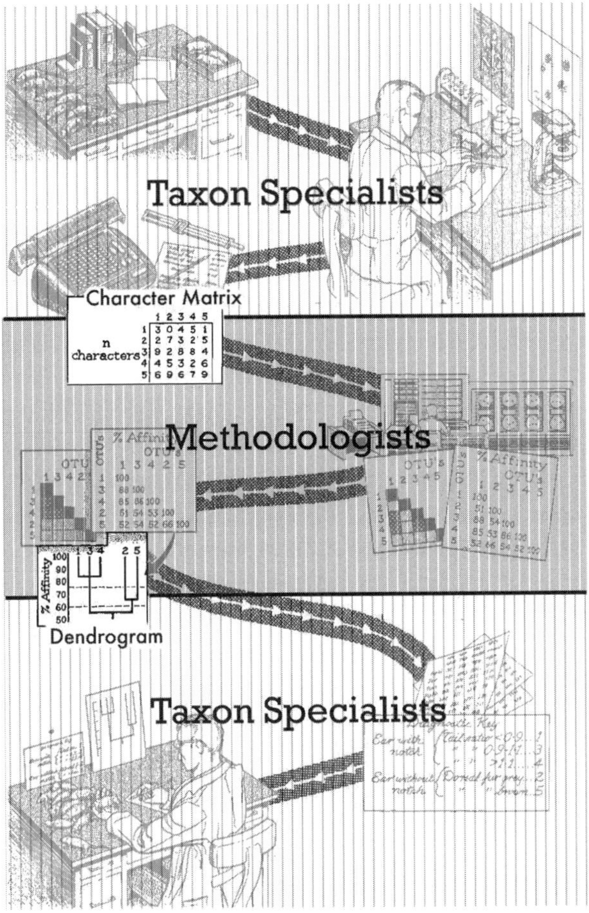

FIGURE 9.2. The workflow of numerical taxonomy edited to emphasize its reorganization of work for methodologists and taxon specialists. Methodologists focus their expertise on the middle stages, while taxon specialists are experts in the first and last stages. These subspecialties are bracketed off from each other by the character matrix and dendrogram, which are highlighted as well.

away aspects of practice, making the organisms a separate subject of study from the algorithm that analyzes them. In other words, they consciously set out a strategy in *Principles* for how to re-organize research work to separate out what is a good method vs. what is a good classification of the arthropods, for example. Figure 9.2 superimposes this reorganization of work I have described onto figure 9.1.

JUSTIFYING THE AUTONOMY OF METHOD

Sokal and Sneath took seriously the issue of formally defining a character in such a way that its nature and representation does not depend on the particular organism from which it came. In other words, their approach effectively aimed to establish that *in principle* no taxon-specific expertise was needed to classify the taxa once a character matrix had been produced, although they acknowledged "only specialists in the various groups will be in a position to define and describe unit taxonomic characters in the organisms they are studying" (Sokal and Sneath 1963, 92).[15] How, then, did they ground their method in the nature of characters while avoiding the "speculative" hypotheses of evolutionary theory and expertise of taxon-specific researchers? Sokal and Sneath brought their own distinctive set of conceptual resources from outside systematics, and together they succeeded at reposing the question of how to judge the similarity of two organisms in a well-structured fashion as a problem of statistical estimation.[16]

From the perspective of NT, character measurements are simply instances of matches or mismatches in the overall similarity function, and neither the quantitative nor the qualitative nature of what the characters *are* is important to their value in the classification process. Indeed, the work of the taxon specialist is effectively recast as one of developing and improving new ways of measuring the genome through phenotypic traits, at least insofar as the methodology of classification was concerned. The meaning of resemblance in NT is independent from the results and concepts of most of systematics: it is ultimately a function of matches between genomes that Sokal and Sneath have defined to be unaffected by any theories about homology or evolution. The problem of classification therefore becomes an essentially statistical one: devise methods for accurately estimating the parameter indicating the proportion of matching traits and then find ways to best represent those resemblances in a branching classification diagram. This way of posing the problem of classification made it well-structured yet complex enough to involve multiple subproblems that were nonetheless all solvable together using standard procedural algorithms on a computer.

Returning to my characterization of the problem situation facing any nonevolutionary taxonomist, we can now see how Sokal and Sneath sought to resolve aspects 1, 2, and 3 of Simon's list of the properties of a well-structured problem. Briefly summarized, these are (1) a definite criterion for testing proposed solutions; (2) a problem space with which they could represent the start, end, and middle states of the classification process; and (3) state changes ("legal moves" in the game), which can also be represented in the problem space. Commensurate with their initial status as outsiders to systematics, Sokal and Sneath drew on external resources to reorganize classificatory work, including mathematical statistics, logical positivism, genetic determinism, and information theory.[17] Gilmour's positivist ideas about the naturalness of classifications would prove important for item 1, along with an adaptation of the correlation coefficient from statistics, and the combination of information theory with genetic determinism was critical for supporting the algebraic coding of characters and the matrix representation of the units of classification in a multidimensional trait space, fulfilling items 2 and 3.

One major source of inspiration came from Sneath's interactions with Gilmour and Gilmour's 1940 essay, "Taxonomy and Philosophy," in *The New Systematics* (Vernon 1988; Gilmour 1940). Gilmour's intention was to bring the epistemological ideas of logical positivism to taxonomy "for an examination of the purpose and method" of the field (Gilmour 1940, 463). Under this interpretation, the work of taxonomy "consists in clipping together the mass of sense-data collectively classed as 'living things' into a logically coherent pattern for the purpose of making inductive generalizations concerning those data" (Gilmour 1940, 465). He goes on to define a "natural" classification as "that grouping which endeavors to utilize *all* the attributes of the individuals under consideration, and is hence useful for a very wide range of purposes," in contrast to an "artificial" classification that uses few attributes and fits only a narrow range of purposes (Gilmour 1940, 472).

Sokal and Sneath fleshed out this perspective in greater detail by offering operationalized definitions of character and homology. They initially defined a unit character "as an attribute possessed by an organism about which one statement can be made," with the understanding that "attributes are formally logical constructs, since they will change if the technique of observation changes" (Sokal and Sneath 1963, 63). The states of characters can be recoded as binary bits according to information theory, and Sokal and Sneath proceed to give a survey of calculations one can make about how many bits of information an organism contains in its genome. However, since gene sequences were rarely observable directly at the time, Sokal and Sneath give a "working definition" of a unit character that uses only phenotypic data: "we may define a unit character as *a taxonomic character of two or more states, which within the study at hand cannot be subdivided logically, except for subdivision brought about by changes in the method of coding*" (Sokal and Sneath 1963, 65; emphasis in original). Since logical divisions depend on the availability of instruments to measure states, a unit character is a set of possible states that no measurement is able to further refine, perhaps because we do not have the capability to observe it in fine enough detail.

The information view of characters therefore sees them as observations ultimately representable in the form of binary yes or no states. Sokal and Sneath believed this view was also consonant with the genetic theory of the time: they claimed that "we may tentatively identify our taxonomic bits with the genetic code." In addition, "it is generally considered that only genetically determined characters should be used in orthodox taxonomy, and with this we concur" (Sokal and Sneath 1963, 92).[18] (What remains unclear, though, is precisely how they proposed to select only genetically determined characters.) Nonetheless, the focus on genes as the sole (or perhaps primary) cause of taxonomic characters was central to their statistical theory of why clustering based on affinity functions was a valid approach for taxonomy.

Given this causal identification of observable taxonomic characters with genomes, however, there arises a problem of characterizing the worth of nongenetic characters as indicators of the similarity of different genomes. In response, Sokal and Sneath advanced several basic hypotheses that essentially justified why traditional characters in systematics could be treated as offering an unbiased sample of the genomes. The first two "fundamental hypotheses" (Sokal and Sneath 1963, 84) are the nexus and nonspecificity hypotheses. The nexus hypothesis states that "every taxonomic character is likely to be affected by more than one genetic factor and that, conversely, most genes affect more than one character." The nonspecificity hypothesis assumes "that there are no distinct large *classes* of genes affecting exclusively one class of characters such as morphological, physiological, or ethological, or affecting special regions of the organism such as head, skeleton, leaves" (Sokal and Sneath 1963, 84–85; emphasis in original). The main point here is that no character is particularly special: all have complex relations to genes, and there are no groupings of genes that solely affect certain groupings of characters. As Sokal and Sneath put it, "There are no a priori grounds for favoring one character over another." Alternatively, one could say that there is effectively no relevant structure to an organism's genotype-phenotype map: detailed knowledge of that structure is not necessary to correctly classify the organism based on its genetic nature while using only phenotypic traits. While they devote a substantial discussion to the evidence for and against these hypotheses, the relevant outcome for us here is that they believed both hypotheses were largely correct for the majority of cases (Sokal and Sneath 1963, 85, 91).

Only given the result that characters are generally of equal worth as reflections of the genomes could Sokal and Sneath proceed to lay the fundamental ground for why NT should work. What they argued is that although it appears even one hundred characters are unlikely to represent a sample of more than 15 percent of a given genome under a simple model, what matters is not "the absolute magnitudes and qualitative nature of the characters sampled but [the] relative degrees of difference between them" (Sokal and Sneath 1963, 112–113).[19] The reason is that NT treats "resemblance as a proportion of characters agreeing (matching) out of the total number being compared" (114). This proportion is then taken as an estimate of the actual proportion of matching traits that exists between two operational taxonomic units. (The issue of how to treat unmatchable characters is not discussed.)

This leads Sokal and Sneath to the matches asymptote hypothesis: "as the number of characters increases, the value of the similarity coefficient becomes more stable," because as the sample size increases, the proportion better reflects that found in the total population of traits being sampled (Sokal and Sneath 1963, 114). The reasons that numerical taxonomic classification produces a true representation of the relation between organisms are therefore that there exists an objective set of pairwise matching relations between the operational taxonomic units' genomes, that resemblance is solely an increasing function of the proportion of matches, and that a (large enough) set of taxonomic characters provides an effectively unbiased estimate of that proportion. The nexus and nonspecificity hypotheses are crucial to establishing that almost any sufficiently large sample of characters should be adequate to producing an unbiased estimate of the true proportion of the total population of characters.[20]

An additional, key factor supporting the autonomy of NT's method was Sokal and Sneath's belief in its ability to produce maximally general-purpose classifications. Compared to their treatment of the statistical properties of characters, however, their justifications for why one should expect this benefit were more vague and intuitive. Following Gilmour's idea of natural classifications, Sokal and Sneath noted that "the nature of a taxonomy depends on its purpose . . . and we choose one way because we think it is best for some purpose. If the purpose is restricted, then the classification conveys less information than a general or 'natural' one" (Sokal and Sneath 1963, 12). Hence they held "the view with Gilmour that a 'natural' taxonomy is a general arrangement intended for general use by all scientists" (12). However, their concept of purpose was the somewhat vague notion of how many things a classification can predict: "a system of classification is more natural the more propositions there are that can be made regarding its constituent classes" (19).

Broadly speaking, NT would maximize this measure of naturalness because it used the equal weighting of many characters, which was more "informative" than unequal weighting (akin to Gilmour's sense of "artificial") or the use of few characters. In addition, the use of mathematical techniques on a computer was critical: as Sneath put it in his early paper, "A classification based on the number of similar features [i.e., equally weighted] and their correlation, in other words the classification which would emerge if one fed one's data into a calculating machine, would in theory be the best classification which could be made with the available data" (Sneath 1957, 198).[21] This very general picture of what was required to make a good classification was a crucial piece of support for the autonomy of methodology, precisely because it meant that the problem of evaluating competing methods could be separated from the questions and nature of particular taxa.

CONCLUSION

The separability of methodological questions from taxon-specific research that Sokal and Sneath articulated quickly ran into problems. Numerical taxonomy faced challenges to its claims to superiority as a methodology both on its own terms and on the broader grounds that it did not in fact serve all purposes equally well.[22] Numerical taxonomy's fate in fact differed from one group of organisms to another: it was highly appreciated within the microbiology community, moderately so in botany, and only occasionally in zoology.[23] Emerging in part from that last group, systematists who wanted their classifications to represent evolutionary history intrinsically rather than post hoc began to develop their own mathematical and computational approaches. In the two decades after the 1963 publication of *Principles,* a cladistic school of classification (and phylogenetics) coalesced and disagreed strongly with some of the tenets of NT. Although numerical taxonomists and cladists often initially mutually tolerated each other, over time their relations soured to the extent of giving rise to a period known as the "systematics wars" (Hagen 2001). However, despite this vitriol and the varying judgments on NT's merits across taxonomic specialties, key aspects of NT's methodology and its autonomy survived.

The impact of NT's new methodology was not limited simply to its direct success or failure with taxon specialists, however, because its methodological framework—that is, the boundary objects distinguishing methodology's proper domain as lying between the character matrix and dendrogram in figure 9.2—has remained largely fixed. In other words, the post-NT systematics wars have been concerned with the substance and interpretation of method, not its general organization and distinction from taxon-specific research as Sokal and Sneath first articulated. In this way, the battle for dominance operated within a framework that depended intrinsically on the use and concepts of mathematics, and therefore the integration of mathematics proceeded even as the battle between research programs intensified.

To be clear, it is obvious that cladistics has departed significantly from the steps indicated by figure 9.1; however, the form of the boundary objects between methodology and taxon-specific research, a data matrix and a branching diagram, have remained constant. As Brent Mishler has recently pointed out, "Paradoxically, despite the logical preeminence of *data matrix construction* in phylogenetic analysis, by far the greatest effort in phylogenetic theory has been directed at the second phase of analysis, the question of how to *turn a data matrix into a tree*" (Mishler 2005, 57; my emphasis). Mishler's point exactly parallels the distinction Sokal and Sneath drew between methodology and taxon-specific work, as I argued above. Indeed, Hagen has documented that early work by James Farris shared a number of methodological tenets with NT: when Farris proposed Wagner trees and networks instead of the Camin-Sokal technique for making cladograms, "Farris did not justify it on philosophical distinctions between phenetic and cladistic techniques. He was in fundamental agreement that the solution to Hennig's dilemma lay in the mathematical analysis of large numbers of characteristics and alternative phylogenies. . . . His was simply a better solution to a computational problem established by Camin and Sokal's earlier paper" (Hagen 2001, 300). Although a full account of NT's contribution to an entrenched division of work within systematics requires much further investigation, Farris's agreement with NT about the importance of large numbers of characteristics hints at a narrative of the "statistical view" of morphological characters in systematics that crosses continuously over ideological divides.

The theoretical foundation of methodology as articulated by Sokal and Sneath and its weak points remain relevant to contemporary problems in systematics: the issue of how to distinguish methodology from taxon-specific research in a functional way is *still* a pressing problem.[24] Despite the fact that the tree is ultimately only as good as the characters that go into it, the steps that transform character matrices into dendrograms have continued to receive the near-exclusive attention of methodologists. Indeed, at the heart of the issue is what constitutes a good character, including properties such as independence from other characters and the absence of too much noisy variation. Since "a systematist spends 95% of his/her time gathering and analyzing character data and less than 5% time turning the assembled data matrix into a tree," Mishler argues, the problem of what makes a good character should be paramount for any taxon specialist interested in methodology (Mishler 2005, 59).

Sokal and Sneath provided the issue with a careful and deep conceptual treatment, and it seems that any revision or reform of how they reenvisioned the process of classification needs to have worked through the merits and weaknesses of their approach. In particular, Sokal and Sneath introduced the idea that sets of characters represent an unbiased "sample" of some underlying nature of the organism, and that a large enough sample should provide statistical convergence in the resulting tree—fifty years later, what if anything justifies this "statistical view" of characters?

Indeed, our attention to the reorganization of classificatory work carried out by Sokal and Sneath has illustrated how important the relationship is between the actions of scientists and their concepts. Just as the initial challenge facing NT was to take its broad "philosophy" of classification and demonstrate its effectiveness in practice, over time foundational concepts can become unmoored from their practical grounding. Sokal and Sneath at least had an argument for what in the nature of morphological characters (i.e., their origins in a determinative, simple genome) should justify an expectation of convergence on a single tree, yet as their argument proved insufficient historically, there does not seem to be a clear alternative to replace it.

From the perspective of epistemic visions, the historical trajectory of NT illuminates the need for an extra dimension to Agar's question, "What difference did computers make?" Agar focused on the consequences of computers for single laboratories or institutions, but the fate of NT's methodological innovations occurred properly at the scope of the entire field. My concern here has been to trace how Sokal and Sneath managed to generalize their local innovations into a universalist methodology, and it is commensurately important to ask what difference computers made to *this* process, that is, beyond their impact on the prehistory of NT as a revolutionary effort. I can only point toward an answer here, but the obvious direction is to examine computers not as tools of calculation but as infrastructure and communication media. Sokal and Sneath spent considerable effort developing the computing infrastructure necessary to support the distribution of numerical taxonomic techniques, including the publication of algorithms in a specialized journal and the sale of magnetic tapes containing these programs across the United States.[25] To my knowledge, the existence of an (at least putatively) universal pool of shared, objective methods that could be published as contributions to the field and exchanged as commodities had no precedent in systematics. Only with the ever-growing accessibility of computers in the 1960s and an exchange network for software programs could NT methodologists make their work usable by taxon specialists, who typically did not have the mathematical and technical background to develop NT's techniques for themselves. Computers provided essential social infrastructure for the spread of the numerical taxonomic revolution, indeed also for its enduring methodological legacy.

Acknowledgments

I would like to thank Andrew Hamilton for his support in pursuing this project, and I also want to acknowledge my appreciation of the extensive comments and suggestions I received from Chris Diteresi, Trevor Pearce, Eli Gerson, Marcie Holmes, Zoe Nyssa, Jennifer Karlin, William Wimsatt, William Sterner, Margot Browning, Erin Barringer, and the CHSS Graduate Workshop. This work was funded by an NSF Graduate Research Fellowship.

NOTES

1. I should briefly clarify a historical ambiguity contained in referring to the practice of systematics over the twentieth century. For the sake of simplicity, and because I do not believe it directly affects the argument here, I am going to project backward the label "systematics" to a diverse field of heterogeneous composition and name at the turn of the century. At the time, the work of

natural history was being redescribed or transformed into taxonomy or systematics, and the people doing this work were "amateurs" more often than what we would recognize as professional scientists today. Although these transformations of name, work, and profession are of great importance to the history of biology, they must remain outside the scope of this chapter. Instead, I will, as do Sokal and Sneath, use taxonomy and systematics broadly in the same sense as Simpson in 1961: taxonomy is "the theoretical study of classification, including its bases, principles, procedures and rules," while systematics "is the scientific study of the kinds and diversity of organisms and of any and all relations among them" (Simpson 1961, 7–9). Taxonomy is therefore generally an activity within systematics, although for exactly what reason is contested.

2. The term *phenetic* is A.J. Cain's coinage, and here I use it to mean any method of classification that insists on making judgments using similarity alone at first and then potentially involving evolutionary theories or interpretations.

3. Of course, Sokal and Sneath also had a hand in preparing this ground, as they collaborated with A.J. Cain and others early on to develop the phenetic perspective (Vernon 1988).

4. My account below is based on Vernon (1988). Also, although Gilmour influenced Sneath considerably, Sneath published separately, in contrast to the collaborations of Cain and Harrison and Sokal and Michener, and therefore I follow Vernon in listing him as a separate individual.

5. Simon in fact wants to argue that AI can also deal with overlapping and backtracking problems because those ill-structured problems are not different in kind from well-structured ones, only different in degree. His solution, however, involves positing several notoriously difficult to realize capacities of computers, such as a "noticing and evoking mechanism" that is crucial to successfully redefining the problem space as new difficulties and information crop up. As these capacities are not required for solving a well-structured problem, one might wonder if he needs a qualitative leap in computing functionality in order to show how ill-structured problems differ only in degree.

6. Of course, one reason for the similarity in phenetic perspectives articulated in the 1960s was that Sokal, Sneath, and Camin began to collaborate, yet the point stands that they embarked on closely parallel trajectories before interacting (Vernon 1988). Also, it will prove relevant to my later discussion to note that the resolution of what were initially specific taxonomic problems proved to be of more interest to Sokal and Sneath than the original problems themselves.

7. William Wimsatt's (2007) take on theories as heuristics all the way down is particularly relevant here to understanding the *positive* relationships that different revolutionary camps have to the discipline as a whole. Heuristics by nature are flawed, and when they fail in systematic fashion, we have the potential to learn from the errors (i.e., "metabolize" them). If this chapter succeeds, it is because revolutions in systematics illuminate major functional failings of their historical predecessors for the historian and philosopher in a way that reflects not only conceptual changes but also how concepts are implicated in the daily practical needs of the scientists.

8. I mean "know" in a generally pragmatic sense here. As Wittgenstein put it, "'But is a blurred concept a concept at all?'—Is an indistinct photograph a

picture of a person at all? Is it even always an advantage to replace an indistinct picture by a sharp one? Isn't the indistinct one often exactly what we need?" (Wittgenstein 2001, 34e).

9. Mark Wilson's (2006) discussion of Heaviside's adventurous use of calculus to model electrical currents in circuits, and the applied mathematicians who made sense of it only decades later, makes a similar point, though Wilson focuses on the "seasonality" of concepts as abstracted from their part in organizing the research practices of a scientific community.

10. Sokal and Sneath in fact cited Simpson, Hennig, and Remane as having provided notable efforts at "analytic studies of systematic principles," but none of them resolved the problem of "circular reasoning" in a satisfying fashion for Sokal and Sneath (Sokal and Sneath 1963, 2).

11. For a later review emphasizing plants, see Pankhurst 1978.

12. Vernon (2001) discusses some of the second generation of NT.

13. Gerson (1998) has a related distinction between taxon specialties and analytic specialties, e.g., population genetics. The main difference between analytic specialties and methodologists is that methodologists are strictly speaking a subspecialization within taxonomy rather than an outside specialty.

14. While these particular terms are not actor's categories, similar distinctions had existed long before the publication of *Principles*. For example, see Simpson's discussion of Gregg's distinction between "taxonomy proper," which is analogous to taxon-specific research, and "methodological taxonomy" (Gregg 1954; Simpson 1961).

15. They had also earlier on written that "it is possible to formulate a list of characters and their states that do not presuppose a prior knowledge of homologies" (74), thereby bypassing a great portion of the relevant expertise of the taxon specialist. In fact, Sokal later went further and attempted to demonstrate that even the discovery and measurement of characters required no specialized knowledge or training. (It did not work as hoped; see Sokal and Rohlf 1966, 1970; Rohlf 1965.)

16. I will not touch here on their emphasis on OTUs as a way of avoiding rank-dependent problems in step 1 of their classification process, but this is an additional, important reason for why NT could separate itself from taxon-specific research.

17. Olivier Rieppel has investigated these issues for cladistics, and my section here is a parallel, preliminary account of similar problems. See Rieppel 2003, 2007.

18. They also added, "however, a study to investigate the influence of environment using numerical taxonomic methods could legitimately include environmentally determined characters."

19. Note that although Sokal and Sneath discard the factor asymptote hypothesis as "not very fruitful," they do not reject the reasoning behind it that treats characters as forming a "random sample of the genome of the operational taxonomic unit."

20. Sokal and Sneath make a few hedges here involving requiring sampling characters across an organism's life cycle and including a broad variety of whatever characters show variation. See pp. 92–93 and others.

21. While Sokal and Sneath did have a way of checking how well the dendrogram diagram represented the initial measurements of affinities (the co-phenetic correlation coefficient; see Sokal and Sneath 1963, 189–194), it would require another step of logic to demonstrate that the whole process maximized the number of possible propositions that could be made using the classification.

22. James S. Farris (1969) found that the co-phenetic correlation coefficient, the primary measure of optimality NT had used, did not in fact measure what Sokal and Sneath thought it did.

23. About thirty years after *Principles* first appeared, approximately 3,000 papers had been published applying NT to bacterial classifications, and the subspecialty had indeed used NT to increase the objectivity and repeatability of its character measurements and classifications. According to a similar estimate for botany, about 500 to 1,000 publications had appeared applying NT to plants in the same time period. Far fewer appeared within zoology. See Vernon 2001; Hagen 2001; Hull 1988; Goodfellow, Jones, and Priest 1985; Duncan 1981.

24. Winther (2009) has set out to address this problem, among others.

25. See, e.g., Rohlf 1969.

REFERENCES

Agar, J. 2006. "What Difference Did Computers Make?" *Social Studies of Science*, 36: 869–907.

Duncan, T. 1981. "Numerical Phenetics: Its Uses in Botanical Systematics." *Annual Review of Ecology and Systematics*, 12: 387–404.

Farris, J.S. 1969. "On the Co-Phenetic Correlation Coefficient." *Systematic Zoology*, 18: 279–285.

———. 1970. "Methods for Computing Wagner Trees." *Systematic Zoology*, 19: 83–92.

———. 1972. "Estimating Phylogenetic Trees from Distance Matrices." *American Naturalist*, 106: 645–668.

Farris, J.S., Kluge, A.G., and Eckardt, M.J. 1970. "A Numerical Approach to Phylogenetic Systematics." *Systematic Zoology*, 19: 172–189.

Felsenstein, J. 2001. "The Troubled Growth of Statistical Phylogenetics." *Systematic Biology*, 50: 465–467.

———. 2004. *Inferring Phylogenies*. Sunderland, MA: Sinauer Associates.

Galison, P. 1997. *Image and Logic: A Material Culture of Microphysics*. Chicago: University of Chicago Press.

Gilmour, J.S.L. 1940. "Taxonomy and Philosophy." In J. Huxley (ed.), *The New Systematics*. New York: Oxford University Press, 461–474.

Goodfellow, M., Jones, D., and Priest, F.G., eds. 1985. *Computer-Assisted Bacterial Systematics*. New York: Academic Press.

Gerson, E.M. 1998. "The American System of Research: Evolutionary Biology, 1890–1950." Ph.D. dissertation, University of Chicago.

Gregg, J.R. 1954. *The Language of Taxonomy*. New York: Columbia University Press.

Hagen, J.B. 1984. "Experimentalists and Naturalists in Twentieth-Century Botany: Experimental Taxonomy, 1920–1950." *Journal of the History of Biology,* 17: 249–270.

———. 2001. "The Introduction of Computers into Systematic Research in the United States during the 1960s." *Studies in the History and Philosophy of Biology and Biomedical Sciences,* 32: 291–314.

———. 2003. "The Statistical Frame of Mind in Systematic Biology from *Quantitative Zoology* to *Biometry.*" *Journal of the History of Biology,* 36: 353–384.

Hamilton, A., and Wheeler, Q.D. 2008. "Taxonomy and Why History of Science Matters for Science: A Case Study." *Isis,* 99: 331–340.

Hennig, W. 1966. *Phylogenetic Systematics.* Trans. D.D. Davis and R. Zangerl. Urbana: University of Illinois Press.

Hine, C. 2008. *Systematics as Cyberscience: Computers, Change, and Continuity in Science.* Cambridge, MA: MIT Press.

Hull, D.L. 1988. *Science as Process: An Evolutionary Account of the Social and Conceptual Development of Science.* Chicago: University of Chicago Press.

Huxley, J., ed. 1940. *The New Systematics.* New York: Oxford University Press.

Johnson, K. 2005. "Ernst Mayr, Karl Jordan, and the History of Systematics." *History of Science,* 43: 1–35.

———. 2009. "The Return of the Phoenix: The 1963 International Congress of Zoology and American Zoologists in the Twentieth Century." *Journal of the History of Biology,* 42: 417–456.

Kuhn, T.S. 1962. *The Structure of Scientific Revolutions.* Chicago: University of Chicago Press.

Mayr, E. 1942. *Systematics and the Origin of Species: From the Viewpoint of a Zoologist.* Cambridge, MA: Harvard University Press.

Mayr, E., and Provine, W.B. 1980. *The Evolutionary Synthesis: Perspectives on the Unification of Biology.* Cambridge, MA: Harvard University Press.

Mishler, B.D. 2005. "The Logic of the Data Matrix in Phylogenetic Analysis." In V.A. Albert (ed.), *Parsimony, Phylogeny, and Genomics.* Oxford: Oxford University Press, 57–70.

Pankhurst, R.J. 1978. *Biological Identification: The Principles and Practice of Identification Methods in Biology.* Baltimore, MD: University Park Press.

Rainger, R., Benson, K.R., and Maienschein. J., eds. 1988. *The American Development of Biology.* Philadelphia: University of Pennsylvania Press.

Rieppel, O. 2003. "Semaphoronts, Cladograms, and the Roots of Total Evidence." *Biological Journal of the Linnean Society,* 80: 167–186.

———. 2007. "The Nature of Parsimony and Instrumentalism in Systematics." *Journal of Zoological Systematics and Evolutionary Research,* 45 (3): 177–183.

Rohlf, J.F. 1965. "A Randomization Test of the Nonspecificity Hypothesis in Numerical Taxonomy." *Taxon,* 14: 262–267.

———. 1969. "GRAFPAC, Graphic Output Subroutines for the GE 635 Computer." *Kansas Geological Society Computer Contributions,* 36.

Simon, H. 1962. "The Architecture of Complexity." *Proceedings of the American Philosophical Society,* 106: 467–482.

———. 1973. "The Structure of Ill Structured Problems." *Artificial Intelligence*, 4: 181–201.

Simpson, G.G. 1961. *Principles of Animal Taxonomy*. New York: Columbia University Press.

Sneath, P.H.A. 1957. "Some Thoughts on Bacterial Classification." *Journal of General Microbiology*, 17: 184–200.

Sokal, R.R., and Rohlf, J.F. 1966. "Random Scanning of Taxonomic Characters." *Nature*, 210: 461–462.

———. 1970. "The Intelligent Ignoramus: An Experiment in Numerical Taxonomy." *Taxon*, 19: 305–319.

Sokal, R.R., and Sneath, P.H.A. 1963. *Principles of Numerical Taxonomy*. San Francisco: W.H. Freeman.

Star, S.L., and Griesemer, J. "Institutional Ecology, 'Translations' and Boundary Objects: Amateurs and Professionals in Berkeley's Museum of Vertebrate Zoology, 1907–39." *Social Studies of Science*, 19 (3): 387–420.

Stebbins, G.L. 1950. *Variation and Evolution in Plants*. New York: Columbia University Press.

Strasser, B.J. 2010. "Collecting, Comparing, and Computing Sequences: The Making of Margaret O. Dayhoff's *Atlas of Protein Sequence and Structure*, 1954–1965." *Journal of the History of Biology*, 43: 623–660.

Vernon, K. 1988. "The Founding of Numerical Taxonomy." *British Journal of the History of Science*, 21: 143–159.

———. 1993. "Desperately Seeking Status: Evolutionary Systematics and the Taxonomists' Search for Respectability, 1940–1960." *British Journal of the History of Science*, 26: 207–227.

———. 2001. "A Truly Taxonomic Revolution? Numerical Taxonomy, 1957–1970." *Studies in the History and Philosophy of Biology and Biomedical Sciences*, 32: 315–341.

Wimsatt, W.C. 2007. *Re-Engineering Philosophy for Limited Beings: Piecewise Approximations to Reality*. Cambridge, MA: Harvard University Press.

Winther, R.G. 2009. "Character Analysis in Cladistics: Abstraction, Reification, and the Search for Objectivity." *Acta Biotheoretica*, 57: 129–162.

10

A Comparison of Alternative Form-Characterization

Approaches to the Automated Identification of Biological Species

NORMAN MACLEOD

In American folklore John Henry was a railway laborer whose job was to hammer down the steel spikes that hold railroad tracks in place. Incensed by the railroad company buying a steam-driven hammer to do his job, Henry challenged the hammer driver to a race to determine whether a man or a machine could do the job more quickly. Henry won the race but collapsed and died of exhaustion immediately thereafter "with his hammer in his hand." The machine went on to replace Henry and his coworkers because, despite Henry's personal reservations and heroic performance, it *could* do the job better and faster than humans.

Taxonomists stand at the heart of many, if not most, research programs in applied biology, ecology, and biodiversity conservation. This painstaking work, often conducted while peering through a microscope, is essential because identification and classification are critical steps to understanding how living systems function. Like John Henry, they pursue their work, for the most part, using traditional methods essentially unchanged in over a hundred years. To be sure, contemporary taxonomists have access to, and readily employ, an amazing variety of advanced technologies to supply them with images, genetic sequences, sounds, and in some cases even smells and tastes of the organisms presented to them for identification. But in the end, most routine identifications are performed by a relatively small number of experts using the same diagnostic data, methods, and types of reference materials that their teachers used, that the teachers of their teachers used, the teachers of those teachers used, and so on down the line.

It is a matter of concern that this comparatively small community of expert taxonomists with the requisite skills to perform the task of delivering authoritative identifications is declining. That universities and other institutions have discontinued many taxonomic research and training programs is unsettling. That the scope and accuracy of many biological research projects is limited by the lack of access to reliable and consistent taxonomic identifications is worrying. But all these issues pale in the face of the fact that there is no tradition of—much less a requirement for—independent testing and verification of the accuracy and consistency of taxonomic identifications that are made by the experts engaged to perform this vital service. If evidence existed to reassure the scientific community that most taxonomic identifications are accurate and consistent the situation might be (barely) tolerable. There is no such evidence.

The production of consistent results from the same input is the hallmark of science. In most physical sciences, few analytical or experimental results are accepted until or unless they have been replicated by independent investigators working in independent laboratories. For a variety of reasons a comparable system of cross-checking and verification never developed in taxonomy or in the myriad different fields that rely on the identification of various object categories. As a consequence, the old ecologist's quip that a species is "whatever an expert in the group says it is," is as true today as it was a hundred, or even a thousand, years ago.

Even more troubling are the results of the few blind verification studies that, for one reason or another, have been performed by taxonomists (Colquboun 1959; Zachariasse et al. 1978; Simpson et al. 1991; Ginsburg 1997 [see MacLeod 1998 for a summary]; Kelly 2001; Gobalet 2001; Culverhouse et al. 2003). These studies suggest that, despite reassurances to the contrary by those whose livelihoods depend on their reputations for making accurate identifications, humans are generally poor at visual discrimination tasks. Empirical cognitive studies explain why (Culverhouse et al. 2003; Culverhouse 2007). We grow tired of repetitive identification tasks easily and so frequently miss items present in scenes, count some objects more than once, and misclassify others. Other sources of error in faunal and floral lists include issues arising from nomenclature (e.g., use of syntypic names) and access to current taxonomic literature. These errors are often magnified by the fact that most taxonomic identifications are performed by single specialists working alone.

Generally speaking, human taxonomists working alone and in isolation from one another are doing well to deliver identifications that are 60 to 75 percent consistent with their peers (Culverhouse et al. 2003; MacLeod 1998). Often a comparison of faunal lists generated during taxonomic blind tests of specimens collected from splits of the same sample (Ginsburg 1997 and references therein) and/or the same specimens (Zachariasse et al. 1978) yield far lower consistency results. Only when taxonomists have frequent collaborative contact with one another do consistency ratios rise to the 95 percent mark that many taxonomists regard as the standard for all taxonomic identifications performed by qualified experts.

Fortunately, appropriate use of computer systems can not only speed and diversify the types of data taxonomists can use to identify groups of objects, but also improve assessment of any sort of digital data, more objectively analyze features, and, once trained, perform repetitive tasks more quickly, accurately, and consistently. These systems are the taxonomic equivalent of John Henry's nemesis. But how good are such systems at present, and what is their potential? Even more important, why are they not playing a larger role in taxonomy at present, and what role are they likely to play in the future?

At present two generalized approaches are considered relevant to the problem of automated species identification. Biometric (or morphometric) systems are based on a set of a priori–designated measurements or assessments of the state of a character thought, or assumed, to be useful for species discrimination. Geometric data are typically extracted from digital images of specimens via automated, semiautomated, and/or manual image segmentation procedures and then used to define a set of linear or nonlinear mathematical discriminant functions based on training sets of authoritatively identified example specimens. Once defined, these functions can be used to place unknown specimens into groups the identification system has been trained to recognize or, in some cases, to flag specimens whose features are not consistent with any of the training datasets that reside in the identification system.

The other major category of approaches to solving this problem are machine learning algorithms designed to accept unfiltered input—usually in the form of generalized digital files or, in some cases, digital images themselves—that devise their own rule-based systems for categorizing the differences between a priori–designated species groups. As with biometric systems, machine learning systems are trained via presentation of a set of authoritatively identified data to the algorithms. However, rather than adjust a specific class of mathematical functions to facilitate between-group discrimination, machine learning algorithms are allowed to model interspecies differences in a far more generic manner, by boosting or diminishing aspects of the input data using a system of recursively adjusted numerical weights. As a result, machine learning algorithms have access to a much greater range of data and of mathematical functions that can be used to accomplish identification. Owing to the recursive nature of the rule-finding process, machine learning systems are usually not able to provide insight into the nature of the rules developed to facilitate discrimination, regions of the forms in question particularly pertinent to discrimination, or other types of information they use to perform their identification tasks. Although this inability to describe the rules by which unknown specimen placement decisions are made has been a cause of concern to some analysts, it does not compromise the practical utility of the rule sets themselves, which can be assessed objectively through empirical testing. It is also the case that human experts are often unable, or unwilling, to provide the same information on request.

Both of these approaches have been used extensively on a variety of small, "bench-top" or "demonstration" datasets (Gaston and O'Neill 2004). Nevertheless, direct comparisons of the performance of different types of data that can be used with these different approaches have been exceedingly rare (see MacLeod, O'Neill, and Walsh 2007 for another example). In addition, distinctions between the performance of these approaches when applied to two-dimensional (2D) and three-dimensional (3D) geometric data have not been assessed by any published report to date. This report compares the performance of five different types of morphometric data-form factors, Procrustes-superposed landmark configurations, Procrustes-superposed outline semilandmark coordinates, Procrustes-superposed 3D semilandmark grid coordinates, and processed versions of raw digital images collected from a test set of planktonic foraminiferal specimen images representing seven recent species. Performance criteria were judged based on reference to the number of correct post hoc identifications achieved by the datasets when submitted to either a jackknifed canonical variates analysis (CVA; see Manly 1997) for the morphometric datasets, or a jackknifed plastic self-organizing map-based artificial neural network (PSOM-ANN) discrimination system for the processed image dataset (Lang 2007; O'Neill 2007).

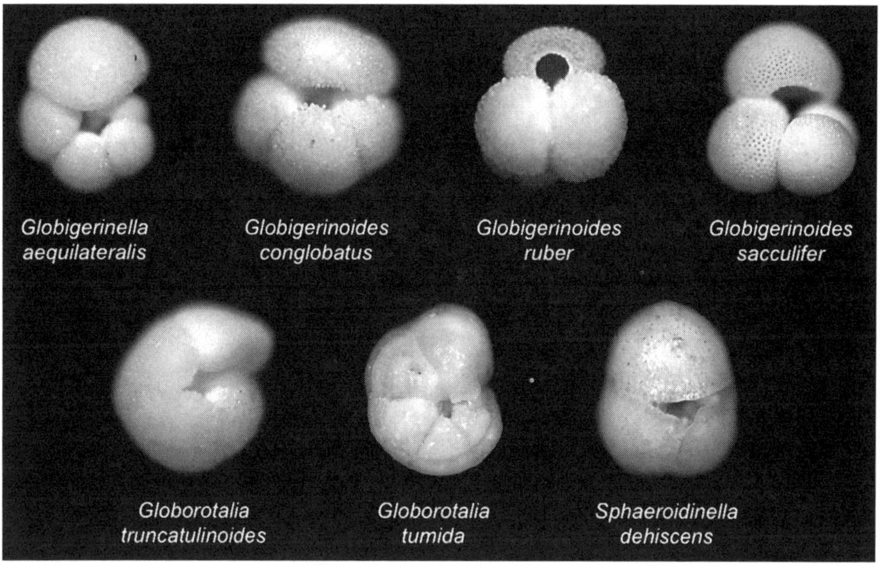

FIGURE 10.1. Plate of representative planktonic foraminifer specimens from each of the seven species considered in this investigation.

MORPHOMETRICS, HOMOLOGIES, AND LANDMARKS: AUTO-READING AN IMAGE

A small sample of 203 color, digital images of randomly selected specimens of seven recent planktonic foraminifer species (fig. 10.1) served as the set of test subjects for this study. These were picked from a Vema Cruise core-top sample (Sample No. V24–99 50) collected from the Baltimore Canyon, off the coast of the U.S. state of New Jersey and oriented in the standard umbilical view. All images were taken with a digital video camera at relatively low resolution (72 dpi).

2D Data

Most quantitative summaries of form variation that have been published in the taxonomic literature have been drawn from two-dimensional representations of organismal morphology. These representations have been collected usually as sets of scaled distances between landmark points, the 2D coordinate values of landmark points themselves, or strings of semilandmark coordinate points placed or otherwise located along a specimen's boundary. In this context it is important to appreciate the difference between landmark and semilandmark data. Landmarks are the locations of individual features of the form that can be located uniquely, without reference to other landmarks. Bookstein (1990b, 1991) identified three types of landmarks: unique point values at intersections of bones or tissues (Type 1), maxima or minima of curvature along a continuous boundary (Type 2), and extreme points (Type III, e.g., tip of a claw, crest of a cusp, or center of an orbit or eye). Mardia and Dryden (1998) proposed an alternative landmark classification, distinguishing between anatomical landmarks defined via reference to some aspect of an organism's biological morphology (= Bookstein's Type I landmarks) and mathematical landmarks that are defined via reference only to a structure's geometry (= Bookstein's Type II and III landmarks).

While many practitioners of geometric morphometrics assume all true landmarks are biological homologues, this is not actually the case as the homology concept pertains to entire structures (e.g., humerus bone) or regional features of structures (e.g., humerus proximate articular surface) but not to the mathematical points placed—for the most part arbitrarily by ad hoc criteria—on such structures (see MacLeod 1999). These individual point locations correspond to a concept of topological homology that, when used together with the unitary structure-based formalism of biological homology, allows detailed, point-by-point anatomical comparisons to be made. The significance of these distinctions between biological homology and topological homology, however, must be kept in mind when planning, and especially when interpreting, the results of a morphometric investigation.

Geometric morphometricians also recognize another broad category of landmarks, termed semilandmarks by Bookstein (1990a, 1990b, 1991) and pseudolandmarks by Mardia and Dryden (1998). These are geometrically constructed points such as the internal coordinate positions that subdivide a boundary whose ends are defined by either anatomical or mathematical landmarks into *n* segments of equal length. Conceptually, pseudo/semilandmarks differ from true landmarks in that the latter conform to the idea of a single, unique point that can be located on a form in some position that corresponds to a unique, topologically homologous feature. In the case of semilandmarks, though, it is the entire pseudo/semilandmark string that represents the unique, topologically homologous feature. Once defined and located, landmarks and/or semilandmarks can be used to define, represent, and characterize a wide variety of systematically important aspects of species' form, size, and shape (see MacLeod 2002a for examples).

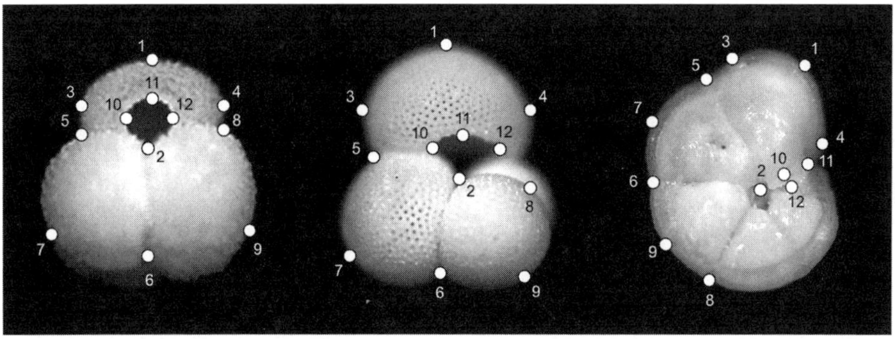

FIGURE 10.2. Landmarks used to quantify shape variation in the planktonic foraminifer dataset: 1—ultimate chamber peripheral midpoint, 2—primary aperture midpoint (umbilical side), 3—ultimate chamber max. chord (left), 4—ultimate chamber max. chord (right), 5—peripheral intersection between ultimate and penultimate chambers, 6—peripheral intersection between penultimate and antepenultimate chambers, 7—penultimate chamber peripheral midpoint, 8—peripheral terminus of antepenultimate chamber, 9—antepenultimate chamber peripheral midpoint, 10—primary aperture max. chord (left), 11—primary aperture midpoint (peripheral side) , 12—primary aperture max. chord (right). Species illustrated are *Gd. ruber* (left), *Gd. sacculifer* (middle), and *Gr. tumida* (right).

For the 2D morphometric analyses, three sets of 2D data were collected. The first of these expressed specimen shape via reference to a series of six standard shape or form factors (Folk 1968; Moelering and Rayner 1979). Form factors are simple ratios of standard geometric properties of closed forms (e.g., area, perimeter, Feret diameter). The specific form factors used in this study were the form index (l/w), form ratio (A/l^2), grain size index (p/l), circularity index ($((w/l^2)^{0.5})$), ellipticity index ($\pi(0.5l)l/A$), and the thinness ratio ($4\pi(A/p^2)$), where A = area, l = length of long axis, p = perimeter, w = length perpendicular to long axis.

The second 2D morphometric dataset consisted of the x,y coordinate locations of twelve discrete landmarks representing, variously, the intersections of chambers, extremes of curvature, and semilandmarks (fig. 10.2). These points expressed the size, relative location, and shape of the last three chambers of the foraminiferan shell (or test) and of the primary aperture.

The third 2D morphometric data type represented foraminiferal test shape by a string of 53 evenly spaced semilandmarks located along the specimen's periphery, with each string beginning at the midpoint of the ultimate chamber and with digitization proceeding in a counterclockwise manner (fig. 10.3). The number of semilandmarks used to quantify

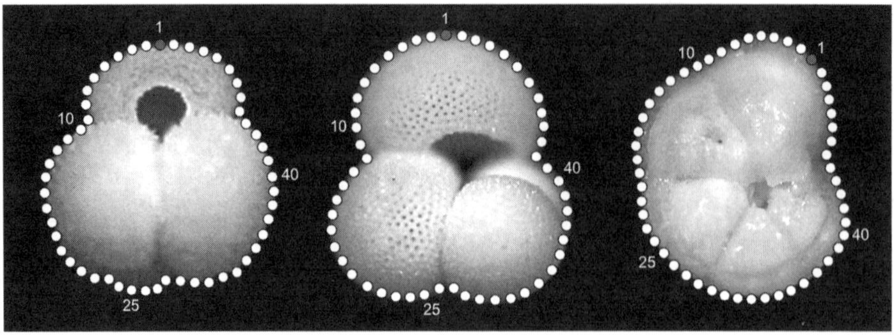

FIGURE 10.3. Outline semilandmarks used to quantify shape variation in the planktonic foraminifer dataset. Note the starting point for digitization was a landmark (ultimate chamber peripheral midpoint), and digitization proceeded in a clockwise direction. Note also that comparable points along the outline sequence (points 10, 25, and 40 are marked) do not fall at topologically or biologically corresponding positions relative to the specimens' structure. Whereas this might induce a source of error for some biological applications, this lack of correspondence is irrelevant from the standpoint of outline characterization for the purpose of group separation via differences in outline geometry. Species illustrated as in figure 10.2.

the specimen outlines derived from an empirical calculation of the minimum number of equally spaced semilandmarks needed to represent all outlines in the dataset to a consistent value of 97.5 percent of the original image outline length (see MacLeod 1999).

3D Data

Qualitative taxonomists have access to, and routinely make use of, assessments of 3D comparisons among specimens to inform their identifications. To determine the importance of treating these 3D objects as 3D objects, two datasets were collected. The first of these was a set of 3D semilandmark points arranged into a rectilinear grid that spatially adapts to each specimen's form.

As a first step in this process, an Alicona Infinite Focus Microscope (www.alicona.com/home/products/InfiniteFocus/InfiniteFocus-Standard-System.en.php) was used to collect 3D scans of the umbilical surfaces of each foraminiferan specimen. Scan resolutions varied but typically exceeded 10,000 vertices (fig. 10.4). These scans were used as a basis for representation of each specimen for a topologically homologous set of 3D semilandmark points arranged into an adaptive sampling grid.

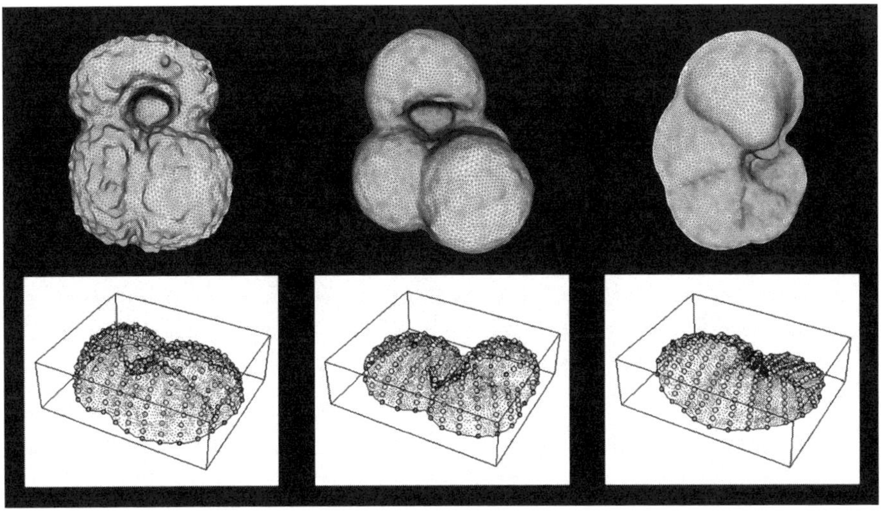

FIGURE 10.4. Eigensurface semilandmark sampling grids used to quantify 3D umbilical surface variation in the planktonic foraminifer dataset. Upper row: meshed 3D scans of representative *Gd. ruber* (left), *Gd. sacculifer* (middle), and *Gr. tumida* (right) specimens. Lower row: umbilical side half-scans of same specimens with a 15-point eigensurface adaptive semilandmark sampling grid superimposed on the virtual specimen model. For each specimen the set of semilandmark points is the same size (245 points), and each point bears a consistent topological position with respect to every other point in the sampling grid but not necessarily with respect to the to the specimens' structure. As with the outline semilandmark data (see fig. 10.2), this lack of correspondence is irrelevant from the standpoint of outline characterization for the purpose of group separation via differences in surface geometry. Indeed, in many, if not most, cases there is no way to judge the biological correspondence between semilandmark point locations as the notion of biological homology refers to unified structure (e.g., ultimate chamber), not point locations within structures.

This sampling grid adaptation proceeded in four steps (see MacLeod 2008 for a more complete discussion) that were applied consistently to each specimen. First, two landmarks were selected at the coordinate locations where a linear chord that divides the form into two halves in the umbilical view intersect the specimen periphery (see fig. 10.4). These landmarks divide the outline into two halves. Next, a series of equally spaced semilandmarks were located along the specimen peripheries. In this analysis, a resolution value of fifteen semilandmarks along each half-outline was used. This was an arbitrary selection designed to minimize the number of semilandmark points used to represent the overall specimen form while at the same time capturing taxonomically important aspects of gross morphology. However, it would be a simple matter to allow the complexity of the outline shapes themselves to determine the resolution required to meet a user-specified tolerance criterion (see MacLeod 1999 for an example of this procedure).

The next step in the 3D grid adaptation process involved projection of a chord joining the two boundary landmarks out of the *x,y* plane onto the surface of the scan (i.e., the virtual specimen), interpolating the positions of the chord coordinates from the positions of the scan vertices using a moving window that was passed down the chord. This is the "backbone" or midline chord of the network. Once this interpolation had been accomplished, the backbone chord was further interpolated into a set of fifteen semilandmark points to match the fifteen semilandmark coordinate positions located along the specimen periphery (see fig. 10.4). Together, these three chords specify the resolution and basic geometry of the adaptive 3D surface-sampling grid.

In a final step of the grid-specification process a series of rib chords were specified that joined the semilandmark locations along the backbone to those on the peripheries on either side. As with the backbone chord itself, these rib chords were projected out of the *x,y* plane onto the surface of the scan (i.e., the virtual specimen), interpolating the positions of the chord coordinates from the positions of the scan vertices within a moving window that was passed down each rib. The number of interpolated grid semilandmarks was set to conform to the average interlandmark spacings found along the periphery and backbone chords. In a final step, the number of rib semilandmark points specified along each topologically homologous rib segment was compared and each grid semilandmark sequence reinterpolated to a constant number of semilandmark points that equaled the maximum number of rib semilandmarks specified for the collection of corresponding rib sequences across the entire dataset.

At the end of this interpolation procedure, each specimen's 3D surface was represented by a set of semilandmark points equal in number and comparable in topological arrangement across all specimens. As can be appreciated from figures 10.1 and 10.4, in some species the primary aperture is a prominent feature of the test morphology in umbilical view. So that each grid network would contain the same number of points, the lumens of the apertures of these specimens were "floored" by forcing the space to be represented by a flat membrane drawn normal to the specimen's periphery at the level of the minimum *z*-coordinate level of the aperture's periphery. For the dataset used in this investigation, the procedure described above resulted in the specification of 245 3D semilandmarks per specimen.

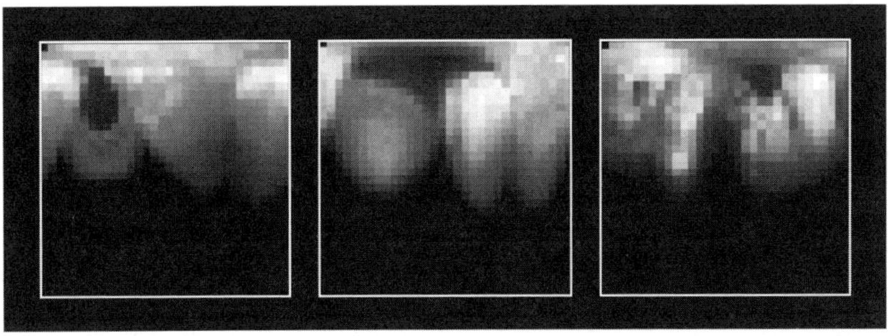

FIGURE 10.5. Processed image-based representations of umbilical surface morphology used to quantify shape variation in the planktonic foraminifer dataset. See text for a description of the processing steps involved in the transformation between the representations of the specimens shown in figure 10.2 and those of the same specimens shown here. For these data the RGB (color images) or grayscale (monochrome images) brightness values were used as quasi-3D shape variable descriptors for PSOM-ANN analysis.

Processed Images

In addition to these data the raw digital images from which the 2D morphometric data were obtained were processed and the processed images submitted to a plastic self-organizing map algorithm (PSOM; see Lang and Warwick 2002; Lang 2007). The PSOM is a relatively new variant of the classic Kohonen self-organizing map (Kohonen 1982) that has highly attractive scalability characteristics (Lang 2007; O'Neill 2007). Prior to PSOM analysis, image processing involved a three-stage transformation: (1) subsampling of the original images to a 32 × 32 square pixel grid, (2) equalization of the image histogram, and (3) conversion of the digital pixel grid from a Cartesian to a polar coordinate format (fig. 10.5). For these data, the pixel location and brightness values constituted the "variables" employed in the PSOM analysis. Of course, brightness values assigned to each pixel in the processed image dataset include information derived from the third (z) axis of morphological variation. But this information is complexly confounded with many other factors that affect the pixel brightness values exhibited by any region of the form under any set of local conditions (e.g., lighting and orientation).

Data Analysis Methods

Multivariate data processing for the 2D landmark and outline semilandmark datasets and the 3D surface semilandmark datasets consisted of Procrustes superposition of the landmark/semilandmark point clouds (Rohlf 1990; Rohlf and Slice 1990) to standardize for position, size, and rotation with subsequent submission of the superposed data to a canonical variates analysis (CVA; see MacLeod 2007). The form factor dataset was submitted directly to CVA. Two-dimensional outline and 3D semilandmark data were first processed through principal components analysis (PCA; see MacLeod 2005a, 2010) to reduce the dimensionality of the data prior to CVA analysis. For these datasets, scores and the number of principal components sufficient to summarize 95 percent of the observed shape variation were retained and used as CVA input. Owing to sample size limitations, all specimens were used to train the CVA and PSOM algorithms. However, identification performance was assessed using an identical jackknife sampling procedure (Manly 1997) for both morphometric and machine learning trials.

RESULTS

Form Factor Data

In the case of the example planktonic foraminifer dataset each image was read into MediaCybernetics' *Image-Pro Plus* software (vers. 7.0, www.mediacy.com/index.aspx?page = IPP) and transformed from a color (RGB) format to a binary (0/1) format by using a threshold value to map the range of pixel brightness values to black (0) or white (1). A boundary-mapping algorithm was then used to identify pixel coordinate locations that represented each specimen's closed-form boundary. After each boundary had been established standard *Image-Pro Plus* algorithms were used to estimate each specimen's area, perimeter, length of the long axis, and length perpendicular to the long axis. From these values the six form factors listed in the "Morphometrics, Homologies, and Landmarks" section above were calculated. These ratios served as the basis for a CVA of within-group dispersion relative to between-groups separation (see MacLeod 2007).

The first three axes of the CVA discriminant space formed as a result of the analysis of the form factor data are shown in figure 10.6. Together these axes represent 96.13 percent of the variation between group centroids within a total space that contains six dimensions. Obviously some species were able to be better discriminated by the form factor data (e.g., *Gr. tumida, Gn. aequilateralis, S. dehiscens*) than others (e.g., *Gd. conglobatus, Gd. ruber, Gd. sacculifer*). Nevertheless, the φ

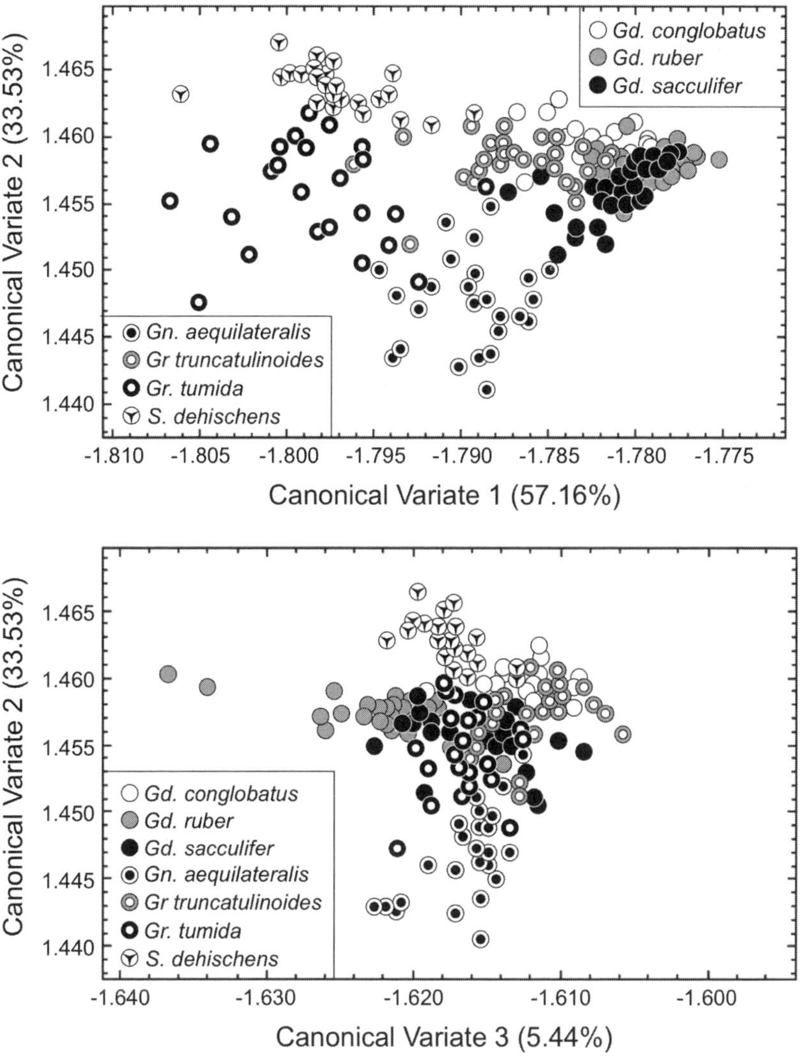

FIGURE 10.6. Species-group ordinations in the subspace created from the first three CVA axes derived from the form factor dataset. See text for discussion.

log-likelihood test (Manly 1994) of group discrimination rejects the null hypothesis of no difference between groups in terms of their form factor data at a very high level of confidence ($\varphi = 851.7$; dof = 32; $\rho = 100\%$). Inspection of the confusion matrix for the raw CVA results (table 10.1) indicates that *Gd. ruber* exhibited the lowest correct

TABLE 10.1 CONFUSION MATRICES FOR RAW AND JACKKNIFED (IN PARENTHESES) ESTIMATES OF FORAMINIFER IDENTIFICATIONS BASED ON FORM FACTOR DATASET

Group	*Globigerinoides conglobatus*	*Globigerinoides ruber*	*Globigerinoides sacculifer*	*Globigerinella aequilateralis*	*Globorotalia truncatulinoides*	*Globorotalia tumida*	*Sphaerodinella dehiscens*	Total Estimated	Total Correct	Percent Correct
Globigerinoides conglobatus	23 ([illegible]3)	3 (3)	2 (3)	—	3 (2)	—	—	31	23 (23)	74.19 (74.19)
Globigerinoides ruber	5 ([illegible])	24 (22)	10 (10)	—	—	—	—	39	24 (22)	61.54 (56.41)
Globigerinoides sacculifer	3 ([illegible])	4 (4)	25 (24)	– (1)	1 (1)	—	—	33	25 (24)	75.76 (72.73)
Globigerinella aequilateralis	—	—	—	25 (24)	1 (2)	1 (1)	—	27	25 (24)	92.59 (88.89)
Globorotalia truncatulinoides	2 (3)	—	2 (4)	—	20 (17)	2 (2)	—	26	20 (17)	76.92 (65.38)
Globorotalia tumida	—	—	—	2 (2)	1 (1)	17 (17)	4 (4)	24	17 (17)	70.83 (70.83)
Sphaerodinella dehiscens	—	—	—	—	2 (2)	—	21 (21)	23	21 (21)	91.30 (91.30)
Total Estimated	33 (36)	31 (29)	39 (41)	27 (27)	28 (25)	20 (20)	25 (25)	203	155 (148)	76.35 (72.91)
Total Correct	23 (23)	24 (22)	25 (24)	25 (24)	20 (17)	17 (17)	21 (21)	155 (148)		
Percent Correct	69.70 ([illegible]3.89)	77.42 (75.86)	64.10 (58.54)	92.59 (88.89)	71.42 (68.00)	85.00 (85.00)	84.00 (84.00)	76.35 (72.91)		

identification consistency (61.54%, as assessed by proximity of the specimen locations within the discriminant space relative to the seven group centroids) and *Gn. aequilateralis* the highest (92.59%). Overall the apparent discrimination efficiency for these discriminant functions when applied to the entire form factor training set is 76.35 percent.

Despite this encouraging result, the performance of form factors as morphological indices would fall off rapidly if more species were included in the analysis. Because form factors ignore all aspects of the underlying biology of the morphologies they are used to quantify, they often force comparisons to be made between directions, distances, and landmarks that make no sense biologically and that no systematist would be comfortable relying on. This problem is signaled in the raw CVA results by the comparison of *Gd. conglobatus*, *Gd. ruber*, and *Gd. sacculifer* identifications. All three of these congeneric species exhibit broadly similar globigerine morphologies (see fig. 10.1). Of the 31 total misidentifications of specimens belonging to these three species, over 90 percent were misassigned to one or the other globigerinoid species. Together, misidentifications just among these three globigerinoid species accounted for almost 60 percent of the total misidentifications.

It would be overly optimistic to imagine that the true performance of these discriminant functions can be assessed by applying to the same data used to estimate their orientation. Accordingly, a "leave one out" jackknife test of the stability and performance of these data at achieving robust between-group discriminations was also undertaken. These jackknife results are also shown in table 10.1 (in parentheses). In this case the jackknife estimate suggests a general lowering of identification performance estimates by a small amount (approx. 3.5%). Here again we see the shadow of the lack of geometric specificity with respect to this data type insofar as the identification accuracy for all the globigerinoid species declined under the jackknife protocol, whereas that of all other species in the training set either increased or remained the same.

Procrustes Superposed Landmark Data

The landmark dataset was much more labor intensive to collect, requiring that each landmark location be marked on each image by hand (using tpsDIG software, http://life.bio.sunysb.edu/morph/soft-dataacq.html), based on the analyst's knowledge of foraminiferal morphology along with a set of a priori decisions regarding the locations of topologically accurate and biologically meaningful correspondences across all specimens in the sample. Once these data had been collected, the raw form coordinate configurations were transformed into shape coordinates using the Procrustes (GLS) method (see above) with the shape coordinates being used as input into a CVA analysis.

A much better result was achieved using the landmark dataset, the first three axes of the discriminant space formed from which are shown in figure 10.7. Together these axes represent 85.28 percent of the variation between group centroids within a discriminant space that, again, contains a total of six dimensions. Groups are well separated by these data with only *Gn. aequilateralis*—*Gr. tumida* and *Gd. conglobatus*—*Gd. sacculifer* exhibiting any appreciable apparent shape overlap. Reflecting this improved between-groups discrimination, the value of the φ log-likelihood test statistic is more than double that of the form factor dataset ($\varphi = 1{,}791.0$; dof = 48; $\rho = 100\%$), providing excellent statistical confidence that the observed differences between species-specific landmark configurations are real. When all six discriminant dimensions are added and identifications determined on the the basis of proximity to group centroids over 95 percent of the training set is classified correctly on the basis of landmark shape configuration similarity (table 10.2), a figure that also stands up well when the more robust "leave one out" jackknife test is applied to the training set data.

In its own way this level of performance is even more remarkable than the form factor result. Given the greater reliance on biological information that is required by the landmark approach and given the larger number of variables used to characterize between-group variation, a better result was expected than the more generic form factor approach. The achievement of a near-perfect result was unanticipated given that these landmarks play little or no role in the formal definition of these species or in the qualitative recommendations for their discrimination (Saito, Thompson, and Breger 1981; Kennett and Srinivasan 1983). In this case it would appear that morphometric analysis has discovered a set of diagnostic shape-based characters that have been overlooked by qualitative taxonomists, albeit ones that can only be assessed in the context of a quantitative shape analysis. However, there is also reason to be suspicious of the degree to which this result can be generalized. Selection of the species to be included in this trial was made from the subset of planktonic foraminiferal species that exhibit multiple chambers in the final whorl that are visible in umbilical view.

TABLE 10.2 CONFUSION MATRICES FOR RAW AND JACKKNIFED (IN PARENTHESES) ESTIMATES OF FORAMINIFER IDENTIFICATIONS BASED ON *PROCRUSTES* SUPERPOSED LANDMARK DATASET

Group	*0*	*0*	*0*	*0*	*0*	*0*	*0*	Total Estimated	Total Correct	Percent Correct
Globigerinoides conglobatus	29 (29)	1 (2)	1 (–)	—	—	—	—	31	29 (29)	93.55 (93.55)
Globigerinoides ruber	—	39 (39)	—	—	—	—	—	39	39 (39)	100.00 (100.00)
Globigerinoides sacculifer	1 (1)	—	32 (32)	—	—	—	—	33	32 (32)	96.97 (96.97)
Globigerinella aequilateralis	—	—	—	24 (24)	—	3 (3)	—	27	23 (23)	85.19 (85.19)
Globorotalia truncatulinoides	—	—	—	—	26 (26)	—	—	26	26 (26)	100.00 (100.00)
Globorotalia tumida	—	—	—	1 (1)	—	23 (23)	—	24	23 (23)	95.83 (95.83)
Sphaerodinella dehiscens	—	—	—	—	—	—	23 (23)	23	23 (23)	100.00 (100.00)
Total Estimated	30 (30)	40 (41)	33 (32)	25 (25)	26 (26)	26 (26)	23 (23)	203	195 (195)	96.06 (96.06)
Total Correct	29 (29)	39 (39)	32 (32)	24 (24)	26 (26)	23 (23)	23 (23)	195 (195)		
Percent Correct	96.67 (63.89)	97.50 (75.86)	96.97 (58.54)	96.00 (96.00)	100.00 (100.00)	88.46 (88.46)	100.00 (100.00)	96.06 (96.06)		

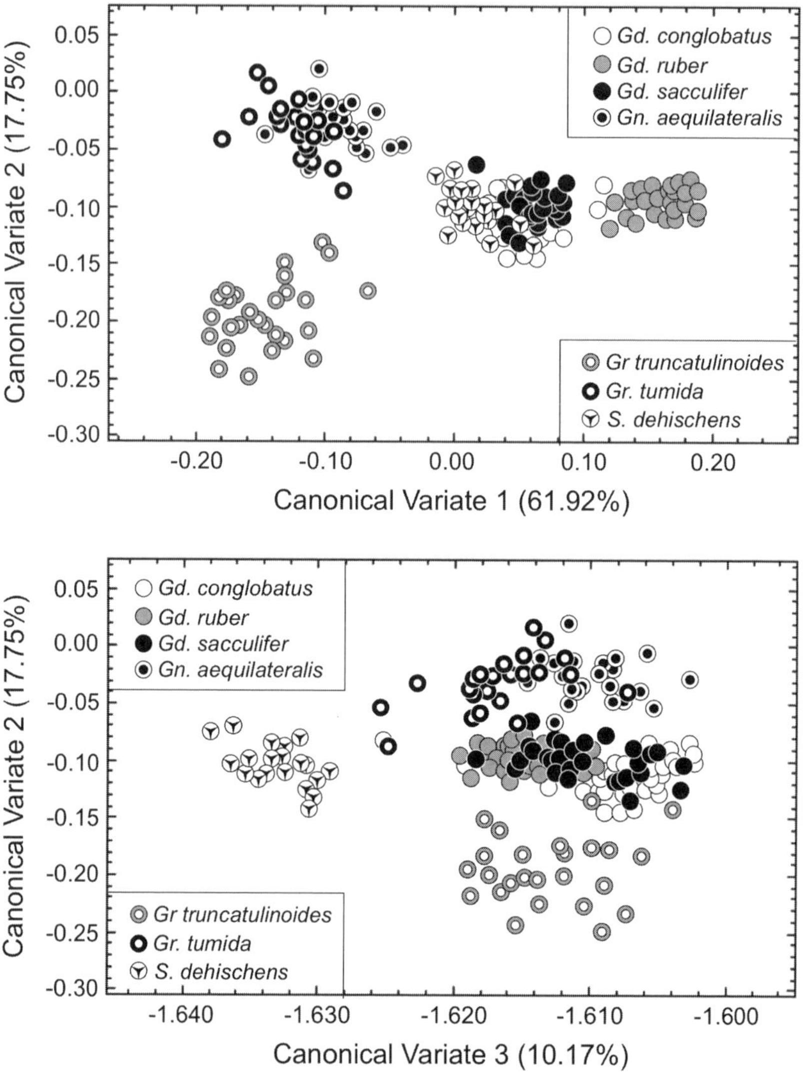

FIGURE 10.7. Species-group ordinations in the subspace created from the first three CVA axes derived from the Procrustes superposed landmark dataset. See text for discussion.

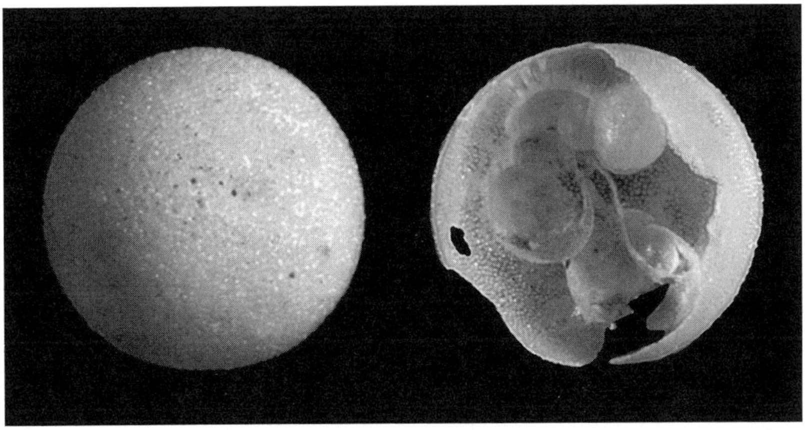

FIGURE 10.8. *Orbulina universa*, a planktonic foraminifer species in which the ultimate chamber envelops all previous chambers completely, thus preventing assessment of the aspects of the test morphology used in the landmark analysis to quantify within-species morphological variation and between-species morphological differences. Complete adult specimen (left); Broken adult test revealing trochospirally coiled juvenile and subadult chambers (right).

Unfortunately, not all planktonic foraminifer species exhibit test morphologies that conform to these standards. An obvious example is *Orbulina universa* (fig. 10.8), in which the ultimate chamber envelops completely all previous chambers that are arranged in a typical globigerinoid trochospiral coil. Because this species does not exhibit the requisite three chambers in the final whorl that are visible in umbilical view—indeed adult *O. universa* specimens lack an umbilicus—the landmarks required for comparison with the other species cannot be located. In such cases either *O. universa* would need to be excluded from consideration in a landmark-based analysis (in which case a full discrimination between planktonic species could not be achieved) or a smaller number of landmarks restricted only to the ultimate adult test chamber would need to be used to quantify geometric distinctions across all species in the training set irrespective of the forms of the majority of species represented therein. If this latter sampling strategy were followed it is very likely that the outstanding result achieved by the full, three-chamber, twelve-landmark dataset would be compromised. It should also be noted that the time and training it takes analysts to locate and record the position of each landmark individually is considerable. If a completely automated identification system was required, devising the image segmentation and point location algorithms required to automatically establish these landmark locations would present a formidable computer vision-style programming challenge in its own right.

Procrustes Superposed Outline Semilandmark Data

Form representation based on equally spaced outline semilandmarks, in a sense, represents the best of both the worlds explored by the previous datasets. Outlines can be located and quantified in either semiautomated or fully automated ways, and, so long as the analyst is careful about placing the specimens on a contrasting background, there are a variety of tools and approaches that can be used to locate and quantify specimen outlines. Also, once quantified, outline semilandmark coordinates can be superposed, scaled, and rotated to orientations of maximum geometric correspondence easily, again using either semiautomated or fully automated algorithms (e.g., Lohmann 1983; MacLeod 1999).

For the example planktonic foraminifer dataset, outlines were captured from the raw images in a semiautomated mode by loading the image into tpsDIG placing the mouse cursor just to the right of the topologically homologous point that was used as the starting point for outline representation and clicking the mouse button. The software automatically drew a chord from the cursor across the frame until it encountered the specimen boundary, which was signaled by a spatially abrupt transition from dark to light pixels. It then followed this boundary around the specimen's image counterclockwise, collecting the coordinate positions of the outline pixels as it went. These coordinate locations were later used to interpolate each outline to a constant number of equally spaced boundary semilandmark points. Once the entire dataset had been collected, the outlines were further aligned using the Procrustes (GLS) method and then submitted to a PCA to reduce the dimensionality of the data. For this dataset the PCA results showed that 95 percent of the observed shape variation was represented in the first 23 Procrustes PCA axes. Scores of individual outline configurations for each of these 23 latent shape vectors were used as the basis for a Procrustes CVA analysis.

The first three axes of the discriminant space formed for this dataset is shown in figure 10.9. Together this subspace represents 86.68 percent of the total variation between group centroids. While the outline data ordination of training set specimens in this portion of the discriminant subspace might not exhibit quite the level of intergroup distinctiveness as the landmark result (compare with fig. 10.7), performance analysis of the complete six-dimensional space indicates this dataset is fully as efficient as the landmark dataset. The pattern of correct and misidentifications differs but only slightly and only for five of the seven species (compare with table 10.3). The φ log-likelihood test statistic for these outline semilandmark data (φ = 1,480.0; dof = 138; ρ = 100%) is a bit lower than that of the superposed landmark dataset but still much higher than the comparable form factor dataset value. This slight, and statistically insignificant, drop in the overall confidence of group discrimination most likely reflects the higher dimensionality of the outline data. A jackknife test of discriminant function efficiency and robustness suggest that these raw results are accurate estimates of the proportions of correct identifications that would be expected in standard application of the discriminant space to sets of unknown specimens.

While it is true the landmark dataset returned a slightly more significant result than the outline dataset, the discrepancy is minor. Both results are well above that which could reasonably be expected from human operators under even the best of circumstances and the ease with which outline data collection can be automated would likely render it the preferred option of the two for all but the most critical discrimination tasks. Moreover, unlike the landmark dataset, all foraminifer specimens, no matter how morphologically complex, possess an outline. Integration of *O. universa* (see fig. 10.8) into the discriminant system using an outline sampling approach poses no particularly difficult challenges for outline analysis.

In the context of evaluating the identification accuracy of the landmark and outline results, it is interesting to note that that biological conformance of the outline was provided by the location of a (in this case) single landmark. It is a much easier task to develop a fully automated means of matching outlines, either by developing an image processing algorithm that will pick out a particular class of landmarks reliably, or by relying on brute force rotation of outlines to positions of maximal correspondence with a reference specimen. Regardless, it is remarkable that such a high-quality result was achieved despite the fact that, unlike the landmark dataset, the degree of biological correspondence between particular regions of the outline diminishes for semilandmarks located at positions remote from that of the single reference landmark (see fig. 10.3). This provides further support for the idea that precise conformance to the strictures of formal biological homology is not as important in terms of solving the species identification problem as some authors have suggested (Bookstein 1991; Zelditch et al. 2004).

TABLE 10.3 CONFUSION MATRICES FOR RAW AND JACKKNIFED (IN PARENTHESES) ESTIMATES OF FORAMINIFER IDENTIFICATIONS BASED ON *PROCRUSTES* SUPERPOSED OUTLINE SEMILANDMARK DATASET

Group	*Globigerinoides conglobatus*	*Globigerinoides ruber*	*Globigerinoides sacculifer*	*Globigerinella aequilateralis*	*Globorotalia truncatulinoides*	*Globorotalia tumida*	*Sphaerodinella dehiscens*	Total Estimated	Total Correct	Percent Correct
Globigerinoides conglobatus	29 (29)	2 (2)	—	—	—	—	—	31	29 (29)	93.55 (93.55)
Globigerinoides ruber	1 [illegible]	38 (38)	—	—	—	—	—	39	38 (38)	97.44 (100.00)
Globigerinoides sacculifer	– (1)	—	32 (32)	—	—	—	—	33	32 (32)	96.97 (96.97)
Globigerinella aequilateralis	—	—	—	26 (26)	—	—	1 (1)	27	26 (26)	96.30 (96.30)
Globorotalia truncatulinoides	—	—	—	1 (1)	24 (24)	—	1 (1)	26	24 (24)	92.31 (92.31)
Globorotalia tumida	—	—	—	—	—	24 (24)	—	24	24 (24)	100.00 (100.00)
Sphaerodinella dehiscens	—	—	—	—	—	—	23 (23)	23	23 (23)	100.00 (100.00)
Total Estimated	30 (30)	40 (40)	32 (32)	27 (27)	24 (24)	24 (24)	25 (25)	203	196 (196)	96.55 (96.55)
Total Correct	29 (29)	38 (38)	32 (32)	26 (26)	24 (24)	24 (24)	23 (23)	196 (195)		
Percent Correct	96.67 (96.67)	95.00 (95.00)	100.00 (100.00)	96.30 (96.30)	100.00 (100.00)	100.00 (100.00)	100.00 (92.00)	96.55 (96.55)		

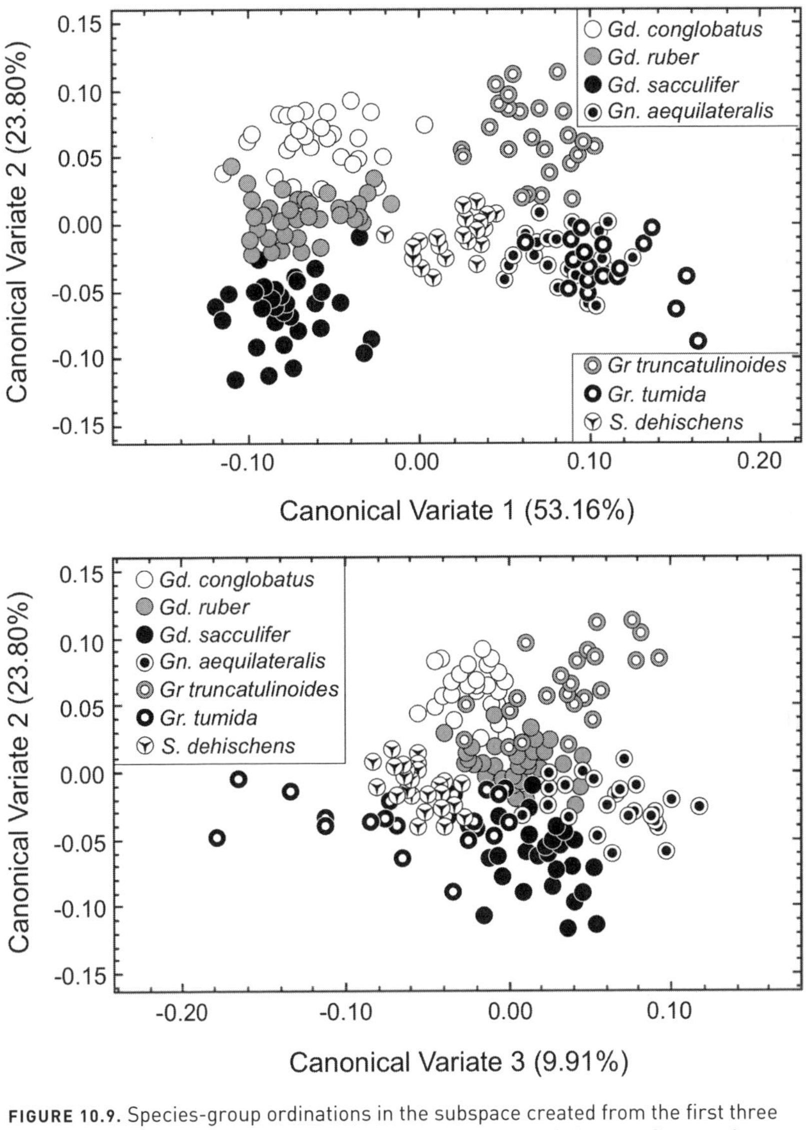

FIGURE 10.9. Species-group ordinations in the subspace created from the first three CVA axes derived from the Procrustes outline semilandmark dataset. See text for discussion.

Moreover, methods are available to take advantage of multiple landmarks located on (MacLeod 1999) or off (MacLeod 2001; Figueirido et al. 2011) the outline in order to achieve a biologically more reasonable cross-specimen morphological mapping. Typically improvement of the outline matching improves the quality and stability of outline-based group discriminations, often dramatically.

Eigensurface (3D) Data

In a final morphometric analysis, 3D eigensurface semilandmark grids were used to determine whether the addition of the third dimension of geometric information would result in improvement of performance of morphometric data. As noted above, a total of 245 semilandmarks distributed over the umbilical surface were used to quantify both within-species and between-species morphological variation (see fig. 10.4).

Figure 10.10 shows a discriminant space formed by the first three axes of a six-axis CVA solution. The subspace formed by these three axes represent 75.51 percent of the between-group centroid variation. Visually, intergroup discrimination appears excellent, with all groups exhibiting discrete and unified ranges of variation within the ordination space. This ordination conforms to the implicit image most systematists have of interspecies distinctions, but rarely see within the data they collect and analyze.

The ordination of specimens within this subspace is comparable to that of the landmark dataset (compare with fig. 10.7), though interestingly the structure of similarity relations between these datasets differs in certain key respects. In addition to a greater general level of intergroup distinctiveness, note the change in the relative position of *Gn. aequilateralis* in the two spaces. The 2D landmark data suggest the shape of this species is most similar to that of *Gr. tumida* and distinct from the central pair of morphologically similar globigerinoid species (*Gd. conglobatus, Gd. sacculifer*). However, the 3D dataset suggests *Gn. aequilateralis* and *S. dehiscens* are more similar morphologically and occupy a central position in the shape space relative to the five other species in this dataset. This discrepancy between results obtained from different sampling schemes applied to the same set of specimens illustrates the degree to which ordinations obtained by numerical analysis are strongly dependent on the data being analysed and so emphasizes the need to take care both in designing quantitative hypothesis tests and interpreting the results of those tests. Both results are correct for the

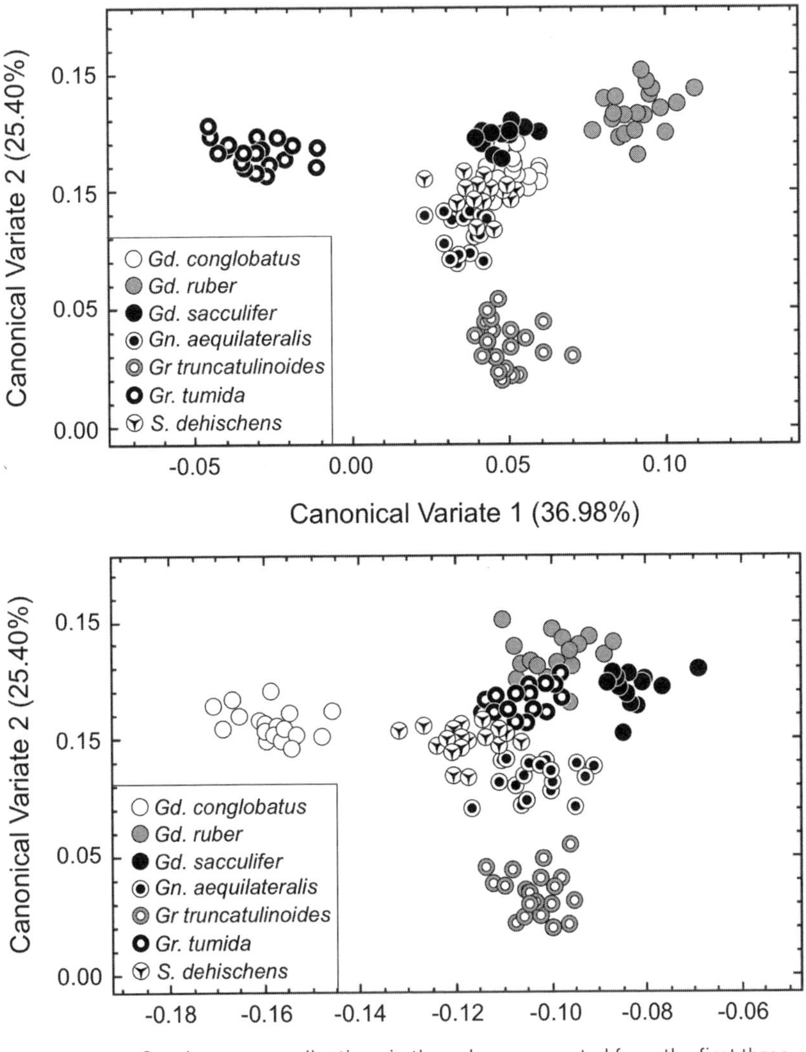

FIGURE 10.10. Species-group ordinations in the subspace created from the first three CVA axes derived from the Procrustes superposed eigensurface semilandmark dataset. See text for discussion.

data to which they apply despite the fact they are mutually incompatible when compared to each other. As for which is a better representation of the shape similarities among these species, this question is, in a sense, unanswerable. Both approaches—along with the results obtained from analysis of the specimen outlines and, to some extent, the form factor data—have inherent advantages and disadvantages that could be regarded as either more or less important in a wide variety of biological contexts. The judgment of appropriateness in every instance lies in the match between the data being collected and the purpose for which the data are required, not in any a priori estimation of what type of data is "best" in any abstract sense.

What can be said unambiguously, however, is that of any of the data collected so far the 3D eigensurface grid data have returned the discriminant ordination that best represents the shape-based distinctions among the a priori groups within which these specimens have been placed. The φ log-likelihood test of intergroup centroid distinctiveness relative to intragroup dispersion (φ = 1,603.0;, dof = 156; ρ = 100%) is somewhat less than the comparable value obtained from the landmark dataset owing mostly to the larger number of variables needed to summarize 95 percent of the observed 3D shape variation (26 as opposed to 8 for the landmark data or 23 for the outline data). Regardless, in all three cases the confidence level of the test statistic is extremely high.

When all six discriminant dimensions are used to place training set specimens into groups based on proximity to group centroids 100 percent of the training set is classified correctly on the basis of 3D surface shape (table 10.4). This result remains unchanged when the more robust "leave one out" jackknife test is applied. In addition, it is worth noting that this result was achieved in much the same way as the outline semilandmark result, with minimal input of biological information that would improve the alignment of each specimen's 3D grids with each specimen's intrinsic biology. While there is no barrier in principle to the use of additional landmark points to help align the 3D grids MacLeod (1999), algorithms for realizing this remain the subject of active research and development.

Image Data

In a final analysis of the planktonic foraminifer image set a collection of down-sampled, processed versions of the same raw images that served as the basis for the 2D morphometric analyses were used to train an unsuperized PSOM-based artificial neural network (PSOM-uANN). These images contain information derived from the 3D aspects of the specimens themselves in the form of shading geometries that are expressed as spatial structures in the pixel brightness values across the image. This signal is complexly confounded with many other sources of variation that affect these brightness values. The separation of the

TABLE 10.4 CONFUSION MATRICES FOR RAW AND JACKKNIFED (IN PARENTHESES) ESTIMATES OF FORAMINIFER IDENTIFICATIONS BASED ON *PROCRUSTES* SUPERPOSED EIGENSURFACE SEMILANDMARK DATASET

Group	*Globigerinoides conglobatus*	*Globigerinoides ruber*	*Globigerinoides sacculifer*	*Globigerinella aequilateralis*	*Globorotalia truncatulinoides*	*Globorotalia tumida*	*Sphaerodinella dehiscens*	Total Estimated	Total Correct	Percent Correct
Globigerinoides conglobatus	17 (17)	—	—	—	—	—	—	17	17 (17)	100.00 (100.00)
Globigerinoides ruber	—	20 (20)	—	—	—	—	—	20	20 (20)	100.00 (100.00)
Globigerinoides sacculifer	—	—	13 (13)	—	—	—	—	13	13 (13)	100.00 (100.00)
Globigerinella aequilateralis	—	—	—	19 (19)	—	—	—	19	19 (19)	100.00 (100.00)
Globorotalia truncatulinoides	—	—	—	—	20 (20)	—	—	20	20 (20)	100.00 (100.00)
Globorotalia tumida	—	—	—	—	—	23 (23)	—	23	23 (23)	100.00 (100.00)
Sphaerodinella dehiscens	—	—	—	—	—	—	20 (20)	20	20 (20)	100.00 (100.00)
Total Estimated	17 (17)	20 (20)	13 (13)	19 (19)	20 (20)	23 (23)	20 (20)	132	132 (132)	100.00 (100.00)
Total Correct	17 (17)	20 (20)	13 (13)	19 (19)	20 (20)	23 (23)	20 (20)	132 (132)		
Percent Correct	100.00 (100.00)	100.00 (100.00)	100.00 (100.00)	100.00 (100.00)	100.00 (100.00)	100.00 (100.00)	100.00 (92.00)	100.00 (100.00)		

z-relief signal from other sources of brightness variation is a task accomplished routinely by traditional taxonomists during their visual inspection of specimens, albeit one that usually requires inspection of the specimen in multiple orientations.

Owing to the nature of artificial neural networks, no ordination of specimens within a linear space can be produced to facilitate visual inspection of the discriminant space (Bishop 1996; Schalkoff 1997). Instead, an internal assessment of the confidence in the identifications made by the PSOM-uANN under the jackknife sampling protocol was produced and is summarized in figure 10.11 in the form of a frequency histogram. In this case the identification quality criterion was a "coordination number" that expresses the degree of "embeddedness" of the unknown image within a feature (map) space populated by the "locations" of training set images. If the *n* nearest neighbors to the position of the unknown data point within the PSOM have all been assigned to the same a priori group, this results in coordination value of *n* for the image. A large *n* (e.g., 10) is interpreted as a high-confidence identification as the unknown image occupies a region of the map space in which a single group predominates. Similarly a small *n* (e.g., 2) is interpreted as a low-confidence identification as the unknown image occupies a region of the map space in which multiple groups are present. For this analysis the coordination statistic was generated by assessing the identity of the 10 nearest training set neighbors. Under the settings employed in the foraminifer investigation, a coordination value of 10 represents a "perfect" score. This identification quality test also employs a threshold function that prevents images that fall outside group boundaries established by the analysis of the training set from being allocated to any group.

Results of this identification confidence test indicate that the overwhelming majority of PSOM-uANN identifications are of the highest quality (coordination = 10). Unsurprisingly given the previous morphometric results, the three globigerine species exhibit the largest numbers of relatively uncertain identifications for this dataset. Only three misidentifications (1.5% of the sample) were recorded in this trial—one for *Gd. conglobatus* and two from *Gd. sacculifer*—yielding an overall correct identification score of 98.52 percent. This result is comparable to the best 2D morphometric result reported thus far. Indeed, this is the most remarkable result of the investigation insofar as neither biological nor geometric information was provided to the PSOM-uANN to guide its development of its discrimination functions.

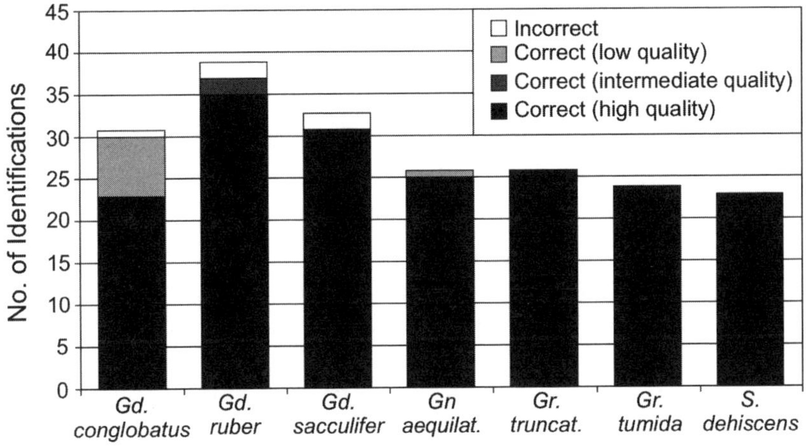

FIGURE 10.11. Histogram of PSOM-uANN identification corrections and confidence estimates for the planktonic foraminifer dataset. High-confidence identifications: coordination values 8–10. Intermediate-confidence identifications: coordination values 5–7. Low-confidence identifications: coordination values < 5. Note large proportion of very high likelihood identifications present for each species.

Because of this lack of guidance based on experience—and manifested in most morphometric analyses by the analysts' selection of what data to collect (e.g., landmarks, outlines, surfaces)—uANNs have long been known to require relatively large training sets into order to achieve maximum discrimination efficiencies (Bishop 1996; Schalkoff 1997). A crude estimate of what effect an increase in training set size might have on the correctness and the confidence of identifications produced by the uANN can be obtained by randomly morphing the training set images so they retain their basic identify but vary from one another in their spatial proportions and then using the morphed images as the uANN training set (see MacLeod 2007 for a detailed description of this morphing technique). Under such a morphing protocol pseudo–training sets of any desired size can be generated quickly and easily. Once trained, the original images—from which the morphed training set images were ultimately derived but none of which are actually included in the pseudo–training sets—can be submitted to the ANN as "known unknowns" and the performance of the simulated uANN determined.

For this analysis a spatial distortion factor of 5.0 percent in both the x and y image axes and a brightness distortion factor of 0.5 were used to produce five randomly distorted pseudo-specimen images for each of the 203 original training set images. This operation increased the total size of the (pseudo)training set to 1,015 images. When the original 203 test images were then submitted to the PSOM-uANN as unknowns all were identified correctly and with a coordination value of 10. These identifications took on the order of 0.5 seconds each and all are perfectly reproducible. While I am making no claim that this extremely high level of performance would be delivered by a system of 1,015 actual training set specimens, this simulation exercise does provide an indication of the fineness with which group discriminations can, in principle, be made for the foraminiferal taxa when recourse to large training sets is (or becomes) a possibility. As use of the same pseudo-specimen images for the morphometric analyses would only create denser clouds of training set points about nearly the same group means, repetition of this exercise would not alter those results appreciably.

DISCUSSION: AUTOMATED IDENTIFICATION?

There are several ways in which this technology can be applied in the context of phylogenetic systematics. Species are the ontological subjects of phylogenetic systematics. Collections of phena assumed to represent species are the points where most phylogenetic systematics investigations begin. The very act of naming and/or identifying species presumes that some type of discontinuity exists that allows members of the species to be unambiguously distinguished from all other species. But how often is this assumption tested in the context of a phylogenetic analysis beyond the immediate subjects of a particular investigation? In principle, automated species identification methods could provide a means of addressing this issue, of objectively determining the kinds of patterns that exist in living organisms and so the scope of subjects suitable for phylogenetic analysis. Moreover, as the automated identification systems are nothing more (and nothing less) than pattern recognition engines, the same techniques used to identify morphological species can be applied at all taxonomic levels—from the character state to family members to local populations through to kingdoms—to different data—from molecular sequences to gross anatomy through to the characteristic vocalizations made by some species and, in the case of humans, linguistic styles, and in a variety of biological and nonbiological contexts—from biogeography to art history (see MacLeod 2009). Humans cannot match the performance, scope, or speed of machine-based analyses in any of these contexts.

Results obtained by this investigation of different types of data and different analytic approaches that might be employed in the construction of automated biological species identification systems have confirmed some published opinions regarding the utility of such techniques and refuted others. Generally speaking, all approaches tested in this study returned good levels of performance, especially when considered against the—lamentably few—investigations of the performance of human taxonomists in delivering consistent and correct identifications of biological species (see above). These studies suggest that the worst result, obtained from the quantitative analysis (76.35%), was considerably better than average levels of consistency and accuracy than could be expected by groups of taxonomists working under normal conditions for the same groups of objects. The overall results of this investigation are also consistent with those obtained from performances reported for a variety of different quantitative discrimination algorithms applied to various types of data collected and analysed from a wide variety of biological species groups (Gaston and O'Neill 2004).

The description of biological species via geometric form factors is easy to implement in an automated species identification system, can be analyzed via reference to a wide range of numerical approaches, and deliverers consistently high levels of identification accuracy when the number of groups being analyzed is relatively small and the between-group geometric differences are large. These types of systems have implemented successfully to discriminate between broad categories of suspended marine particles including biological taxa (e.g., copepods, chaetognaths), taxonomically indeterminate products of organic activity (e.g., plankton eggs), and sedimentary detritus (e.g., fibers and aggregates; see Gorsky et al. 2010). In this study, form factors delivered the lowest the accuracy rates of the systems and combinations tested. Nevertheless, this comparatively low level of performance may have been due, in part, to the small number of form factor indices used to characterize within-group and between-group morphological variation. The fact that the form factors used here to distinguish between globigerine species that are similar in gross form, but readily distinguishable by qualitative visual inspection, suggests this approach to morphological discrimination may be limited if the purpose is to make fine distinctions between morphologically similar biological species. This potential limitation should be subjected to further exploration. While this issue by no means excludes form factors from being used for biological species identification, the inherent performance limitations of this data type when applied in biological contexts should be taken into consideration when making a selection of morphological descriptors to use in the construction of automated species identification systems.

Shape characterization using Procrustes superposed landmarks delivered a far better result than did the form factor data. This is consistent with the results of many other morphometric studies that have been undertaken using landmarks to characterize biological species (Bookstein 1991; Zelditch et al. 2004 and references therein). It is usually recommended that care be taken to define landmarks explicitly and, to the maximum extent possible, avoid use of semilandmarks in an overall landmark-based approach (Bookstein 1991, but see below). Regardless, the most serious problems with the use of landmarks as a general source of data with which to identify biological species are (1) the difficulty of devising automated image processing algorithms to locate landmarks quickly and accurately across a wide range of forms and (2) the difficulty of locating a sufficient number of common landmarks that can be used to characterize forms across a broad spectrum of species morphologies. The latter is a well-known limitation of landmark-based analysis (Bookstein 1991; Zelditch et al. 2004; MacLeod 1999). Landmark-based approaches will likely be most useful in the designs of systems intended to make fine distinctions between species whose characteristic morphologies are very similar to one another but of limited utility for use in general-purpose automated systems (e.g., Van Bocxlaer and Schultheiß 2010).

Outline semilandmarks delivered a result that was comparable to those of, and arguably slightly better than, the landmark dataset, and also has the highly desirable property of being far easier to automate and of far more general applicability (see also MacLeod 1999, 2002, 2005a). It is important to note here that the strength of this approach lies in the complete characterization of outlines by evenly spaced sequences of semilandmarks that accurately and exhaustively capture the geometry of the object's outline. Mixing a few semilandmarks into a sparse set of true landmarks often introduces error at the sites of semilandmark location and degrades the overall quality of the landmark result. This characteristic derives directly from the dependency of semilandmark locations of information derived from other aspects of the form and the typically high level of morphological abstraction inherent in the landmark-based approach to shape characterization (Bookstein 1991). Nevertheless, dense strings of semilandmarks minimize this problem, because the simple presence of closely spaced, adjacent semilandmarks in the sequence exerts a constraint on the placement of individual semilandmarks that helps ensure the accuracy of the entire sequence. In this sense the deficiency of semilandmarks originally—and correctly—understood by

Bookstein (1991) in the context of the characterization of shape via sparse sets of landmarks may be turned into a strength when dense sets of semilandmarks globally constrained to lie on a well-defined track are used to represent the shape of that track. Semilandmarks are also susceptible to becoming misaligned with respect to well-defined structures within the form, but this misalignment can be compensated for by inclusion of a small number of true landmarks that can be used to further constrain semilandmark placement (MacLeod 1999).

Neural networks in general, and plastic self-organizing maps in particular, are relatively new concepts in the fields of biometry/morphometrics and automated species identification. They are known to be susceptible to several problems (e.g., dependence of large sample sized, failure to converge, and overtraining; see Bishop 1996; Schalkoff 1997). These issues need to be explored further in the context of biological species identification from digital images, as does the issue of the scalability of such systems.

The PSOM-uANN test performed as part of this study was the most radical departure from traditional biometric/morphometric form analysis considered. Here no particular measurements were taken, but processed versions of the digital images themselves were used as the subjects of analysis. These data, in turn, were not subjected to a data analysis method that applies any particular geometric model to the separation of groups but rather was free to devise its own model of within-group morphological variance and between-group discrimination. This is a supremely generalized and very flexible approach to automated species identification system design (see Reed 2010). That this approach delivered levels of performance as good as the best result obtained from any morphometric dataset/data analysis method, and was clearly superior to the results obtained from some (e.g., form factor data), is not only impressive, but suggests that systems designed along similar lines have genuine potential to serve as the basis for highly accurate, generalized, large-scale, automated species identification systems.

Finally, the performance of 3D surface data as the source material for morphological discrimination was also quite encouraging. This approach represents an extension of the geometric morphometric analysis of outline semilandmark data. An analysis of the morphological character of object surfaces is what qualitative taxonomists use when they evaluate the forms of biological species and compare unknowns to the species concepts they have formulated in their minds based on direct observation of hundreds or thousands of specimens over the course of a professional lifetime. The ability to capture this sort of information and then analyze it quantitatively (see MacLeod 2008; Polly and MacLeod 2008; Polly 2008) represents a major innovation in morphometrics with great potential for the design of automated biological species identification systems.

The time required to obtain 3D scans of biological objects, unfortunately, limits the practical utility of this approach at present. This situation will change as data-capture technology matures and as desktop computers grow ever more powerful. It is interesting to note that the PSOM-uANN approach represents an attempt to incorporate an aspect of 3D morphological information into the analysis of 2D biological data insofar as it utilizes whole digital images the patterns of pixel shading within which reflect, in part, the third (*z*) dimension of morphological variation. This is the same type of 3D information that is present in a single view or digital image of a specimen presented to a qualitative taxonomist for identification. However, it is a degraded and highly complex representation of 3D surface morphology. Direct geometric measurement of 3D surfaces would be clearly preferable to their qualitative assessment via imaging for the purposes of morphological analysis. It would be a simple matter to include information about the RGB color or grayscale brightness values for the locations of the semilandmark grid points and, in so doing, achieve a hybrid spatio-geometric and textuo-visual dataset that could be analyzed by either multivariate or machine learning approaches. Such a dataset would provide hyper-realistic information of a type that few systematists could ever access for the purpose of making species identifications from morphological data.

At the moment most scientists regard the technologies required to acquire and process the data needed to automate identification tasks as the stuff of science fiction, like the Star Trek tricorder. Based on these and other results published over the course of the past two decades, it can be concluded with confidence that this assumption is mistaken. Engineers and scientists—including systematists—interested in this problem have made outstanding progress in constructing precisely such devices, albeit in forms far less elegant and compact than the tricorder. Systems now exist that can now deliver better, faster, more accurate, and more consistent semiautomated or fully automated identifications than can be performed by any human taxonomist for small sets of groups (approx. 2–25 categories).

Interestingly, the performance of various approaches to the design of these automated identification systems appear to all work more or less equally well for problems in this size range. But in order to handle more general and more practical problems in this area, especially in the context of applied taxonomic analysis, systems capable of identifying hundreds and even thousands of different groups will be needed. The conceptual and technological challenges involved in scaling up the current generation of automated identification systems to this level are formidable. Judging by the performance of current systems, though, they cannot be regarded as impossible. It is unlikely, however, that such scaling leaps can be made successfully without an integrated interdisciplinary approach that brings experienced taxonomists together with information technology engineers, software designers, mathematicians, artificial intelligence specialists, and others. It is also unlikely that scaling up the technology to address more complex classification problems can be accomplished without appropriate interdisciplinary funding programs.

The foraminifer identification data and results presented here are just one example of a far larger requirement that spans many—arguably all—scientific disciplines: the need to accurately and remotely image objects of all sizes and then automatically categorize the contents into scientifically useful groups. Atmospheric scientists need to measure the size and shape of snowflakes and water droplets in the atmosphere for reasons similar to those of ocean modelers trying to predict the dispersal patterns of particulate pollutants. Ecologists need to collect data on the abundance of morphologically defined object categories that include both organic and nonorganic constituents. Agriculturalists and border security staff need to identify potential pest species quickly, in situ, and at the time they encounter them. Ecologists need to identify some species by the morphology of the sounds they make (e.g., bats, crickets, and frogs). Molecular systematists need to quickly and accurate associate unknown gene sequence data with sequences collected from authoritatively identified and vouchered specimens based on the pattern of their nucleotides. Parataxonomists ranging from members of biodiversity assessment teams conducting surveys of remote habitats to lone biostratigraphers sitting at petroleum exploration and production wells on offshore drilling platforms need to produce accurate species lists for hundreds of different species from myriad different groups for biodiversity and biostratigraphic investigations. All these example applications require that group identification results be delivered quickly, and with high levels of consistency, often across large geographic areas and, in some cases, vast spans of geological time. Given the high level of demand for the consistent and accurate identification of natural objects, we can no longer expect that these identifications continue to be carried out manually by trained experts using qualitative forms of visual assessment. In all these instances and many others, the most cost-effective approach for producing the data necessary to rapidly test scientific hypotheses is to obtain automated identifications based on digital images, audio records, and/or chemical data.

Both my own and others' experience working in this field over the past decade is that, like John Henry, many taxonomists are apprehensive about the idea of turning over the unquestionably important task of taxonomic identification to a "mere" machine. I believe this fear is misplaced. As has been demonstrated time and time again since the beginnings of the industrial revolution, scientific progress lies, in part, in constructing machines that do what machines are better able to do than humans in order to allow humans to do what humans are better able to do than machines. Scientific and technological developments have placed many of the tools necessary to automate aspects of one of the oldest human activities at our disposal. Systems already exist that can make substantial contributions to solving the taxon identification problem. With modest increases in funding and incentive, and with the encouragement of the relevant professional and stakeholder communities, such systems can grow rapidly in sophistication, accuracy, and generalizability. Such systems should—indeed must—be developed and used if taxonomy is to fulfill its potential as a twenty-first-century science.

Once the systematics community comes to recognize and accept the need for a partnership with automated systems in order to gain the speed, objectivity, and consistency that is, at present, missing from systematics' portfolio of goods and services, what role(s) will humans fill? The simple answer is everything other than routine identification and, especially, the "science bit." Automated identification systems will perform only as well as the taxonomic concepts and the authoritatively identified data they must use as the basis for their identifications. Systematists will, at the very least, need to participate in, and hopefully lead, the interdisciplinary teams of computer scientists, mathematicians, interface designers, and engineers who will build, develop, and integrate these systems with other parts of the biodiversity informatics infrastructure. Taxonomic specialists will be able to partner with automated identification systems to test and refine species concepts as well as explore new types of data that can be used to deliver identifications. But most important, once they are freed from the time-consuming and repetitive task of delivering large numbers of routine identifications, systematists will be able to be more active and vital partners in collaborations created to describe the patterns that exist among organisms and between organisms and their environment in the natural world and so to test hypotheses and understand the processes that brought those patterns into existence in the past, that maintain them in the present, and that will need to be conserved into the future if humanity chooses to accept the responsibility associated with the power to disrupt these processes and to recognize that its own survival depends on this understanding, maintenance, and conservation. Very little of the progress we must make in these areas will be possible unless we improve our ability to identify the constituents of the natural world.

CONCLUSIONS

The automated identification of biological objects (organisms) and groups (e.g., species, guilds, characters) has been a dream among systematists for centuries. The goal of some of the first multivariate biometric methods was to address the perennial problem of group discrimination and intergroup characterization. Despite much preliminary work in the 1950s and 1960s, progress in designing and implementing practical systems for fully automated object identification has proven frustratingly slow.

Part of the problem in making more rapid progress in this area has been a lack of any consensus regarding what types of data should be collected from organisms in order to support generalized species identification. Typically, automated species identification systems have been developed to handle specific types of data (e.g., form factors and linear distances between landmarks) in the absence of a knowledge that other types of geometric data exist and are being employed routinely in many other biological data analysis contexts. Moreover, new developments in machine learning and 3D surface characterization have further increased the number of options available to those who seek to characterize species form for the purposes of taxonomy.

In this investigation, five types of data that may be used to characterize aspects of species' form, and two alternative data-analysis procedures, were evaluated for their potential to distinguish between a test set of seven recent planktonic foraminifer species. The performance of all these data types and data analysis procedures delivered better performance than would be expected of human taxonomists working under standard conditions. Among the different data types, only form factor data delivered markedly lower identification accuracies, especially in terms of the discrimination of superficially similar species. The remaining data types did differ in terms of the generalizability for the purposes of discrimination of a wide variety of forms and ease with which they could be collected. For the purpose of maximizing identification accuracy and algorithmic generalizability, 2D landmark-registered semilandmark outlines analyzed via CVA of Procrustes-aligned shape coordinates (with or without size information included) and processed digital images analyzed using the PSOM-uANN delivered the best overall performance in these trials. The 3D eigensurface (semilandmark) grid data delivered marginally better discrimination accuracies than either of these two data types but was not considered an attractive option at present owing to the time needed to collect such data. The attractiveness of 3D semilandmark data will likely improve with time as the design of the digitizers needed to collect such data and the software required to process the raw scans improve.

Acknowledgments

I would like to acknowledge and thank the following for their contributions to this investigation: Stig Walsh (for collecting the 3D scans that were used as the basis for the eigensurface analysis, Jonathan Krieger (for collaborating on the programming of the adaptive gridding routines used for eigensurface analysis), and Mark ONeill (for providing access to and technical assistance with the DAISY PSOM-ANN program). Andrew Hamilton and the reviewers of the original submission also made comments and corrections that improved this chapter substantially. The research described in this report was made possible through a generous grant from the Petroleum Research Fund of the American Chemical Society.

REFERENCES

Bishop, C.M. 1996. *Neural Networks for Pattern Recognition*. Oxford: Oxford University Press.

Bookstein, F.L. 1990a. "Analytic Methods: Introduction and Overview." In F.J. Rohlf and F.L. Bookstein (eds.), *Proceedings of the Michigan Morphometrics Workshop: Ann Arbor, MI, 1990, the University of Michigan Museum of Zoology*, Special Publication 2: 61–74.

———. 1990b. "Higher-Order Features of Shape Change for Landmark Data." In F.J. Rohlf and F.L. Bookstein (eds.), *Proceedings of the Michigan Mor-*

phometrics Workshop, Ann Arbor, MI, 1990, the University of Michigan Museum of Zoology, Special Publication 2: 237–250.
———. 1991. *Morphometric Tools for Landmark Data: Geometry and Biology.* Cambridge: Cambridge University Press.
Coulbourn, W.T., Parker, F.L., and Berger, W.H. 1980. "Faunal and Solution Patterns of Planktonic Foraminifera in Surface Sediments of the North Pacific." *Marine Micropaleontology,* 5: 329–399.
Culverhouse, P. 2007. "Natural Object Categorization: Man vs. Machine." In N. MacLeod (ed.), *Automated Taxon Recognition in Systematics: Theory, Approaches and Applications.* Boca Raton, FL: CRC Press, Taylor & Francis Group.
Culverhouse, P.F., Williams, R., Requera, B., Herry, V., and González-Gil, S. 2003. "Do Experts Make Mistakes?" *Marine Ecology Progress Series,* 247: 17–25.
Figueirido, B., MacLeod, N., Krieger, J., De Renzi, M., Pérez-Claros, J.A., and Palmqvist, P. 2011. "Constraint and Adaptation in the Evolution of Carnivoran Skull Morphology." *Paleobiology,* 37: 490–518.
Folk, R.L. 1968. *Petrology of Sedimentary Rocks.* Austin: Hemphill's.
Gaston, K.J., and O'Neill, M.A. 2004. "Automated Species Identification—Why Not?" *Philosophical Transactions of the Royal Society of London, Series B,* 359: 655–667.
Ginsburg, R.N. 1997. "An Attempt to Resolve the Controversy over the End-Cretaceous Extinction of Planktic Foraminifera at El Kef, Tunisia Using a Blind Test. Introduction: Background and Procedures." *Marine Micropaleontology,* 29: 67–68.
Gobalet, K.W. 2001. "A Critique of Faunal Analysis; Inconsistency among Experts in Blind Tests." *Journal of Archaeological Science,* 28: 377–386.
Gorsky, G., Ohman, M.D., Picheral, M., Gasparini, S., Stemmann, L., Romagnan, J., Cawood, A., Pesant, S., Garcia-Comas, C., and Prejger, F. 2010. "Digital Zooplankton Image Analysis Using the Zooscan Integrated System." *Journal of Plankton Research,* 32: 285–303.
Kelly, M.G. 2001. "Use of Similarity Measures for Quality Control of Benthic Diatom Samples." *Water Research,* 35: 2784–2788.
Kennett, J.P., and Srinivasan, S. 1983. *Neogene Planktonic Foraminifera: A Phylogenetic Atlas.* Stroudsbourg: Hutchinson Ross.
Kohonen, T. 1982. "Self-Organized Formation of Topologically Correct Feature Maps." *Biological Cybernetics,* 43: 59–69.
Lang, R. 2007. "Neural Networks in Brief." In N. MacLeod (ed.), *Automated Taxon Recognition in Systematics: Theory, Approaches and Applications.* Boca Raton, FL: CRC Press, Taylor & Francis Group.
Lang, R., and Warwick, K. 2002. "The Plastic Self-Organizing Map." *In Proceedings of the International Joint Conference on Neural Networks.* Honolulu: Institute of Electrical & Electronics Engineers.
Lohmann, G.P. 1983. "Eigenshape Analysis of Microfossils: A General Morphometric Method for Describing Changes in Shape." *Mathematical Geology,* 15: 659–672.
MacLeod, N. 1998. "Impacts and Marine Invertebrate Extinctions." In M.M. Grady, R. Hutchinson, G.J.H. McCall, and D.A. Rotherby (eds.),

Meteorites: Flux with Time and Impact Effects. London: Geological Society of London.

———. 1999. "Generalizing and Extending the Eigenshape Method of Shape Visualization and Analysis." *Paleobiology*, 25: 107–138.

———. 2001. "Landmarks, Localization, and the Use of Morphometrics in Phylogenetic Analysis." In G. Edgecombe, J. Adrain, and B. Lieberman (eds.), *Fossils, Phylogeny, and Form: An Analytical Approach*. New York: Kluwer Academic/Plenum.

———. 2002a. "Geometric Morphometrics and Geological Form-Classification Systems." *Earth-Science Reviews*, 59: 27–47.

———. 2002b. "Phylogenetic Signals in Morphometric Data." In N. MacLeod and P.L. Forey (eds.), *Morphology, Shape and Phylogeny*. London: Taylor & Francis.

———. 2005a. "Principal Components Analysis (Eigenanalysis & Regression 5)." *Palaeontological Association Newsletter*, 59: 42–54.

———. 2005b. "Shape Models as a Basis for Morphological Analysis in Paleobiological Systematics: Dicotyledonous Leaf Physiography." *Bulletins of American Paleontology*, 369: 219–238.

———. 2007. "Groups II." *Palaeontological Association Newsletter*, 65: 36–49.

———. 2008. "Understanding Morphology in Systematic Contexts: 3D Specimen Ordination and 3D Specimen Recognition." In Q. Wheeler (ed.), *The New Taxonomy*. London: CRC Press, Taylor & Francis Group.

———. 2009. "Who Is Procrustes, and What Has He Done with My Data?" *Palaeontological Association Newsletter*, 70: 21–36.

———. 2010. "Principal Warps, Relative Warps, and Procrustes PCA." Palaeontological Association Newsletter, 75: 22–33.

MacLeod, N., O'Neill, M.A., and Walsh, S.A. 2007. "A Comparison between Morphometric and Artificial Neural Net Approaches to the Automated Species-Recognition Problem in Systematics." In G. Curry and C. Humphries (eds.), *Biodiversity Databases: Techniques, Politics, and Applications*. Boca Raton, FL: CRC Press, Taylor & Francis Group.

Manly, B.F.J. 1997. *Randomization, Bootstrap and Monte Carlo Methods in Biology*. London: Chapman Hall.

Mardia, K.V., and Dryden, I. 1998. "The Statistical Analysis of Shape Data." Biometrika, 76: 271–282.

Moelering, H., and Rayner, J.N. 1979. "Measurement of Shape in Geography and Cartography." *Reports of the Numerical Cartography Laboratory, Ohio State University*, NSF Report #SOC77–11318.

ONeill, M. 2007. "DAISY: A Practical Tool for Semi-Automated Species Identification." In N. MacLeod (ed.), *Automated Taxon Recognition in Systematics: Theory, Approaches and Applications*. London: Taylor & Francis.

Polly, P.D. 2008. "Adaptive Zones and the Pinniped Ankle: A 3D Quantitative Analysis of Carnivoran Tarsal Evolution." In E. Sargis and M. Dagosto (eds.), *Mammalian Evolutionary Morphology: A Tribute to Frederick S. Szalay*. Dordrecht: Springer.

Polly, P.D., and MacLeod, N. 2008. "Locomotion in Fossil Carnivora: An Application of the Eigensurface Method for Morphometric Analysis Of 3D Surfaces." *Palaeontologia Electronica,* 11: 10A.

Reed, S. 2010. "Pushing DAISY." *Science,* 328: 1628–1629.

Rohlf, F.J. 1990. "Rotational Fit (Procrustes) Methods." In F.J. Rohlf and F.L. Bookstein (eds.), *Proceedings of the Michigan Morphometrics Workshop: Ann Arbor, The University of Michigan Museum of Zoology,* Special Publication No. 2.

Rohlf, F.J., and Slice, D. 1990. "Extensions of the Procrustes Method for Optimal Superposition of Landmarks." *Systematic Zoology,* 39: 40–59.

Saito, T., Thompson, P.R., and Breger, D. 1981. *Systematic Index of Recent and Pleistocene Planktonic Foraminifera.* Tokyo: University of Tokyo Press.

Schalkoff, R.J. 1997. *Artificial Neural Networks.* New York: MIT Press and McGraw-Hill.

Simpson, R., Culverhouse, P., Ellis, R., and Williams, B. 1991. "Classification of *Euceratium gran.* in Neural Networks." *IEEE International Conference on Neural Networks in Ocean Engineering.* Washington, DC: IEEE.

Van Bocxlaer, B., and Schultheiß, R. 2010. "Comparison of Morphometric Techniques for Shapes with Few Homologous Landmarks Based on Machine-Learning Approaches to Biological Discrimination." *Paleobiology,* 36: 497–515.

Zachariasse, W.J., Riedel, W.R., Sanfilippo, A., Schmidt, R.R., Brolsma, M.J., Schrader, H.J., Gersonde, R., Drooger, M.M., and Broekman, J.A. 1978. "Micropaleontological Counting Methods And Techniques—An Exercise on an Eight Meters Section of the Lower Pliocene of Capo Rossello, Sicily." *Utrecht Micropaleontological Bulletins,* 17: 1–265.

Zelditch, M.L., Swiderski, D.L., Sheets, H.D., and Fink, W.L. 2004. *Geometric Morphometrics for Biologists: A Primer.* Amsterdam: Elsevier/Academic Press.

11

The New Systematics, the New Taxonomy, and the Future of Biodiversity Studies

QUENTIN WHEELER AND ANDREW HAMILTON

THEORY, PRACTICE, AND TECHNOLOGY-ENABLED TAXONOMY

This final chapter attempts to bring together several themes in this volume by placing the future practice of taxonomy in historical and conceptual context and asking what kinds of technological investments make sense given taxonomy's theoretical structure and the need for accelerating taxonomic studies. The view we defend here is that the theoretical and epistemological foundations of taxonomy point toward, rather than away from, the continued development of a new cyberinfrastructure that importantly includes morphology, even as molecular techniques become more common and more sophisticated.

The second half of this chapter describes efforts in cybertaxonomy that are currently under way and provides a rationale for new innovations, including large-scale e-typification and the building of virtual spaces for collaboration. Our reasons for looking toward these particular changes in taxonomists' ways of working are supported by arguments in the first half of the chapter about the way we understand what the science of taxonomy is and does. While we are excited and enthusiastic about developments in molecular systematics, we think comparative morphology should retain a central place in our collective attempt to understand life more fully. This is not just because of the pride of place morphology has in the history of natural history, but because of what we understand the foundational ideas of contemporary taxonomy to require. We also think morphology presents the complexity and richness that the biodiversity community at large most wants to have described and explained.

REINVENTING (PARTS OF) TAXONOMY, AGAIN

Seventy years ago, John Scott Lennox Gilmour led a large effort that was aimed at understanding systematics and its place among the biological sciences. Published in 1940, *The New Systematics* represented four years of effort by biologists in what was then called the Committee on Systematics in Relation to General Biology to come to grips with the emerging focus on phylogeny as an organizing principle for systematics, as well as with the importance for systematics of new data and techniques in developmental biology, paleontology, cytology, ecology, and other new and newly significant fields of study, most particularly population genetics (Association for the Study of Systematics 1937; Winsor 1995; Forey 2000). The volume, edited by J.S. Huxley, is an overt attempt to synthesize a great deal of knowledge, come to consensus over what it all means for the theory and practice of systematics, and, reaching even further, possibly unify biology. The Committee on Systematics, by the way, is now known as the Systematics Association.

As with all honest intellectual enterprises, *The New Systematics* offers more questions than answers. The arguments it contains for and against phylogenetics as a basis for classification are one example: contributors disagreed in fundamental ways about whether a phylogenetic systematics is possible or even desirable, and offered either praise or condemnation of the idea in equally enthusiastic terms (Winsor 1995).

Over the course of the past forty-five years, phylogenetics has increasingly become the basis of biological classification. This change has been importantly underwritten by the efforts of many theorists—principally among them Willi Hennig—who were able to show just how phylogenies could be rendered conceptually compatible with the previously existing classificatory system (Hennig 1966; Nelson and Platnick 1980; Wiley 1981; Schuh 2000; Hamilton this volume). As the systematics wars of the late 1960s to the early 1980s, described in vivid detail by Hull (1988), have resolved in favor of the phylogenetic approach, new difficulties have arisen. There is a case to be made that these are as fundamental in nature as the old arguments about what the best basis and method for systematics ought to be and that they go right to the heart of what systematics is and how it relates to the rest of biology, including concerns about how best to understand and manage biodiversity.

In the face of large-scale anthropogenic species loss, some have argued that systematics should be reinvented. Old-fashioned alpha taxonomy, they argue, has far outlived its usefulness in an age in which putatively objective, high-quality genetic information is plentiful and cheap. Alpha taxonomy requires the kind of specialist knowledge of morphologies and literatures that we can scarcely afford the time and financial resources to cultivate. It also requires years of advanced-degree-level work on particular taxonomic groups, and therefore also requires many taxonomists, since each taxonomist can only hope to master one or a few related groups. Because alpha taxonomy is strongly comparative and morphological in nature, doing it well also means huge investments in the curation of collections of billions of specimens that have been built up worldwide over a few hundred years. Classificatory work that is based on molecules, by contrast, requires much less by way of physical infrastructure and little or nothing by way of taxon-specific training, can be approached in a high-throughput way, and is orders of magnitude faster and cheaper. On this line of reasoning, we just do not have the luxuries of time and money that taxonomy based in comparative morphology asks for: species loss is happening too quickly, and we need to use new technologies to accelerate our inventory and understanding of the biological world.

We think this argument is half right. It is time to reinvent the tools of systematics. It would be a mistake, however, to abandon the fundamental conceptual approach of successful alpha taxonomy in favor of new methods that appeal mostly or only because they are efficient. Efficiency is a virtue but only inasmuch as it gets us where we want to go. A maximally fuel-efficient automobile that only has the capacity to complete the first leg of the journey to a far-distant destination may be an engineering marvel, but it is not sufficient for arriving. The chief problem with making taxonomy molecular is not the emphasis on genes over morphology—genes *are* phylogenetically relevant morphology, too—but that too often researchers use the latest tools to go only part of the distance.

As part of our work at Arizona State University's International Institute for Species Exploration, we have reconceptualized morphological aspects of integrative taxonomy as cyber-enabled taxonomy. Our vision of cybertaxonomy is a taxonomy that preserves the conceptual structure of taxonomic reasoning, with goals that include deep knowledge of the origin and diversification of species and their evolutionary novelties, and practices that have taken so many workers so long to build. It is also a taxonomy that embraces cyberinfrastructure and digital tools that can accelerate species exploration. With apologies to Huxley, and playing on slight differences in the terms *taxonomy* and *systematics*, we are pushing toward the New Taxonomy (Wheeler 2008): a new systematics that emphasizes descriptive taxonomy and the modernization of species exploration as a scientific enterprise.

WHY SPECIES "EXPLORATION"?

We describe taxonomy in terms of species exploration because the initial description of a species new to science is a beginning of research and hypothesis testing rather than a static end. And whether that initial characterization of a species is based on morphological or molecular data, or both, it, too, is a starting point. By making a species known and identifiable, one opens that species as a frontier for exploration. Over time, more and more information about that species accumulates and is critically tested and either improved or rejected. The ultimate aim is to know as much as we can about a species. As observations and information accretes, we slowly come to understand the geographic and seasonal distribution of the species, the genetic variation within and among its populations, the unique and shared aspects of its morphology and genome, its phylogenetic relationships, and on and on. In a perfect world we would have every species as well explored as the laboratory rat, fruit fly, or human. In practice most species will be less well known, as progress is governed by curiosity, user demands, and available resources.

One foundational aspect of species to be explored is morphology. Why should this be so when molecular data might save money and time? Our reasons are epistemological. Morphology is one of the most interesting and direct reflections of millions of years of natural selection and is therefore so central to the phylogenetic questions asked by taxonomists that it is literally and practically impossible to separate species (or clades) from characters (or homologues, if you prefer). The only unequivocal evidence of species status is character transformation (including molecular characters but most often also conspicuous morphological evolution); the only unequivocal evidence of monophyletic status is synapomorphy (again including any and all character sources, as embraced by the holomorphology concept of Hennig [1966]).

Even a simple description of the morphology of a species is a set of precise hypotheses and predictions about the distribution of attributes among organisms, as David Hull pointed out decades ago (Hull 1970).

Some features are claimed, based on available specimens, to be constantly expressed as a set that makes that species unique. These characters are described in explicit distinction from traits that are observed or theorized to vary within the species (see definitions in Nixon and Wheeler 1992). Each character is conceived of as an original state and all its subsequent modifications (see Platnick 1979 for discussion). Thus the presence of a character does not assume lack of variation but rather variation only with the predicted parameters of that character, which is itself a theoretical construct, a testable hypothesis. Taxonomists predict which features are characters and what the distribution of those characters will be as new specimens are collected and studied. Similarly, taxonomists predict that newly discovered characters, including molecular ones, will be distributed in a way that is consistent with the predicted pattern of species status and monophyly among species. For this reason, species are not merely described. They are *theorized* in terms of patterns of characters presented as explicitly testable predictions. Testing those predictions is a central part of the exploration of species and continues as long as new evidence or specimens are collected.

WHY MORPHOLOGY MATTERS: A PHYLOGENETIC PERSPECTIVE

Before moving on to discuss and describe innovations and ideas for taxonomic cyberinfrastructure, it will be important to be clear about what kind of taxonomic enterprise we are trying to support and why it makes sense to continue an emphasis on morphology. In the previous section we offered a snapshot of an epistemology for descriptive taxonomy but did not offer evidence to the conclusion that the epistemic strategy we discuss is the best or most appropriate one. That is, we have not yet said why morphology as we understand it is central to the *right* taxonomic enterprise.

In its most basic form, our argument is that taxonomy is not in the same position now as it was in 1940. Though many conceptual challenges remain, there is at least now general agreement that phylogeny is the basic organizing principle for classifications, that monophyly is the appropriate grouping concept, and that shared, derived characters, along with some considerations about parsimony or likelihood, or both, are the best avenue toward diagnosing particular lineage relationships. Further, species are (or at least can be made) objectively testable hypotheses that make predictions about character distributions (see Wheeler and Platnick 2000; Wheeler 2010).

This way of understanding the conceptual structure of taxonomy puts a heavy emphasis on characters and their patterns and distributions. A great deal of the intellectual work of modern taxonomy lies in discovering good characters—those that reveal the pattern of relationships between putatively related taxa and frame testable hypotheses about these relationships. Without good characters, even the most sophisticated analyses of phylogenetic data are not of much use. Homology retains its central conceptual position in taxonomy (Williams 2004), and synapomorphy, not overall similarity, is the necessary indicia of monophyly (Williams, Ebach, and Wheeler 2010).

This is not an argument for organism-level morphology and against using sequence data in taxonomy. On the contrary, by our lights sequence data *is* morphological information—no more, no less—and should be treated in exactly the same fashion as morphology. The problem is that it often is not. The notion of homology has not been well thought out at the micro level, at least in the absence of functional information about nucleic acid sequences (de Carvalho and Ebach 2009), and metrics of similarity and dissimilarity stand in for careful attention to homlogy. Sequence alignments, made either "by eye" or by a computer using algorithms, are often not subjected to the same careful scrutiny as claims about the information content of characters at the macro level (Mooi and Gill 2010). There are all kinds of reasons for this, ranging from the naive perceived objectivity and unproblematic nature of sequence data (Nelson 2004) to an understandable, but regrettable, reliance on technology at the cost of close attention to the biological details of the organisms and groups under study. We don't deny that there are good micro-level characters and that the conceptual machinery of phylogenetic systematics can operate on them to resolve phylogenies and help make classifications. We are just concerned that approaching molecules in this way is fairly rare.

Molecular data are, at best, just one of several sources of information about evolutionary history and the relationships between groups, just as morphology is just one other source. Morphological information, however, is largely the basis of our current understanding of the biological world, mostly for better and sometimes for worse, but it remains important enough for most taxonomic groups that access to it is crucial for doing taxonomic work. More important, understanding the evolution of morphological characters is a worthy intellectual and scientific goal in itself, quite apart from species identification or phylogeny for its own sake. The phylogenetic patterns of characters are inseparably linked to our understanding of phylogeny itself, and are no less valuable or interesting.

Those who think that distinguishing and identifying species is sufficient in the absence of an informative description of species, such as some advocates of DNA barcodes, should consider how silly their argument sounds when applied to any other field of inquiry. For example, would astronomers be content with simply knowing how many "stars" exist by making each quickly and uniquely identifiable as a distinct spot of light in the night sky by virtue of some set of coordinates alone? The reason that astronomers want to chart heavenly light sources is not merely to count them or make them identifiable, but rather to understand each in terms of its uniqueness. Astronomers care whether a particular star is a red dwarf or a hyper giant, whether it is indeed a near star or a distant galaxy, and what position each occupies in relation to various gravitational pulls and a predicted expansion of the universe as a consequence of the Big Bang. Taxonomists care about the morphology of each species, its position in the tree of life, and all the ways in which its biology and natural history are unique or interesting.

Initial hypotheses about homology, species limits, and relationships (the "description") must be tested time and again. Even when the species hypothesis is highly corroborated, accumulation of reliable information about ecological, seasonal, and geographic distribution, phylogenetic relationships, and behaviors is a prolonged effort. Given new evidence, even long-standing species hypotheses may be refuted or significantly modified. Every newly collected specimen or discovered character has the potential to falsify hypothesized characters and character distributions (Platnick 1979; Nelson and Platnick 1980; Nixon and Wheeler 1992; Wheeler and Platnick 2000). Describing and understanding the origins of organism-level morphology remain crucial for this work and an ultimate goal of taxonomy. Let us be clear: even if DNA approaches made possible recognizing all species and their phylogenetic relationships, we should still want to study morphology to witness the structural results of millions of years of natural selection. Those who think otherwise only reflect upon applied taxonomy for purposes like making species identifications, for instance, not the practice of taxonomy as a science. Needless to say, the best applications flow from the best fundamental science.

We want to comment in more specificity on currently fashionable DNA barcoding. As Will, Mishler, and Wheeler (2005) quipped, aspects of DNA barcoding are both new and good; however, that which is good is not new, and that which is new is not good, at least as far as classification goes. Some major proponents of barcoding confuse species identification and classification, underestimate the number of specimens necessary to assess infraspecific genetic variation across the full range of a species, and are content to simply arrive at an identification for a species about which they know nothing of real interest (Hamilton and Wheeler 2009).

Let us illustrate with a simple analogous "bar code":

> statelyplumpbuckmulligancamefromthestairheadbearingabowloflatheronwhichamirrorandarazorlaycrossedayellowdressinggownungirdledwassustainedgentlybehindhimbythemildmorningairheheldthebowlaloftandintonedintroiboadaltaredeihaltedhepeereddownthedarkwindingstairsandcalledupcoarselycomeupkinchcomeupyoufearfuljesuitsolemnlyhecameforwardandmountedtheroundgunresthefacedaboutandblessedgravelythricethetowerthesurroundingcountryandtheawakingmountainsthencatchingsightofstephendedalushebenttowardshimandmaderapidcrossesintheairgurglinginhisthroatandshakinghisheadstephendedalusdispleasedandsleepyleanedhisarmsonthetopofthestaircaeandlookedcoldlyattheshakinggug
>
> hereisedwardbearcomingdownstairsnowbumpbumpbumponthebackofhisheadbehindchristopherrobinitisasfarasheknowstheonlywayofcomingdownstairsbutsomegtimeshefeelsthattherereallyisanotherwayifonlyhecouldstopbumpingforamomentandthinkofitandthenhefeelsthatperhapsthereisntanyhowhereheisatthebottomandreadytobeintroducedtoyouwinniethepoohwhenifirstheardhisnameisaidjustasyouaregoingtosaybutithougthhewasaboysodidisaidchristopherrobinthenyoucantcallhimwinnieidontbutyousaidheswinnetherpoohdontyouknowwhathermeansahyesnowidoisaidquicklyandihopeyoudotoobecauseitisalltheexplanationyouaregoingtogetsometimeswinniethepoohlikesagameofsomesortwhenhecomesdownstairsandsom

These two sequences are the first 650 letters in the opening paragraphs of two books. These unique sequences of letters are, like DNA bar codes (comprised of about 650 base pairs of the COI gene), capable of uniquely identifying these two titles from among millions of books. Fed into a computer or parsed by eye, our two examples would rapidly and accurately distinguish any copy of the two books from one another. Of course, they could not identify any other book positively; that would require learning its first 650 letters, too. More to the point, because we have not, at this stage of this literary analysis, read either of the books, we would be at a loss to guess which might be more interesting or useful for any particular purpose, or which we might wish to purchase. The two 650 letter "sequences" are, by the way, from James Joyce's *Ulysses* and A.A. Milne's *Winnie-the-Pooh*. We favor reading both books and morphology rather than simply identifying them.

ACCESSING MORPHOLOGY: TWO STRATEGIES FOR CYBER-ENABLED TAXONOMY

There is much to the criticism from those outside the science of taxonomy that taxonomists are using eighteenth-century tools to solve a twenty-first-century problem (Hebert et al. 2003). These tools include the library, the light microscope, the pitfall trap, a well-stocked and carefully curated natural history collection, and an office with a door that can be closed. It is no exaggeration to say that taxonomy begins and ends at the collection drawer. Access to these drawers is surely the greatest remaining obstacle to a robust infrastructure for taxonomy, as many of the other tools of post–field taxonomy have been rendered digital.

In just the past few years, great strides have been made to create a new infrastructure for taxonomy. The Global Biodiversity Information Facility (GBIF) now provides free access to nearly 200,000,000 natural history museum records. These contain primary information including label data, associated observations, and notes on specimens. The Encyclopedia of Life (EoL) is now in the process of building a webpage for each species (Wilson 2003), and is on track to complete the first million pages in five years. EoL will eventually provide basic information on every species to everyone who needs it. The Biodiversity Heritage Library (BHL) has digitized and recognized the text on more than 32,000,000 pages from more than 86,000 volumes, making even previously obscure and hard-to-find primary documents, including species descriptions, available under Creative Commons licensing at no charge to anyone with an Internet connection.

In addition to these efforts, a great deal of work has been done to track taxon names, find synonyms, and build a name-based cyberinfrastructure for biodiversity studies and management. These include the Universal Biological Indexer and Organizer (uBio), which has cataloged nearly ten million biological names, and similar efforts by the International Plant Names Index, FishBase, the World Register of Marine Species, the Index Fungorum, and other name repositories. The Global Names Architecture project and its associated Global Names Index attempt to provide access to all repositories of scientific names as part of an effort to make sense of these names and their uses across multiple large databases. The Global Names Architecture has as its goals the anchoring of all species data in a unique identifier such that one can find all the existing data for particular species in many databases using that identifier (which is associated with a taxonomic name).

Sorting out the confusion surrounding millions of names and providing digital access to the taxonomic literature are critical steps toward a new infrastructure for taxonomy. Indeed, both are sine qua non of for a twenty-first- century discipline that answers the needs of science and society. Taxonomy, however, is ultimately neither library nor name based. Scientific names have served as identifiers for species in botany and zoology for more than 250 years and are the basis for storage and retrieval of biological information in contemporary databases, from GBIF to GenBank. Names, however, are ephemeral: they reflect current understanding for what is a valid taxon. As our understanding of species advances, the orderly application of scientific names depends upon an objective basis of their meanings. That continues to be provided by typification, that is, the designation of single specimens to which scientific names are anchored. As our theories and available evidence change over time, it is possible to allow the name to follow the type specimen; whichever currently recognized species the type specimen falls within, there follows the name, whether valid or introduced into synonymy.

Taxon scholars frequently examine type specimens to assure nomenclatural accuracy, particularly in the context of preparation of revisions, monographs, floras, and faunas. Frequently a cursory examination of the type specimen is sufficient to affirm or exclude its conspecificity. Other times, more detailed observation or specimen manipulation may be required. Digital images can never replace the physical specimen, yet they can answer a large number of questions and avoid wear and tear on these delicate and irreplaceable specimens, which are often a century old or more. Enormous expense and inefficiency are involved in traveling to dozens of museums to examine types in the course of otherwise routine work. Making type specimens accessible online will eliminate much travel and expense for taxonomists, inform other biodiversity scientists for whom species identifications are important, and also make a significant part of the world's biological heritage approachable in a new and compelling way. With the comparatively—and we stress comparatively—easy task of digitizing the literature and building large-scale name databases well under way, it is time to turn to digitizing morphology with a special urgency for electronic typification.

The process of creating digital representations of type specimens is also well under way, with certain taxa and institutional collections more or less complete (e.g., New York Botanical Garden type sheets). At Arizona State, working with partners at several academic institutions and in industry, we are exploring ways in which 3D images might be rapidly compiled, including ideas for automating much of the process. In addition, we are working on a system that would link specialists with type specimens by means of remotely operable instruments. A combination of such technologies is required: what is needed is a standardized process automated as fully as possible for quickly creating a set of images where none or few exist, including 3D imaging if possible, *and* opening remote access for real-time study and imaging of type specimens (Wheeler et al. 2012). As experts access specimens to examine specific characters, images and annotations are added to the initial 3D image so that a full picture of the type gradually emerges and the need to examine it, in person or remotely, diminishes over time.

The backlog is formidable. Obvious are the type specimens for the approximately two million known species (Chapman 2009). Less obvious are the types for species that have been synonymized, raising the total to perhaps five or six million specimens. For many species, of course, there are a few to many paratypes that also should be imaged, raising the number even higher. Fortunately, there are probably half a million or more types already digitized to some degree. In addition to making realization of a rapid, semiautomated instrument a top priority, there is no excuse for adding to this already daunting backlog. Imaging types for all newly described species should become absolutely routine. The importance of creating such "e-types" can be greatly enhanced if their access is addressed through open knowledge of their existence, as through ZooBank (Polaszek et al. 2005).

It will not be necessary to build and distribute thousands or even hundreds of these instruments in order to complete the set of taxonomically informative e-types; a small network will suffice for most purposes, because more than 80 percent of all the relevant collections are held at three or four large institutions.

Electronic types will complement other digitization projects to open access to research resources. Not only does this digital revolution accelerate the work of taxonomy and nomenclature, but it has enormous implications for the democratization of taxonomy. Access to archival literature and rare specimens was once the province of a few lucky taxonomists in major museums in Europe and the United States. Soon, students at smaller institutions, scientists in developing (and, importantly, biodiversity-rich) nations, and amateurs will share access to 250 years of accumulated taxonomic knowledge and specimens. This will unleash an unprecedented renaissance in species exploration, partly and significantly by making it possible for hobbyists and pro-ams to contribute as never before (Pearson, Hamilton, and Erwin 2011). As important as access to specimens is, other tools are needed to fuel this revolution.

The community needs to harness these resources and deliver them as part of a classification commons. Whether newly created or adapted from EoL's LifeDesks, MX, Scratchpads, or other existing software, it is critically important that we have a virtual taxonomic platform where imaging, character data, historical literature, phylogenetic analysis, video communications, online publishing, the research resources mentioned above, and software for creating e-monographs converge seamlessly. Exciting online publications like *ZooKeys* are, through hyperlinks, giving us a glimpse of the future of taxonomic publication. This evolution will ultimately, we predict, go far beyond such enhanced "papers." Rather, we foresee in the classification commons a shared core knowledge base that includes all and up-to-the-minute data and information for any given taxon. Taxonomists will test hypotheses about characters, species, and clades as rapidly as new specimens and data are analyzed, and their new, improved, or rejected hypotheses will be instantly reflected in name databases and, in turn, in keys, maps, checklists, field guides, labels for images, and every other form and permutation of taxonomic information consumed by a diverse and growing number of user communities.

Our vision also hinges on a sociological ground shift. NSF-funded Planetary Biodiversity Inventories (PBI) projects have demonstrated dramatic acceleration of description and redescription of species through collaborative international teams (Page 2008; Knapp 2008). Providing such teams with efficient videoconferencing and telemicroscopy facilities, combined with online access to literature (BHL), specimen data (GBIF), virtual type specimens, and other resources has the capacity to enable groups of researchers that are focused on large, complex taxa to generate and analyze data that are important to taxonomists as well as to ecologists, conservationists, and natural resource managers.

CONCLUSION

We have here tried to outline a way forward for a twenty-first-century taxonomy to meet the needs of science and society. We are conceptualizing digital instrumentation that can bridge the gap between specimens and taxon experts, and we have projects under way to engineer prototypes. Our goal is to modernize the study of morphology so it advances as quickly as knowledge of genomic, fossil, and developmental phenomena; to leverage 250 years of specimens and taxonomic knowledge to expedite our understanding of evolutionary history and ecosystem function and status; to conserve and advance the parts of theory that taxonomists have gotten right while encouraging advancements and improvements in other theoretical aspects of character analysis and phylogeny reconstruction; and to open the wonder of taxonomy and biodiversity to curious students, citizen scientists, and specialists everywhere.

Beyond new technology, we believe that several parallel transformations are necessary for cybertaxonomy to succeed. Specifically, we must migrate taxonomic work to a cyber-enabled "collaboratory" that links experts and institutions around the world. We must study and understand the recent history of taxonomy, in order to change the perception of this rigorous but nonexperimental science, and we must reform the sociology of taxonomy to both maximize collaboration and change the ways in which team science is recognized and rewarded by administrators and funding agencies.

Because Linnaean binomials are uniquely important in informatics, and because we are convinced that access to images and types themselves is easily within reach by adapting existing technology, we regard this as an urgent priority. Nomenclatural decisions have huge implications for information retrieval and analysis, from genomics to ecosystem science. By making type specimens and their images openly available, the speed and quality of nomenclatural decisions will be enormously improved. This demonstration of the potential transformative impact of cyber tools for taxonomy will be a large step in the emergence of a rejuvenated taxonomy—the New Taxonomy—necessary to the exploration and conservation of biodiversity.

Acknowledgments

Andrew Hamilton's work in this area is sponsored by the National Science Foundation (SES-09083935).

REFERENCES

Association for the Study of Systematics. 1937. "Association for the Study of Systematics in Relation to General Biology." *Nature*, 140: 163–164.

Chapman, A.D. 2009. *Numbers of Living Species in Australia and the World.* 2nd ed. Canberra: Australian Government, Department of the Environment, Water, Heritage and the Arts.

de Carvalho, M.R., and Ebach, M.C. 2009. "Death of the Specialist, Rise of the Machinist." *History and Philosophy of the Life Sciences,* 31: 467–470.

Forey, P. 2000. "How We Were Born." *Systematics Association Newsletter,* 15: 3–4.

Hamilton, A., and Wheeler, Q.D. 2008. "Taxonomy and Why the History of Science Matters for Science: A Case Study." *Isis,* 99: 331–340.

Hebert, P., Cywinska, A., Ball, S., and deWaard, J. 2003. "Biological Identifications through DNA Barcodes." *Proceedings of the Royal Society of London, Series B,* 70: 313–321.

Hennig, W. 1966. *Phylogenetic Systematics.* Urbana: University of Illinois Press.

Hull, D. 1970."Contemporary Systematic Philosophies." *Annual Review of Ecology and Systematics,* 1: 19–54.

Huxley, J.S. 1940. *The New Systematics.* Oxford: Oxford University Press.

Knapp, S. 2008. "Taxonomy as a Team Sport." In Q.D. Wheeler (ed.), *The New Taxonomy.* Boca Raton, FL: CRC Press, 33–53.

Mooi, R.D., and Gill, A.C. 2010. "Phylogenies without Synapomorphies—A Crisis in Fish Systematics: Time to Show Some Character." *Zootaxa,* 2450: 26–40.

Nelson, G., and Platnick, N.I. 1980. "A Vicariance Approach to Historical Biogeography." *Bioscience,* 30: 339–343.

Nixon, K.C., and Wheeler, Q.D. 1992. "Extinction and the Origin of Species." In M.J. Novacek and Q.D. Wheeler (eds.), *Extinction and Phylogeny.* New York: Columbia University Press, 119–143.

Page, L.M. 2008. "Planetary Biodiversity Inventories as Models for the New Taxonomy." In Q.D. Wheeler (ed.), *The New Taxonomy.* Boca Raton, FL: CRC Press, 55–62.

Pearson, D.L., Hamilton, A.L., and Erwin, T.L. "Recovery Plan for the Endangered Taxonomy Profession." *Bioscience* 61: 58–63.

Platnick, N.I. 1979. "Philosophy and the Transformation of Cladistics." *Systematic Zoology,* 28: 537–546.

Polaszek, A., Agosti, D., Alonso-Zarazaga, M., Beccaloni, G., Bjørn P. dP., Bouchet, P., Brothers, D.J., Cranbrook, G., Evenhuis, N.L., Godfray, H.C.J., Johnson, N.F., Krell, F.-T., Lipscom, D., Lyal, C.H.C., Mace, G.M., Mawatari, S., Miller, S.E., Minelli, A., Morris, S., Ng, P.K.L., Patterson, D.J., Pyle, R.L., Robinson, N.J., Rogo, L., Taverne, J., Thompson, F.C., Tol, J. van, Wheeler, Q.D., Wilson, E.O. 2005. "A Universal Register for Animal Names." *Nature,* 437: 4.

Schuh, R. 2000. *Biological Systematics: Principles and Applications.* Ithaca, NY: Cornell University Press.

Wheeler, Q.D. 2008. "Towards the New Taoxonomy." In Q. Wheeler (ed.), *The New Taxonomy.* Boca Raton: CRC Press, 1–18.

———.D. 2010. "What Can We Learn from 20th Century Concepts of Species? Lessons for a Unified Theory of Species." In I. Jahn and A. Wessel (eds.), *Fur eine Philosophie der Biologie.* Munich: Kleine Verlag, USP Publishing, 43–59.

Wheeler, Q., Bourgain, T., Coddington, J., Gostony, T., Hamilton, A., Larimer, R., Polaszek, M., Schauff, M., and Alma Solis, M. 2012. "Nomenclatural Benchmarking: The Roles of Digital Typification and Benchmarking." *ZooKeys,* 209: 193–202.

Wheeler, Q.D., and Platnick, N.I. 2000. "The Phylogenetic Species Concept (*sensu* Wheeler and Platnick)." In Q.D. Wheeler and R. Meier (eds.), *Species Concepts and Phylogenetic Theory: A Debate.* New York: Columbia University Press, 55–69.

Wiley, E.O. 1981. *Phylogenetics: The Theory and Practice of Phylogenetic Systematics.* New York: Wiley-Interscience.

Will, K.W., Mishler, B.D., and Wheeler, Q.D. 2005. "The Perils of DNA Barcoding and the Need for Integrative Taxonomy." *Systematic Biology,* 54: 844–851.

Williams, D.M. 2004. "Homologues and Homology, Phenetics and Cladistics: 150 Years of Progress." In D.M. Williams and P.L. Forey (eds.), *Milestones in Systematics.* Systematics Association Special Volume 67. Boca Raton, FL: CRC Press.

Williams, D.M., Ebach, M.C., and Wheeler, Q.D. 2010. "Beyond Belief: The Steady Resurrection of Phenetics." In D.M. Williams and S. Knapp (eds.), *Beyond Cladistics: The Branching of a Paradigm.* Berkeley: University of California Press, 169–196.

Winsor, M.P. 1995. "The English Debate on Taxonomy and Phylogeny, 1937–1940." *History and Philosophy of the Life Sciences,* 17: 227–252.

Index

About the Editor

Andrew Hamilton is Associate Dean in the Honors College at the University of Houston, where he also teaches in the Department of Biology and Biochemistry. His research focuses on both the conceptual foundations and the policy ramifications of biological classification. His work has been published in academic journals and discussed in the popular press, including *ScienceNow,* the *Christian Science Monitor,* and *Scientific American.* Hamilton also has a strong interest in the classification of groups and individuals as evolutionary objects and has published widely on this topic. Before taking his current position, he held a faculty appointment in the School of Life Sciences at Arizona State University (ASU) and an appointment as Senior Sustainability Scientist at ASU's Global Institute of Sustainability. Hamilton is a graduate of Berea College, Boston College, and the University of California, San Diego.

Species and Systematics

SPECIES: A HISTORY OF THE IDEA
John S. Wilkins
www.ucpress.edu/9780520260856

COMPARATIVE BIOGEOGRAPHY: DISCOVERING AND CLASSIFYING BIOGEOGRAPHICAL PATTERNS OF A DYNAMIC EARTH
Lynne R. Parenti and Malte C. Ebach
www.ucpress.edu/9780520259454

BEYOND CLADISTICS
Edited by David M. Williams and Sandra Knapp
www.ucpress.edu/9780520267725

MOLECULAR PANBIOGEOGRAPHY OF THE TROPICS
Michael Heads
www.ucpress.edu/9780520271968

THE EVOLUTION OF PHYLOGENETIC SYSTEMATICS
Edited by Andrew Hamilton
www.ucpress.edu/9780520276581